EVOLUTION AND ANIMAL BREEDING

Alan Robertson

EVOLUTION AND ANIMAL BREEDING

Reviews on Molecular and Quantitative Approaches in Honour of
Alan Robertson

Edited by

William G. Hill and Trudy F.C. Mackay

Institute of Animal Genetics, University of Edinburgh
and Department of Genetics, North Carolina State University

C·A·B International

Published by
C·A·B International
Wallingford
Oxon OX10 8DE
UK

Tel: Wallingford (0491) 32111
Telex 847964 (COMAGG G)
Telecom Gold/Dialcom: 84: CAU001
Fax: (0491) 33508

British Library Cataloguing in Publication Data
Evolution and animal breeding.
1. Animals. Genetics
I. Hill, W. G. (William George), *1940–*
II. Mackay, Trudy F. C.
591.1′5

ISBN 0-85198-639-0

First printed 1989
Reprinted 1991

Typeset by Editorial Enterprises, Torquay, Devon and Arrowsmiths, Bristol
Printed in the UK by The Cambrian News Ltd, Aberystwyth

This book is dedicated to the memory of
ALAN and to MEG ROBERTSON
to both of whom we owe so much

Contents

A POPULATION GENETICS AND EVOLUTION

B GENETICS OF METRIC TRAITS

C QUANTITATIVE APPROACHES TO ANIMAL BREEDING

Preface

The idea for this book in honour of Alan Robertson was conceived at the Second International Conference on Quantitative Genetics in Raleigh in May – June, 1987. There it became apparent to a group of us (Stuart Barker, Joe Felsenstein, Dick Frankham, Carlos López-Fanjul, Oliver Mayo, Laurie Piper, Charlie Smith and ourselves) that the important contributions of Alan Robertson to the development of population and quantitative genetics and to animal breeding should be noted in some way. His influence is apparent in his publications, in his training of PhD students, who are currently scattered throughout the world, and in animal breeding practice. For those of us who have studied and worked at the Institute of Animal Genetics in Edinburgh, Alan's influence has perhaps been greatest in the informal morning coffee sessions held daily in his office, which provided a uniquely stimulating and friendly environment in which to voice and discuss new ideas.

The book is divided into four main sectiaons, reflecting the great breadth of Alan's interests and significant contributions: population genetics, quantitative genetics, and both quantitative and molecular approaches to animal improvement. At the beginning of each section is a chapter which reviews Alan's own contributions in the field, followed by a series of minireviews covering up-to-date developments and prospects in topics in which Alan was interested. We hope it will be of value to students and research workers in these areas, as well as being a tribute to Alan Robertson.

The authors of the various chapters have all been associated with Alan in some way, as colleagues, students or visitors at the Institute of Animal Genetics. It is indicative of the high regard and respect in which he is held that, with very few exceptions, everyone asked to contribute to the volume readily agreed to do so. We, the editors, in particular have benefited from our close connections with Alan, both as students and colleagues. About the time of his official retirement in 1985 Alan began to suffer a decline in health. It is very sad that, as the book goes to press, his health had deteriorated to the extent that he will now not be able to derive the pleasure from it that it was intended to give him. We hope, however, that his wife, Meg, who has so enthusiastically supported Alan and his friends, colleagues and students and to whom we also owe much, will approve of the volume.

We have to add another sad note. Our longstanding friend and colleague, Roger Land, died suddenly a few days after giving us what he described as a first draft of his paper. We wanted to include it, and have done so after only minor amendment. We are grateful to John Clark for comments and John Land, Roger's brother, for permission to publish. Roger's paper was not originally destined as a section introduction, but it seemed an appropriate way to mark his foresight and honour his memory. We should like to thank all of the authors for the time and effort put into writing their contributions and delivering them (mostly) on time. We are grateful also to Douglas Falconer for his help in preparing the reference list and advice. We should especially like to thank Jackie Bogie and Anne Brown for expert secretarial assistance – we could not have produced this volume without them. (Thanks also to the telecommunications system and NATO which facilitated correspondence between us.)

Bill Hill (Edinburgh)
Trudy Mackay (Raleigh)
November 1988

We regret that Alan Robertson died on 25th April, 1989, while this book was in proof. —The Editors

Biographical sketch of Alan Robertson

Alan Robertson was born on 21 February, 1920 and educated at Liverpool Institute and Cambridge University, graduating with a BA in Chemistry in 1941. He then undertook postgraduate research in physical chemistry at Cambridge, but did not complete a PhD. Subsequently he was awarded a DSc in 1952 by Edinburgh University for his work in genetics.

Alan worked on Operational Research with C.H. Waddington during the war, and subsequently joined the Agricultural Research Council (ARC) Animal Breeding and Genetics Research Organization initially in Hendon and then, following a period spent in the US with Sewall Wright and with Jay L. Lush, in Edinburgh. He remained in Edinburgh for the rest of his career, in what became the ARC Unit of Animal Genetics directed by Waddington and then by Douglas Falconer. Alan was promoted to Deputy Chief Scientific Officer in 1966, a rare appointment for staff without administrative duties, retiring in 1985. He was appointed an Honorary Professor by Edinburgh University in 1967.

Alan received many honours: he was elected a Fellow of the Royal Society of London and of Edinburgh, a Foreign Associate of the National Academy of Sciences of the USA, and a Foreign Honorary member of the Genetics Society of Japan; was awarded honorary doctorates from the University of Stuttgart–Hohenheim, the Agricultural University of Norway, the State University at Liège and the Danish Agricultural University; and was appointed OBE in 1965.

Meg and Alan were married in 1947 and have three children, Mark, Hilary and Michael and three grandchildren.

After a long period of declining health, Alan died on 25 April, 1989.

A POPULATION GENETICS AND EVOLUTION

Chapter 1
Alan Robertson's Contributions to Population Genetics

Joseph FELSENSTEIN

Department of Genetics, University of Washington SK-50, Seattle, Washington 98195, USA

Alan Robertson has been an influential figure in theoretical population genetics over the past 30 years. One might imagine, as most population genetics theorists do, that in analyzing his contributions one should start with theoretical population genetics, and that his work in quantitative genetics and animal breeding flow from that base. One could not be more wrong. A cursory glance at a list of his papers shows that his work in animal breeding was his first and enduring concern, the population genetics coming from questions arising in animal breeding and quantitative genetics. In almost every case his population genetic work came from practical concerns in animal breeding.

This reflects a view of theoretical population genetics far different from that held by many figures in that field. To many, theoretical population genetics is an abstract discipline, a rigorizable branch of applied mathematics. Any applications are to be considered only after the fundamental principles of the discipline are established and made rigorous. Alan Robertson has taken the opposite approach. His approach has been relentlessly practical, however interested he may have become in some of the theoretical questions along the way. As such it has been far less concerned with rigour than with practical utility.

Other major figures in quantitative genetics have approached population genetics differently. Clark Cockerham and Oscar Kempthorne, to take two notable examples, started from an interest in plant or animal breeding only to become heavily involved with working out the principles of population genetics and quantitative genetics on a rigorous basis. Alan Robertson has scarcely participated at all in that line of development, so influential in the 1970s. If a rough approximation would be sufficiently accurate, then he saw no need to examine the general case more precisely if there were ways to use the approximation to get practical results.

Genetic variation in subdivided populations

Domestic animal populations are highly subdivided, and the individual demes are usually small. Taking that into account leads a quantitative geneticist inevitably into population genetics. Alan Robertson's path into population genetics from breeding can be traced through his paper on the effect of inbreeding on the variation due to recessive genes (Robertson, 1952). Written shortly after Sewall Wright's visit to Edinburgh, this is a generalization to recessive and overdominant cases of Wright's (1951) results on the effect of inbreeding on variation within and between inbred lines, which were valid only for purely additive loci. Robertson used an exact calculation in the case of sibmating, and a general recursion of moments of gene frequencies to establish the more general patterns. Comparing numbers, the two methods give closely similar, but not exactly identical, results. It might be thought that they should be identical, since both are exact calculations, but the general calculation is exact only for the case of a pure Wright–Fisher model with selfing allowed, while the sib mating case has selfing prohibited.

Although Robertson could have given full algebraic expressions for the general case, they would have been so complicated that there seemed no point in doing so. Instead he described the numerical patterns seen in the results and computed exact expressions for the relations between the initial and the ultimate values of within- and between-line variance. Here, as throughout his work, he concentrates on providing a sufficiently accurate approximate description rather than exact algebraic solutions.

Alan Robertson later returned to the problem of population subdivision and obtained a startling result concerning mating structure in regular systems of mating. Kimura and Crow (1963) had shown that a circular half-sib mating system had an extraordinarily low asymptotic rate of inbreeding. This involved half-sib matings, whereas a system which avoided mating relatives whenever possible actually showed a higher ultimate rate of inbreeding. Robertson (1964b) explained this pattern with a general result whose full implications have not yet been assimilated into breeding practice.

For the case in which each mating has exactly two offspring he was able to show that the asymptotic rate of decay of heterozygosity in the line will be equal to the rate of increase of variance between lines and the rate of increase of the inbreeding coefficient toward unity, and will be $[1/(4N)](1-C_{t-1}^*)/(1-C_t)$, where C_t is the coancestry (coefficient of kinship) of random pairs of individuals in the line, and C_{t-1}^* is the coancestry of those individuals actually mated in the previous generation. The value 4 comes from the assumption that all individuals have exactly two offspring. The result does not look useful at first, until we realize that $1-C$ and $1-C^*$ will approach a stable ratio to each other which does not then depend on t.

The immediate consequence is that the larger C_{t-1}^* is, the smaller is the rate of inbreeding of the whole line. If individuals who are as closely related as possible are mated, the overall rate of inbreeding is decreased! This counter-intuitive result does, of course, have an *a posteriori* rationale. With the number of offspring of all individuals equalized, all genetic drift must come from Mendelian segregation as there is none to come from the unequal contribution of offspring to the next generation. The variance due to Mendelian segregation is least when heterozygosity is least, hence Robertson's paradox.

Robertson notes that some of the loss of genetic variability can be entirely prevented by completely dividing a population into isolated inbred lines. The penalty is that individuals become inbred, although for different genes in different sublines. Systems such as circular half-sib mating are an attempt to slow down the rate of loss of genetic variability without permanently forming separate sublines. It will be interesting to see whether Robertson's quite general results will be of renewed interest in an era when there is increasing concern with loss of genetic variability in zoo populations and rare wild species.

Alan Robertson returned yet again to the theory of subdivided populations in 1975–6 in a series of papers which dealt with the use of distributions of gene frequencies among populations as a test of selective neutrality of loci. These papers do not arise directly from animal breeding considerations, but rather are a response to Lewontin and Krakauer's (1973) attempt to use the heterogeneity of variance between populations as a test for selective neutrality of loci, the argument being that genetic drift was expected to cause about the same divergence between different populations for all loci. Lewontin and Krakauer used calculations based on a model of islands equally exchanging migrants to produce a test that would reject neutrality if the heterogeneity among loci were too great. I noticed that, if the migration patterns were not so symmetrical, the number of degrees of freedom in their test would be dramatically affected; but Takeo Maruyama and I were dissuaded from publishing a note on this when it was pointed out that Lewontin and Krakauer had indeed mentioned this problem in their paper, although few readers had noticed it. Later Nei and Maruyama (1975) did publish a note demonstrating the problem by simulation. Robertson (1975a) published a summary of some theoretical calculations accompanying this showing that the problem was a serious, indeed a fatal one. The Lewontin–Krakauer test was then a dead letter, as its authors (Lewontin and Krakauer, 1975) essentially admitted. Subsequently Robertson (1975b) presented the details of his argument.

Genetic drift in small populations

Genetic drift in finite populations was the theme of the paper by Robertson (1961c) in which he analyzed the effect of selection on the rate of inbreeding, a problem motivated by his interests in animal breeding but applicable more generally in that selection induces variation in fitness. An expression in terms of selection intensity and heritability is derived that, for the most straightforward case, is

$$N/N_e = 1 + 2i^2h^2[1-h^2i(i-x)],$$

with h^2, i and x being the heritability, standardized selection intensity and standardized cut-off point for truncation selection. In practice, the effective number should be $2N_e$ because the above formula is established by equating the variance of gene frequencies to that with N full-sib pairs, whereas N usually represents the population size. Robertson also derived formulae for the case in which family sizes have been equalized, and for a combination of individual and family selection. He shows how his approach relates to earlier papers by Crow (1954) and Crow and Morton (1955). Subsequently Nei and Murata (1966) treated the case where fertility is inherited, whereas in Robertson's analysis the character that is the sole determinant of viability is inherited, a slightly different case.

Robertson (1965b) addressed the implications of observing departures from the Hardy–Weinberg equilibrium in animal populations. The small numbers of parents (particularly the small numbers of sires) in each population will lead to random differences in gene frequency between females and males, and thereby cause more heterozygotes to be found than expected. This is related to Levene's (1949) observation that the finite number of offspring scored leads to an inaccurate estimate of the gene frequency, which on average also leads to there being more heterozygotes than expected. Both are small effects, but Robertson's concern was that in animal breeding situations there may be many progeny coming from only a few sires, so that the effect of different gene frequencies in different sexes may be the more important one.

Another study in this area was on the estimation of inbreeding coefficients from departure from Hardy–Weinberg proportions in a population with multiple alleles (Robertson and Hill, 1984). This detailed treatment of the problem established that the departures of the homozygote classes from their expectation contained all the information, and that in estimating gene frequencies from a sample in which there is departure from Hardy–Weinberg proportions, a simple count of alleles is not the maximum likelihood estimate of gene frequencies. This paper should remain an essential reference for anyone dealing with estimation of inbreeding coefficients from genotype frequencies.

Natural selection in small populations

It was inevitable that, having been concerned with the effects of artificial selection on effective population size, Alan Robertson should become concerned with the effectiveness of selection in small populations. He used the technology of diffusion equations to arrive at a startling result about overdominance in small populations (Robertson, 1962a). Using an expression developed by Miller (1962) he examined whether the presence of overdominance in a finite population would retard fixation of the alleles, compared with their rate of fixation if neutral. It should be obvious that it will always retard fixation. The startling fact is that it does not. But like all startling facts, it too is obvious in retrospect. We may recall that when a set of replicate populations start in the same situation and drift toward fixation, the distribution of gene frequencies among the unfixed replicates becomes nearly uniform, the populations with most extreme gene frequencies being most likely to fix next. If instead we had overdominance with an equilibrium frequency near 0 or 1, it would act to pull populations out of the intermediate range, and so might even place them at greater risk of fixation.

For example, where s_1 and s_2 are the selection coefficients against the two homozygotes, if $N(s_1+s_2) = 16$ any equilibrium gene frequency value outside the range 0.2 to 0.8 would be a case in which the selection accelerates the net rate of fixation. If selection is stronger, the range is somewhat larger; the equilibrium toward which selection is pulling must be closer to 0 or 1 to allow the population, once it is pulled into the region of the equilibrium, to have a good chance of fixing the common allele by drift. Similar effects occur for the probability of fixation of rare overdominant mutants, which can in some cases actually be decreased by the presence of the overdominance, and for the mean gene frequency when there is symmetrical overdominant selection and nonsymmetric mutation, which can be more extreme for small populations than for large ones.

Hill and Robertson (1968b) used exact iteration of Markov chains, and a diffusion scaling argument which is discussed below, to examine the retention of genetic variability at a single locus by overdominance. In effect it extended the results of Robertson's (1962a) paper to consider not just the asymptotic rate of loss of variation, but the entire process, when a single overdominant locus with two alleles is acted on by genetic drift. As with the earlier paper, overdominance for alleles with intermediate equilibrium gene frequencies has the effect of retaining genetic variability longer than it otherwise would be found, but overdominance that pulls the gene frequencies toward extreme frequencies can speed the loss of genetic variability.

Later Robertson (1970d) made a simple but useful approximation addressing the reduction in fitness that would be produced by genetic drift acting on an overdominant locus. Using one of Sewall Wright's results on the variance maintained at loci where mutation is balanced by genetic drift, he argued that when selection is strong ($4Ns$ large) it acts as a kind of mutational pressure towards the equilibrium. This simple linearization argument shows that the variance of gene frequencies around the equilibrium will be proportional to $1/[4N(s_1+s_2)]$. The result is startlingly simple: for large enough values of $4Ns$, the reduction in fitness due to drift is simply $1/(4N)$, irrespective of the exact values of selection coefficients against the two homozygotes. He shows that for multiple (n) alleles the reduction is (n-1) times as great if all heterozygotes have equal frequency.

Ohta and Kimura (1970a) extended the argument to a number of different cases of frequency-dependent selection. Lande (1980b) has stated the result more generally, arguing that the population suffers a load of $1/(4N)$ per selective constraint, so that an attempt to maintain m phenotypes at optimum values would suffer a load of $m/(4N)$. He draws the startling implication that if m is greater than $4N$ the population will not be able to maintain these constraints – if the number of phenotypes maintained by natural selection is greater than the effective population size by a large enough factor, either the population or the phenotype will suffer. The result makes me wonder how long complex adaptations can be sustained in species that have a small effective population size. More work needs to be done on this.

Robertson and Narain (1971) made a detailed examination, using exact computations with transition probability matrices, of the time that a partially dominant or overdominant lethal would remain in a finite population, and the dependence of lethal allelism on time. In this problem one cannot use the simulation scaling argument, which uses the fact that the result depends on the product $N_e s$ of the effective population size and the selection coefficient. The selection coefficient against a lethal homozygote is 1 so its product with the effective population size will not be constant as N_e varies. In the calculations in this paper, values above $N=50$ could not be considered because of the size of the matrices involved. This paper will be essential to any consideration of the dynamics and statics of lethal allelism, although observations of lethal allelism will probably be quickly superseded by studies using DNA sequencing to detect relationships among lethals and deleterious mutants.

A similar study with a greater direct application in animal breeding is Robertson's (1978) analysis of the time to detection of a recessive mutant in a finite population. This used both exact matrix methods (computers had improved and some of the matrices were as large as 1000×1000) and simulations, to find out how long it would take to detect a visible recessive mutant in a finite breeding population. The matrix methods enabled the probability that the mutant would ever be detected and the distribution of times-to-detection to be computed,. The simulations allowed the effect of prohibiting selfing to be checked. The results showed that the mean time to detection was about $2N^{1/3}$ over the range of N examined, although many mutants would never be detected.

Normalizing selection and selection for an optimum

In the 1950s, the major alternative to maintenance of variability by overdominance was considered to be maintenance by normalizing or optimum selection. The paper on the effect of selection against extreme deviants (Robertson, 1956b) addresses problems suggested by the experimental quantitative genetic work which he did with Douglas Falconer (Falconer and Robertson, 1956). Robertson considers two patterns of selection. In one, the survivors are those near the population mean, in the other those at the two extremes of the population are retained and negatively assortatively mated with each other. Using approximations which are accurate when each gene contributes a small fraction of the phenotypic variance, and treating the dynamics of individual genes separately without a consideration of linkage disequilibria, he showed that selecting from near the population mean resulted in fixing individual loci with the loci of largest effect changing most rapidly. He approximated the whole process as multiple stages of selection, each of which would be completed before the next began. The ultimate effect is that all loci fix, but that the population mean remains unchanged. With mating of extremes there is the opposite effect – maintenance of variation at the individual loci.

Robertson (1956b) points out that the result can be extended to the case in which fitness is a declining function of the deviation of the phenotype from an optimum. The results are qualitatively the same. He notes that Sewall Wright (1935) obtained similar results for the case of selection for an intermediate optimum. In fact, Wright's results are more precise, in that he mentioned that it was possible that one locus remained polymorphic at the end of the process. This line of work has continued in the papers of Lewontin (1964), Singh and Lewontin (1966), Bulmer (1971) and Barton (1986). The picture painted by Wright and by Robertson has been filled in considerably, but the patterns they saw are still broadly valid, in spite of the approximate treatment.

Robertson (1956b) considers one case that none of these others do. It is inspired by Lerner's (1954) hypothesis of 'genetic homeostasis', according to which the loci affecting a quantitative trait were not selected based on their effect on the trait but because of a separate fitness effect, with the individual loci being overdominant. Robertson showed that if this were the case, there would still be a quadratic relationship between individual phenotype and average fitness, and a quadratric decline in mean population fitness as one changed the mean population phenotype. Thus selection that was independent of the phenotype might appear to be dependent on it.

Robertson has remained a sceptic of the validity of the intermediate optimum model which has enjoyed renewed popularity as quantitative genetics has become popular among ecologists. Just as selection for an intermediate optimum has seemed to most of us a pattern that must be frequent in nature, Lerner's genetic homeostasis has seemed implausible, requiring as it does two separate effects of each locus, and with different dominance patterns at that. Robertson's paper (1968a) in the symposium at Syracuse University is a firmly-stated counterbalance to these views, with unpublished experimental work cited as empirical support for his position, as are his papers in the Cold Spring Harbor volume on quantitative genetics (1955b) and in the Proceedings of the 11th International Congress of Genetics (1965c). I wish those papers were better known to today's evolutionary ecologists.

Experimental population genetics

That Alan's interest in evolution was becoming less strictly dependent on problems in animal breeding is shown by two experimental papers with George Knight (Knight *et al.*, 1956; Knight and Robertson, 1957), one an attempt to induce sexual isolation within *Drosophila melanogaster* by discarding offspring of mating between the two lines, the other using markers to measure the overall fitness of lines. Each was among the earliest uses of these techniques.

In the 1970s he returned to experimental population genetics, in part by supervising students, using perturbations of the environment to look for measurable fitness differences between genotypes at electrophoretically polymorphic loci. Briscoe *et al.* (1975) found that when ethanol was added to the medium not only were there measurable fitness differences between different electrophoretic genotypes at the Alcohol Dehydrogenase (ADH) locus of *D. melanogaster*, but a pattern of dominance was found, the Slow allele being deleterious and recessive to the Fast allele. McKay (1981), a student of Alan Robertson's, undertook a massive survey of second chromosomes from 247 different strains of *D. melanogaster*, isolating them in a standard genetic background and examining their ADH activity and thermostability. Almost all of the variation found could be attributed to the two known electrophoretically detectable alleles. This study seems consistent with Kreitman's (1983) survey of variation at the ADH locus at the DNA level, in which the only variations in the amino acid sequence found were the known ones corresponding to those Fast and Slow alleles.

Frequency-dependent selection has often been invoked as the pattern responsible for maintenance of polymorphism at loci such as ADH. One of the postulated mechanisms for frequency-dependent selection is differential use of limiting resources by different genotypes. If larvae of different strains of drosophila were depleting different resources then having other larvae previously present would always be expected to be deleterious, especially so if they were of the same strain. Dolan and Robertson (1975), however, did not find this expected effect: conditioning by larvae of the same strain was no more deleterious than conditioning by different strains. In fact, conditioning often seemed to actually improve survival of the strain!

Linked loci in finite populations

In two papers with W. G. Hill, Robertson extended his concern with genetic drift effects in animal breeding to consideration of the interaction of genetic drift, selection and linkage (Hill and Robertson, 1966, 1968a). These papers are essential to the understanding of the interaction of these three forces, although they have too often been overlooked.

Hill and Robertson (1968a) deal with the amount of linkage disequilibrium produced by genetic drift in a small population. Considering simple genetic drift at two loci each with two alleles, with gene frequencies respectively p and q, and linkage disequilibrium value D, they derived recursions for the three moments, $E[D^2]$, $E[D(1-2p)(1-2q)]$, and $E[p(1-p)q(1-q)]$, which are related in successive generations by three linear equations whose coefficients depend on the population size N and the recombination fraction c. They were able to find exact expressions for these moments for the case of no recombination. For arbitrary recombination fraction c they iterated the moment equations numerically for the case of $N=16$, then used the principle that populations are well-approximated by diffusion processes whose behaviour as a function of the rescaled time t/N depends on N and c only through their product Nc. Thus a population of size 16 with $c=0.1$ would show about the same expected amount of squared disequilibrium after 16 generations as a population of 1600 individuals would after 1600 generations if the recombination fraction were 0.001. This scaling argument is very powerful, since the approximation of these populations by diffusion processes is very good, and numerical checks verified that it worked. Indeed Alan Robertson's work has frequently involved use of recursion of moments and of this scaling argument applied to simulations or numerical computations.

More simulations using this scaling argument showed that the correlation measure of linkage disequilibrium, $r^2 = D/[p(1-p)q(1-q)]^{1/2}$, had an expectation among lines segregating for both loci that approached an asymptotic value depending on Nc. In fact, although not noted in this paper, it was not far from $1/(1+4Nc)$. (Hill and Robertson did note that it approached $1/(4Nc)$ for large values of Nc.) All results showed that recombination would cut the amount of squared linkage disequilibrium in half when $4Nc$ was in the vicinity of unity. Sved (1971) gave $1/(1+4Nc)$ as an approximation for the squared correlation at two loci a few years later (see also Sved and Feldman, 1973). Ohta and Kimura (1969) gave another derivation of the approximation $1/(4Nc)$, as well as a more detailed approximation, and Weir and Cockerham (1974) compare the various approximations with an exact calculation. Hill and Robertson's paper concludes with a demonstration by simulation that the results are not much affected by overdominance at the two loci.

Much of the impetus for this work came from the observation in computer simulation studies that linkage of genes seemed to retard response to truncation selection. I had suggested (Felsenstein, 1965) that, since truncation selection generates a small amount of departure from multiplicative fitnesses at different loci, it would on average generate repulsion linkage disequilibrium which would interfere with response to selection. This deterministic force seemed an adequate explanation of the simulation results. Hill and Robertson were able to show that there was a stochastic force which was the likely culprit, one much larger than the deterministic effect.

This is developed at length by Hill and Robertson (1966), the importance of whose results is matched only by the difficulty other population geneticists have had understanding and assimilating them. It is evident from their 1968 paper, and the work of others as well, that the effect of genetic drift is to create random linkage disequilibria, sometimes coupling and sometimes repulsion equilibria. My own work (Felsenstein, 1965) showed that these equilibria sometimes speed and sometimes retard the selection of linked loci. One would expect that if possession of a favourable allele tended to be correlated with the possession of others as well, that the apparent selection coefficient acting to fix the allele would be increased, and that is what is found. If there is repulsion disequilibriim so that the possession of a favourable allele at a locus tends to be associated with possession of unfavourable alleles at a second locus, that would reduce the apparent selection coefficient and the probability of fixation. One naturally expects that, if the average disequilibrium is zero, that on average these effects will cancel out and there will be no net effect of the selection at the other locus.

Strangely, this is not what is found. There are nonlinear terms whose net effect is negative. The presence of selection at a nearby locus on average reduces the effect of directional selection at a given locus. The effect is due to the negative disequilibria having on average a greater effect than the positive ones. The 1966 paper devotes much effort to understanding and approximating the effect, which I like to call the 'Hill–Robertson effect'. The effect is large only when selection coefficients $4Ns$ are large at both loci, and the recombination fraction c is less than $1/(4N)$, so that substantial disequilibrium is expected to be generated by genetic drift. Much use is made in this paper of the diffusion scaling argument to allow generalization from a limited number of computer simulations. I have argued (Felsenstein, 1974) that it is the Hill–Robertson effect that underlies Fisher's and Muller's explanations for the evolutionary advantage of recombination, as well as 'Muller's ratchet'. In effect the reason for the Hill-Robertson effect is 'hitch-hiking'. Alleles at nearby loci, as they are selected, randomly perturb the gene frequencies at the locus. This has the effect of increasing the random noise at the locus, and hence decreasing the effect of selection. Hill and Robertson used their simulations to make the practical recommendation that the greatest long-term selection response would be achieved if one selected a bit more than the 50% of parents which was the main recommendation of Robertson's 1960b paper on selection limits. I hope that some day the Hill–Robertson effect will be accorded its proper place in population genetics.

A useful review of Alan Robertson's contributions to the theory of linked loci in finite populations is given by Robertson and Hill (1983b).

Molecular evolution

Robertson and Hill's (1983b) paper is notable for a short section at the end which draws an important conclusion from calculations on random linkage disequilibria. They note that in examining closely linked restriction site polymorphisms, substantial linkage disequilibrium is found over distances of as much as 7.1 kb, and that the theory of disequilibrium generated by genetic drift suggests that it will be substantial only if $c < 1/(4N)$. With a human effective population size assumed to have been about 10^5 over the relevant time period, a value consistent with the degree of restriction site polymorphism under the assumption of selective neutrality with a mutation rate of 5×10^{-9}, we should expect disequilibrium when $c < 2.5 \times 10^{-6}$. The human genome is about 3×10^9 bases long and its map is about 30 Morgans, so that if recombination were uniformly distributed over the DNA we would expect the critical distance for disequilibrium to be 120 bases. 'This is well below the indicated value from the data though that would also suggest that crossing over is not homogeneous along the DNA sequence' (Robertson and Hill, 1983b, p. 263). This is one of the few times I have seen a strong molecular conclusion arrived at from population-genetic arguments.

Another interesting molecular conclusion is given by Robertson and Hill (1983a), in response to work on thalassemia genes. They point out that any class of homogeneous mutant alleles, such as lethals at the same locus, will to a good approximation behave as if neutral towards each other, and the probability that two of them are descendants of the same identical mutation can then be calculated by using neutral mutation theory. The results show that for lethal genes like thalassaemia, there should be alleles originating from many different mutational events segregating in the population.

More recently, Alan Robertson has supervised a research student, Frank Wright, on investigations of codon useage patterns in proteins (Wright, 1988).

Other evolutionary topics

Alan Robertson's excursions into population genetics at times carried him quite far from animal breeding. An extreme example is the paper by Rosado and Robertson (1966), who derived approximate expressions for the change of gene frequency of alleles affecting the sex ratio. They showed that selection for alleles modifying the sex ratio before the end of parental care would be selected in the way predicted by Fisher's (1930) theory of the evolution of the sex ratio. Selection on alleles acting on viability of the sexes after the end of parental care would act so as to maximize the product of female and male mean fitnesses.

Overview

Alan Robertson's population genetic studies have thus carried him from animal breeding to molecular genetics. It is a diverse body of work and it is appropriate to try to identify some themes. In terms of technical methods, his work has shown the fruitfulness of relatively straightforward approximations such as iterations of systems of moments, a technique used more intensively by Hill. The diffusion scaling argument also shows up repeatedly, as a means of generalizing from a limited base of computer simulations or exact matrix iterations.

In terms of subject matter, it is notable how much of his work is concerned with effects of random genetic drift, with the joint effects of drift and linkage and with drift, linkage, and selection. Although it was Sewall Wright, partly through his influence on Jay L. Lush, who first treated the effects of genetic drift in breeding, this had not penetrated deeply by the mid-1950s. Inbreeding coefficients were widely used, and inbreeding depression avoided, but there had been no serious attempt to consider the interaction of drift with selection. Alan Robertson's role has been to fully integrate genetic drift into quantitative genetics and breeding, using the diffusion theory developed by Kimura and others. That this would lead to an interest in evolution and to contributions to that field was inevitable.

In terms of his style of thought and communication, one cannot read any number of his papers without immediately being struck by the conciseness and clarity of his thought and the elegant simplicity with which it is expressed. It has been a real pleasure to reread these papers, for their clarity and conciseness remind me strongly of the many hours of conversation with him in Edinburgh, particularly during the morning coffee hour in his office. I was a postdoctoral fellow at the Institute of Animal Genetics during 1967–68. Coming from a background in population genetic theory and knowing little about quantitative genetics I found myself in the best place in the world to learn about it. There were interesting students and visitors from all over the world, and one could have coffee with the Edinburgh group each morning. Those conversations about animal breeding, quantitative genetics, and evolution were a wonderful way to learn. I have not encountered an atmosphere quite like that elsewhere.

It is therefore with a sense, not only of gratitude for that experience, but of nostalgia for a paradise lost, that I must acknowledge my own indebtedness to Alan Robertson.

Acknowledgements

I wish to thank the Editors of this volume for useful pointers to papers I might have otherwise overlooked. Work on this paper was supported by National Science Foundation grant number BSR-8614807.

Chapter 2
Neutral Theory

Motoo KIMURA

National Institute of Genetics, Mishima, 411 Japan

In contrast to the Darwinian theory of evolution by natural selection, the neutral theory (Kimura, 1968) claims that the great majority of evolutionary changes at the molecular level are caused not by natural selection acting on advantageous mutants, but by random fixation (due to sampling drift) of selectively neutral or very nearly neutral mutants under continued mutation pressure. In other words, the neutral theory emphasizes the predominant role that mutation and random genetic drift play in evolutionary changes at the molecular level.

The theory does not deny the role of natural selection in determining the course of adaptive evolution, but it assumes that only a minute fraction of DNA (or RNA) changes are adaptive in nature. The neutral theory also asserts that most of the intraspecific variability at the molecular level (including protein and DNA polymorphism) is essentially neutral, so that the majority of polymorphic alleles are maintained in the species by the balance between mutational input and random extinction. In other words, it regards protein and DNA polymorphisms as a transient phase of molecular evolution and rejects the notion that the majority of such polymorphisms are adaptive and maintained by some form of balancing selection.

As a scientific hypothesis, the neutral theory has the advantage that its underlying assumptions are sufficiently simple that its population-genetical consequences can be worked out using suitable mathematical methods, particularly the diffusion equation method, or the diffusion model as it is often called (Kimura, 1964). For details, see Kimura (1983).

Some simple formulae

Let us consider an evolutionary process in which mutant genes are substituted sequentially within a species. Each such substitution is made up of a sequence of events in which a rare mutant form, usually represented in single copy, in the population, finally spreads through the whole population. If such fixations are caused by random drift acting on selectively neutral (i.e. selectively equivalent) mutations, we have the following formula for the rate of evolution per generation:

$$k_g = v_T f_0, \qquad (2.1)$$

where v_T is the total mutation rate and f_0 is the fraction of neutral mutations. Under the neutral theory the evolutionary rate is therefore equal to the mutation rate to neutral alleles. Note that k_g represents the rate per generation at which molecular mutants are substituted one after another within the lineage in the course of evolution. Each of these events takes a long time, on average four times the effective population size (Kimura and Ohta, 1969). Advantageous mutations may occur, but the neutral theory assumes that they are so rare that they may be neglected from consideration. Thus $(1 - f_0)$ represents the fraction of definitely deleterious mutants which are eliminated from the population without contributing either to evolution or polymorphism. What is remarkable in the above formulation is that the rate of evolution is independent of population size and, assuming neutrality, of environmental conditions. When we estimate the actual rates of evolution through comparative studies of protein or DNA sequences, we usually express the evolutionary rate by taking one year as the unit length of time, while mutation rate is usually measured per generation. The above formula (2.1) may therefore be modified to give the evolutionary rate per year:

$$k_1 = (v_T/g)f_0, \qquad (2.2)$$

where g is the generation span, and v_T/g is thus the total mutation rate per year.

Let us now consider intraspecific variability and assume the infinite allele model (Kimura and Crow, 1964) in which whenever mutation occurs at a locus, it leads to a new, nonpre-existing allele. At equilibrium in which mutational input and random extinction of neutral alleles balance each other, we have the following formula for the average heterozygosity per locus:

$$\bar{H}_e = 4N_e v_0/(4N_e v_0 + 1) \qquad (2.3)$$

where N_e is the effective population size and v_0 is the mutation rate for selectively neutral alleles, so that $v_0 = v_T f_0$.

Features of molecular evolution and the neutral theory

Ample evidence has now accumulated to show that molecular evolution is distinguished from phenotypic evolution by two remarkable features. They are:

(i)   constancy of the rate of molecular evolution: for each protein or gene region, the rate of amino acid or nucleotide substitutions is approximately constant per site per year (known by the term 'molecular evolutionary clock'); and

(ii) the 'conservative nature' of changes: functionally less important molecules or portions of molecules evolve faster than more important ones.

The first feature, the constancy of the rate, may be explained by the neutral theory by assuming that v_T/g remains the same (constant) among diverse lineages and over time for a given protein or gene, for which f_0 is assumed to be constant. Therefore, if the neutral theory is valid, the rate of production of neutral mutations per year for a given gene must be nearly constant among diverse organisms whose generation spans are very different. Note that mutation here refers to changes that led to DNA base replacements, but not to lethal and 'visible' changes. These latter types of mutation, whose incidence has been known to be generation dependent, are suspected in *Drosophila melanogaster* to be largely caused or controlled by various families of movable genetic elements (Finnegan and Fawcett, 1986). Transposons and insertion sequences are common components of eukaryotic and prokaryotic genomes and may prove to be generally responsible for these classes of mutation. No definite data are available at present, however, to settle the issue of whether the mutation rate for nucleotide substitutions is proportional to year or to generation. Experimental studies on this subject are much needed.

From the standpoint of the neutral theory, a universally valid and exact molecular evolutionary clock would exist only if, for a given molecule, the mutation rate for neutral alleles *per year* (v_0/g) were exactly equal among all organisms at all times. Any deviation from the equality of neutral mutation rate per year makes the molecular clock less exact. Such deviations may be due to two causes: one is the change of the mutation rate per year (such as due to changing generation span), and the other is the alteration of the selective constraint on each molecule (due to change in the internal molecular environment). I must emphasize that departure from exact clockwise progression of molecular evolution by no means invalidates the neutral theory. For more detailed recent discussion on the molecular evolutionary clock and the neutral theory, see Kimura (1987).

The conservative nature of molecular evolution has now been well established. Those mutant substitutions that cause less drastic changes in the existing structure and function of a molecule occur more frequently than those which cause more drastic ones. This is easy to understand from the neutral theory, because the less drastic or more conservative the mutational change, the more likely it is to turn out to be selectively neutral. This means that for more conservative changes the values of f_0 are larger.

More than a decade ago, we enumerated five principles which govern molecular evolution, including constancy in rate and conservatism in mode (Kimura and Ohta, 1974). One of the principles states that functionally less important molecules or parts of a molecule evolve (in terms of mutant substitutions) faster than more important ones. When this was proposed, accompanied by its neutralist explanation, much opposition was voiced by the Neo-Darwinian establishment, but I am glad to note that it has become a part of common knowledge among molecular biologists. It is now routine practice for them to search for various signals by comparing a relevant region of homologous DNA sequences of diverse organisms and to pick out a constant or 'concensus' pattern, but disregard variable parts as unimportant.

Evolution of DNA sequences

The recent explosion of DNA sequence data has brought much new information that lends strong support to the neutral theory. It has now been well established that synonymous base substitutions within codons, which do not cause amino acid changes, occur at a much higher rate in evolution than amino-acid altering substitutions. It has also been found that base substitutions in 'introns' occur at an evolutionary rate comparable with synonymous substitutions or even higher. Considering the fact that natural selection acts on the phenotype of the organism, these observations which show a preponderance of synonymous and other silent substitutions suggest that *molecular changes that are less likely to be subject to natural selection occur more rapidly in evolution*. This can readily be explained by the neutral theory, because such molecular changes must have a higher chance of being selectively neutral (i.e. larger f_0) and therefore neutral evolution occurs at higher rate.

I once predicted (Kimura, 1977) that, based on the neutral theory, the maximum evolutionary rate is set by the mutation rate ($k \leqslant v_T$), and that the maximum rate is attained when all the mutations are selectively neutral ($f_0 = 1$). A dramatic example vindicating this prediction was the discovery of very high evolutionary rates of pseudogenes (or 'dead' genes) which have lost their function. This was first shown clearly by Miyata and Yasunaga (1981) who made a careful analysis of the evolutionary rate of a pseudo alpha-globin gene in the mouse. This was followed by a more elegant statistical analysis of the evolutionary rates of pseudogenes by Li et al. (1981). What is really interesting is that the rates of substitution revealed by these studies are equally high in all three codon positions. The estimated rate in globin pseudogenes is about $k = 5 \times 10^{-9}$ substitutions per nucleotide site per year in mammals. Moreover, pseudogenes accumulate deletions and additions at very high rates, suggesting that they have been liberated from the constraint of negative selection and that they are on the path to disintegration by accumulating various mutational changes.

Intraspecific variability

Space does not permit an extensive discussion on this subject (for details, see Kimura, 1983), but I would like to mention one type of observation which is very important from the standpoint of the neutral theory. As is evident from a comparison of equations (2.2) and (2.3), it is expected that genes (proteins) or portions of DNA that evolve more rapidly must show higher intraspecific variability.

The recent observations on DNA polymorphism in natural populations of fruit flies are consistent with this expectation: on the whole synonymous and other silent sites are much more polymorphic than amino acid altering sites, although more systematic studies are required based on more extensive data than currently available. In this respect, a recent study by Ward and Skibinski (1985) is reassuring. Using electrophoretic data for 42 proteins from over 200 invertebrate and 300 vertebrate species, they found that there is a very strong correlation between protein genetic distance and protein heterozygosity among different proteins, and that the observed relationship between the two can be explained quantitatively by the neutral theory.

Some recent developments

The predominant role played by mutation pressure in molecular evolution (quite in line with the neutral theory) has become increasingly evident from recent studies. One of the most remarkable illustrations is the very rapid evolutionary change observed in RNA viruses which are known to have very high mutation rates, roughly a million times higher per year than DNA organisms. Such high rates of evolution have been revealed by the studies of Hayashida et al. (1985) and Saitou and Nei (1986) on influenza virus genes. Gojobori and Yokoyama (1985) have also compared the evolutionary rates of the retroviral oncogene of the Moloney murine sarcoma virus and its cellular homologues, and found that the former evolves nearly 0.8 million times faster than the latter.

The concept of mutation-driven neutral evolution is also useful in understanding the evolution of the deviant coding system recently discovered by Osawa's group (Muto et al., 1985; Yamao et al., 1985) in Mycoplasma capricolum. In this bacterial species, UGA, a stop codon in the standard code table, codes for tryptophan instead of UGG, the standard codon for tryptophan. This organism has the characteristic of having a very high A + T content (75%) in its genome. Such a high A + T content must have been brought about by A/T directed mutation pressure (A/T pressure), that is by a higher mutation rate from G/C to A/T than the reverse. Presumably, this was caused by modification in the DNA polymerase system. It was pointed out by Jukes (1985) that A/T pressure can lead to the replacement of UGA (as a stop codon) by UAA (another stop codon), followed by a change in the anticodon of one of the duplicated copies of the tRNA gene for tryptophan from CCA to UCA, so that UGA (as a new codon for tryptophan) can replace UGG. Thus this 'capture of a stop codon' would have been brought about by a series of changes, none of which were deleterious and were probably neutral.

I conclude that the most prevalent evolutionary changes that have occurred at the molecular level since the origin of life on Earth are those that have been caused by random genetic drift rather than by positive Darwinian selection.

Acknowledgements

This work was supported in part by a grant-in-aid from the Japanese Ministry of Education, Science and Culture. This is contribution number 1759 from the National Institute of Genetics, Japan.

Chapter 3
Inferences From or About Molecular Evolution

C.H. LANGLEY

NIEHS, PO Box 12233, Research Triangle Park, North Carolina, 27709, USA

The interactions between evolutionary biology and molecular biology (especially molecular genetics) fall into two distinct categories. The first, *DNA palaeontology*, deals descriptively with taxonomically distinct organisms on a geological time scale, and the second, *molecular population genetics*, focuses on the genetic and ecological mechanisms that generate the diversity within and between populations.

In the seminal work in DNA palaeontology, Zuckerkandl and Pauling (1965) observed the positive correlation between amino acid sequence differences among haemoglobins in different species and the estimated times of their most recent common ancestors. This approximate proportionality of divergence of the protein sequences (and, more recently, the DNA sequences) with the time since a common ancestor became known as the *molecular clock*. It remains the singularly outstanding contribution of molecular evolution to general biology.

The accumulated protein sequence data prior to the introduction of recombinant DNA technology have supported several other general conclusions about molecular evolution in addition to the molecular clock. The rate of amino acid substitution has been roughly constant and characteristic of the individual genes (Kimura, 1968; Ohta and Kimura, 1971a). Thus haemoglobins have evolved faster than cytochrome c, and fibrinopeptides have accumulated amino-acid substitutions faster than haemoglobins. These characteristic differences are thought to reflect different levels of functional constraint: the more demanding the characteristic functional requirements of a particular protein, the fewer amino acid substitutions that can be accepted. This view has also been applied to differences in substitution rates among the amino acid sites in given proteins: there have been more substitutions at sites on the outside of proteins than in the inner cores or active sites (Perutz and Lehman, 1968; Dickerson, 1971). Furthermore, many substitutions have been chemically conservative, in that the alternative amino acids have similar charge and size (Miyata *et al.*, 1979).

Recombinant DNA technology has produced a burgeoning amount of data to which the principles of DNA palaeontology can be applied. Perhaps the most common use of DNA palaeontology is as a tool in molecular biology, for those sequences with the slower rates of divergence are often suspected to have greater functional significance. Many exciting discoveries have come from following such leads.

Another area where substantial contributions have been made in DNA palaeontology is the reconstruction and quantification of phylogenies. The use of highly conserved ribosomal RNA sequences to build phylogenetic trees that reflect the historical relationships of kingdoms, phyla and classes is a clear contribution (Kimura and Ohta, 1973; Fox *et al.*, 1977; Hori and Osawa, 1979, 1987). Quantitative application of DNA palaeontology to recent evolution has contributed to significant adjustments in the accepted times of occurrences of the most recent common ancestors of various hominids (Sarich and Wilson, 1966). Thus the tools of DNA palaeontology are of great utility in drawing inferences *from* molecular evolution about different regions of the genome and about qualitative and quantitative phylogenetic relationships.

The model underlying DNA palaeontology is formulated in taxonomic terms. The fundamental event is referred to sometimes as a mutation and at other times as a substitution. This ambiguity in terminology reflects the shallow nature of the concepts underlying this empirical branch of molecular evolution, for mutations occur in genes in individuals, substitutions occur in populations. DNA palaeontology fails to distinguish between genetic mechanisms and population genetic mechanisms and this confusion limits the kinds of inferences that can be drawn. For example, 'rapid divergence' may indicate a less selective constraint or a higher mutation rate, or both. The 'last common ancestor' is assumed to have occurred at the same time for each nucleotide position in a sequence; it is assumed there is no polymorphism or recombination at the time of speciation (Gillespie and Langley, 1979; Hudson, 1983).

The second category of interaction between evolutionary biology and molecular biology is *molecular population genetics*, which is mechanistic and essentially Darwinian, in that it fosters the reductionistic view of evolution as a population phenomenon. Molecular population genetics is based on extensions of the classical population-genetic and demographic models of how new mutations are ultimately lost or fixed (substituted) in the population. The particular population-genetic model that has so far provided an intellectual foundation for both molecular population genetics and DNA palaeontology is the neutral theory of molecular evolution (Kimura, 1983). The conceptual elegance and mathematical tractability of this theory have made it a favourite with most theoretical population geneticists and DNA palaeontologists. It is blessed (or cursed) with the two hallmarks of a 'successful' population-genetic theory. As Lewontin (1974) has pointed out, the only two parameters of the theory (effective population size, N_e (a large unknown number), and the mutation rate to selectively neutral mutants, u (a small unknown number)) are multiplied to give the critical parameter, $_{,,} = 4N_e u$, a very unknown number. The second 'blessing' is that the theory's robust predictions comfortably fit most of the observed data from both the DNA palaeontological and molecular population-genetic perspectives (Kimura, 1983). Indeed, while DNA palaeontology flourished on the meagre protein sequence data of the 1960s and 1970s, population geneticists were attempting to assess the level of protein sequence variation in natural populations utilizing electrophoretic techniques (Lewontin and Hubby, 1966; Harris, 1966) and what little sequence data was available (Robertson, 1968a). Despite much experimental work and theoretical analysis, this approach failed to test the neutral theory to the satisfaction of some (Lewontin, 1974; Lewontin and Krakauer, 1973; Robertson, 1975a). Nevertheless, Kimura has argued persuasively for his view that most molecular evolution does not involve the forces of natural selection, but his contention that this theory has been tested adequately (let alone, severely) is less convincing. In what follows I shall draw attention to several recent molecular evolutionary studies that truly challenge the neutral theory and afford the opportunity to draw inferences *about* molecular evolution, i.e. about the forces and mechanisms determining evolution at the DNA sequence level.

The neutral theory has always needed a few good alternatives

Perhaps the weakest point of the neutral theory has been the failure of population geneticists to formulate simple and testable alternative hypotheses. Ohta and Kimura (1971b) noticed that the rates of molecular evolution of several proteins during the mammalian radiation were not as constant as expected; but Kimura (1983) argued that the deviations were small and no threat to the theory. Using the abundance of recently generated DNA sequences, Gillespie (1986a) re-examined the question of whether the ratio of the variance to the mean of the observed rate of molecular evolution is larger than expected under the neutral theory. He confirmed that the ratio of the variance to the mean is consistently larger than the expected value of one, and in some instances is as large as 35. Gillespie

(1984a) also pointed out that for typical sample sizes the ratio is rather insensitive to large deviations from the neutral theory.

Gillespie (1978) reasoned that the expected values of this ratio under reasonable models of natural selection may be quantitatively indistinguishable from that under selective neutrality. Recently Gillespie (1984b, 1986b) proposed his 'molecular landscape' model under which molecular evolution occurs in episodes, with periods of no substitution followed by rapid, multiple substitutions associated with an environmental change. This model's episodic behaviour accounts for the observed increase in the ratio of the variance to the mean of the rates of molecular evolution and suggests that all substitutions may be selected. The real lesson of Gillespie's analysis is that the excesses observed in the ratio of the variance to the mean in the rate of substitution can signify profound (as opposed to minor) mechanistic departures from selective neutrality. Although the molecular landscape hypothesis (as it has so far been formulated) accounts for observed molecular variation between species, it predicts no variation within populations, a prediction contrary to reported observations.

Drosophila Adh polymorphism is the source of much good theory

The rate of accumulation of data on DNA sequence variation is slower within populations than among species. Two basic approaches have been used to study population genetic variation at the DNA level: restriction mapping and DNA sequencing. Surveys of restriction map variation have yielded information about both nucleotide and insertion/deletion polymorphisms over large segments of the nuclear (e.g. Jeffreys, 1979; Kazazian et al., 1983; Aquadro et al., 1986; Kreitman and Aguade, 1986b) and mitochondrial (e.g. Shah and Langley, 1979; Brown et al., 1979; Avise et al., 1979) genomes of a wide variety of organisms (see also Robertson and Hill, 1983b). There have been several studies of intraspecific variation using DNA sequencing (Aquadro and Greenberg, 1983; Kreitman, 1983) which showed that, as for interspecific comparisons, synonymous polymorphisms are more frequent. Noncoding introns and flanking regions appear to be more variable than coding regions, as expected from the neutral theory. Although it is too early to know how useful these surveys of polymorphism will turn out to be in the study of evolution, several interesting results are already clear. The distribution of repetitive dispersed DNA sequences varies in interesting ways in different parts of the genome and in different species (Langley et al., 1988 and references therein). The overall level of nucleotide sequence polymorphism and linkage disequilibrium in different regions of the drosophila genome and in different drosophila species appear to be qualitatively distinct (Miyashita and Langley, 1988; Aquadro et al., 1988). It is possible that molecular population-genetic studies at the DNA level will be sufficient to discern the primary forces shaping the polymorphism and determining substitution rates in such outbreeding species with large effective population sizes (such as drosophila).

Several recent studies based on DNA sequence variation at the *Adh* locus in drosophila demonstrate the power of such data to stimulate new theoretical approaches that challenge the neutral theory and even support alternative hypotheses (Hudson et al., 1987; Kaplan et al., 1988; Hudson and Kaplan, 1988). Implicit in Darwin's explanation of evolution by natural selection is the view that systematic diversity must arise as the consequence of the genetic and ecological mechanisms operating within populations. As pointed out above, the ability of the neutral theory to mechanistically unify molecular evolution between species and molecular polymorphism within populations has been its strength. Before the study of Kreitman and Aguade (1986b) there had been no data appropriate for a quantitative comparison in the context of the neutral theory of the amounts of variation found within and between species. Kreitman and Aguade (1986b) assayed the levels of polymorphism in the *Adh* transcription unit and in a flanking noncoding region in *Drosophila melanogaster* and found an excess of polymorphism in the coding region (especially exon three) relative to that in the flanking region when they used the relative interspecific divergences as a standard.

Hudson *et al.* (1987) rigorously and conservatively tested these divergence and polymorphism levels against predictions of the neutral theory. Figure 3.1 shows the predicted and observed numbers of segregating sites within populations of *D.melanogaster* and the numbers of site differences between *D.melanogaster* and *D.sechellia* in two regions of DNA, the *Adh* locus and its 5′ flanking region. The predicted values are based on standard neutral theory (Watterson, 1975; Gillespie and Langley, 1979) assuming that :

(i)   each region has a constant mutation rate to selectively neutral mutations (u_1 and u_2);
(ii) there is constant effective population size (N_e) for *D.melanogaster*;
(iii) there is no crossing over within each region and free recombination between them (most conservative parameters); and
(iv) *D.sechellia* and *D.melanogaster* have a common ancestral population (at *t* generations in the past with population size = N_e).

They concluded that the observed data are inconsistent with the predictions of the neutral theory (see Figure 3.1). Because this test has only one degree of freedom, it is not possible to identify which of the observed values is out of line: either the system was not at equilibrium or the fundamental model failed to include an essential mechanism such as balancing selection leading to an excess of variation in the third exon of *Adh* of *D.melanogaster*.

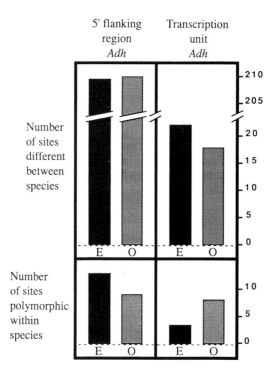

Figure 3.1. Observed numbers (*O*) of sites polymorphic within the *D. melanogaster* population or different between *D. melanogaster* and *D. sechellia* and expected numbers (*E*) obtained from an extension of the infinite sites model in Hudson *et al.* (1987). The statistical significance of the observed values was determined by computing a statistic similar to chi-squared in which $(O-E)^2$ was divided by its theoretical variance, and the critical values were determined by Monte Carlo simulation. The probability is 0.016 of observing a chi-square larger than the observed value of 6.07.

To answer the question of whether the apparent excess of molecular variation in the third exon of *Adh* can be explained by some form of balancing selection acting on the allozyme polymorphism requires the theoretical tools developed by Kaplan *et al.* (1988) and Hudson and Kaplan (1988). They presented a method of computing the means and variances of the numbers of segregating, selectively neutral sites surrounding a selected site (with arbitrary crossing over) and introduced the *sliding window* analysis that measures the level of polymorphism *along* the DNA. As Figure 3.2 shows, the polymorphism in the third exon of *Adh* stands out as grossly in excess of what is predicted by the neutral theory, while a simple model of balancing selection can easily account for the observed distribution. These authors also investigated similar quantities when there is selection-mutation balance for deleterious alleles.

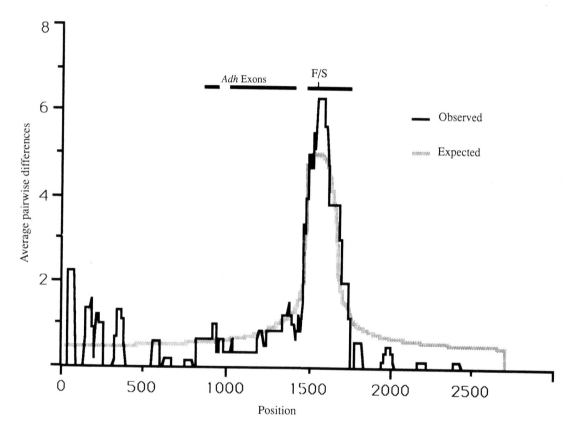

Figure 3.2. Comparison of observed and expected numbers of differences in a 'sliding window' between Adh^F and Adh^S alleles (kindly provided by R. Hudson). The observed values are the average number of pairwise differences between Adh^F and Adh^S alleles within a symmetric window of 150 possible silent changes. In noncoding regions the window is 50 nucleotide sites wide, in coding regions it is somewhat wider. The expected values were calculated assuming: the *Fast/Slow* polymorphism is selectively maintained at a frequency of Adh^S equal to 0.7; the rest of the observed nucleotide sequence variation is selectively neutral (and silent) and maintained by an equilibrium between mutation and genetic drift ($4Nu = 0.006$, where N is population size and u is the mutation rate per nucleotide site); $4Nr = 0.002$ (where r is the rate of crossing-over per nucleotide site); and the product of $2N$ and the mutation rate between the Adh^S and Adh^F alleles is 0.001. The coding regions of the *Adh* gene and the position of the *Fast/Slow* difference are indicated. The analysis depicted in this figure is the same as in Hudson and Kaplan (1988).

Several conclusions arise from these studies. The first seems to be that any inference about the adequacy of the neutral theory (or any other theory) as a mechanistic explanation of molecular evolution is only as good as the alternative theories that challenge it and fail. The approximate agreement of the rate of molecular evolution to the predictions of the neutral theory may make it a convenient theoretical basis for a practical tool, but this is not convincing scientific proof. The recent analysis by Gillespie (1986a,b) suggests that an episodic model may account for the observed variation in rates of substitution, but his molecular landscape model fails to predict the observed level of variation within populations. Hudson, Kaplan and colleagues (Hudson *et al.*, 1987; Kaplan *et al.*, 1988; Hudson and Kaplan, 1988) have shown that the observed distribution of DNA sequence polymorphism and molecular evolution in the *Adh* region fit a model that incorporates some form of balancing selection better than the neutral theory, but the generality of this result must be tested by similar examinations of other regions and other organisms. The observation of excess polymorphism does not require the neutral theory to be abandoned; the model proposed by Hudson and Kaplan (1988) required only one selected site while the rest of the DNA sequence polymorphism in the *Adh* region was assumed to be selectively neutral. Thus Kimura's conjecture that *most* molecular evolution is free from the forces of natural selection is not in conflict with their results.

Acknowledgements

I thank Richard Hudson, John Gillespie and Norman Kaplan for many hours of discussion about their research, and Alan Robertson for his encouragement and intellectual leadership, especially in molecular population genetics.

Chapter 4
Evolution at 'Silent' Sites in DNA

Paul M. SHARP

Department of Genetics, Trinity College, Dublin 2, Ireland

Molecular evolution emerged as a discipline with the characterization of protein sequences in the 1960s, but it is the increased availability, during the last ten years, of DNA sequence data that has allowed the study of molecular evolution to truly flourish. These DNA sequences can be used to infer more protein sequences, but also allow many new questions to be addressed; among these, one of the most intriguing concerns the evolutionary significance of 'silent' sites within the translated regions of genes. These silent sites arise due to the degeneracy of the genetic code – they are positions within codons where the nucleotide is potentially subject to synonymous substitution, i.e. replacement without changing the amino acid encoded. The question is whether equivalence of meaning in the genetic code is reflected in functional equivalence or, in other words, are 'silent' sites neutral?

Silent sites have been subject to two major lines of investigation: (i) comparison among genes, to investigate the frequency of alternative nucleotides at these sites, or 'codon usage', and (ii) comparison of homologous sites in different species to assess their rate of evolution. On the one hand, the pattern of codon usage has been seen to vary extensively among species, and even among genes within species (Grantham *et al.*, 1980; Ikemura, 1985a), an observation that has been taken as evidence that there are selective differences among synonymous codons. On the other hand, it has been suggested that the rate of substitution at silent sites is roughly similar in different genes (Miyata *et al.*, 1980) and even in quite disparate taxa (Ochman and Wilson, 1987; Wilson *et al.*, 1987), which might imply that synonymous changes are neutral, or nearly so. In the past, investigations of these two phenomena, i.e. codon usage frequencies and rates of silent substitution, have largely been conducted separately. Here I examine them in turn, and show how it has recently become possible to interrelate them.

Codon usage patterns

The scope of any attempt to make generalizations about molecular evolution is limited by the small number, as yet, of species which are heavily represented in the DNA sequence databank (Aota *et al.*, 1988). Thus codon usage patterns are well documented for only a handful of different organisms, although it may be possible to extrapolate from these cases to many others. This is most likely to be true if the similarities and differences between species (in codon usage) can be related to their evolutionary distance, and the particular population biology of the different organisms.

Codon usage in prokaryotes

Patterns of codon usage are most clearly understood in *Escherichia coli*. It is known that alternative synonymous codons are translated at different rates (Pedersen, 1984; Varenne *et al.*, 1984); they may also have different proof-reading costs (Ehrenberg *et al.*, 1986). This arises because tRNAs have varied abundances, and because a tRNA may translate alternative codons with differing efficiency. It is observed that genes expressed at high levels have very biased synonymous codon frequencies. Generally, one codon per amino acid is heavily favoured (Table 4.1), and that codon is the one most preferentially recognized by the most abundant tRNA for that amino acid (Ikemura, 1985a). There is a strong positive correlation between expression level and degree of codon usage bias (Gouy and Gautier, 1982; Ikemura, 1985a,b); in genes expressed at low levels, codon usage is more nearly uniform (Table 4.1). In between the two extremes of high and low bias, there are trends in codon bias which occur in parallel across most groups of synonymous codons (Sharp and Li, 1986a; Bulmer, 1988). Thus, there is clear evidence that alternative synonymous codons can be selectively different, although mutation can randomize codon usage when the selective constraints (arising from expression level) are weaker. It is worth noting that even the highly biased pattern of codon usage seen in some genes can arise from quite small selective differences among codons, if (as in *E.coli*) the effective population size is very large (Li, 1987; Bulmer, 1987).

Salmonella typhimurium is an enterobacterium related to, but certainly distinct from, *E.coli*; from 16S rRNA comparisons they have been estimated to have diverged 140 million years (Myr) ago (Ochman and Wilson, 1987). Codon usage patterns in *S.typhimurium* genes are very similar to those in *E.coli* (Ikemura, 1985b; Sharp and Li, 1987b), even though the genes of these two species show an average divergence of nearly one substitution per silent site (see below). The codon usage similarity reflects the lack of divergence (in sequence or abundance) of tRNAs in these two species (Ikemura, 1985b).

In contrast, *Bacillus subtilis* is a gram-positive bacterium very distant from the gram-negative enterobacteria. The divergence time between *E.coli* and *B.subtilis* may be about ten times greater than that between *E.coli* and *S.typhimurium* (Ochman and Wilson, 1987). The pattern of codon usage in *B.subtilis* is rather different from that in *E.coli* (Shields and Sharp, 1987). Again, highly expressed genes utilize only a subset of the genetic code, but for at least 6 of the 18 amino acids where there is a choice the preferred codons are different from those in *E.coli,* and overall the bias is much weaker (Table 4.1). While less is known about tRNAs in *B.subtilis* than in *E.coli*, it seems likely that these differences in codon preference reflect differences in tRNA abundances and anticodon sequences. The lower overall bias suggests that selection on differences among codons is less strong and/or less effective, perhaps indicating that *B.subtilis* has a smaller long-term effective population size than *E.coli*. Interestingly, codon frequencies in lowly expressed *B.subtilis* genes are not uniform (Table 4.1). However, the pattern of bias is such that frequencies of dinucleotides in different codon positions (excluding the first dinucleotide of each codon, since it directly determines the amino acid encoded), and on the sense and antisense strands of the DNA, are highly correlated (Shields and Sharp, 1987). The simplest explanation is that codon frequencies in these genes result entirely from biases in the pattern of mutation, these biases being context-dependent.

The genomes of some bacteria have a highly skewed base composition, either a very high or a very low G+C content, and this dominates codon usage (Muto and Osawa, 1987). For example, the genome of *Mycoplasma capricolum* contains only 25% G+C and over 90% of codons end in either A or T. At the other extreme, *Micrococcus luteus* has a 75% genomic G+C content and codons ending in G or C are used almost exclusively (Muto and Osawa, 1987). It is most likely that these biases reflect mutational biases, rather than direct selection among particular codons. In these examples there is no apparent variability among genes (such as could be attributed to codon selection, varying with gene expression level), but most of the available sequences encode ribosomal proteins and in *E.coli* and *B.subtilis* such genes all have a high codon bias. Of course, in some cases the mutational biases may themselves be adaptive; for example, it has frequently been suggested that a high genomic G+C content is a response to life at high temperature (since G:C hydrogen bonds are more stable than A:T bonds), but counter-examples exist.

Examination of codon frequencies in different bacterial genes has therefore revealed the influences of both translational selection and mutational bias. Variation among species can arise because:

(i) the mutation patterns differ,
(ii) the abundances and anticodons of tRNAs differ (leading to different translationally optimal codons), and
(iii) the strength or efficacy of selection differs.

To the extent that these factors affect all genes within a species equally, there is a common pattern of codon usage within a genome (Grantham *et al.*, 1980). The differences among genes from a single bacterial species are not likely to reflect differences in the pattern of mutation or selection, but rather different points of balance between these two forces, dependent on the strength of selection on a particular gene.

Codon usage in eukaryotes

With regard to codon frequencies, the best understood eukaryotic organism is the budding yeast *Saccharomyces cerevisiae*. As in *E.coli*, a trend across genes in degree of codon usage bias can be correlated with the level of gene expression (Sharp *et al.*, 1986). Highly expressed genes use a very restricted subset of the genetic code (Bennetzen and Hall, 1982); their bias is, if anything, more extreme than that in *E.coli*. The 'optimal' codons in *S.cerevisiae* differ from those in *E.coli* (Table 4.1), in accord with the known tRNA abundances and anticodons of the two species (Ikemura, 1985a). Functional differences between synonymous codons in *S.cerevisiae* have been emphatically demonstrated in experiments where substituting an increasing number of nonoptimal codons into a gene progressively reduced its level of expression (Hoekema *et al.*, 1987). Thus *S.cerevisiae* ranks with *E.coli* as a clear example of the potentially strong effect of selection on codon frequencies. In the absence of direct data, one might deduce that the effective population size of *S.cerevisiae* has been very large.

The fission yeast *Schizosaccharomyces pombe* is very distantly related to *S.cerevisiae*; protein molecular clocks suggest that they diverged between 800 and 1200 Myr ago. Codon usage patterns in *S.pombe* and *S.cerevisiae* differ somewhat (Table 4.1); different codons are favoured for a few amino acids, and overall the degree of bias is less extreme in *S.pombe*. However, the degree of bias in codon usage is very highly correlated between homologous genes in the two species (Sharp and Wright, 1989).

Codon usage in multicellular organisms is less well understood, even though three mammals (human, mouse and rat) are the best represented species in the sequence databank. Human and rodent genes are generally similar. This is not surprising: these two mammalian orders probably diverged about 75 Myr ago, i.e. perhaps more recently than *E.coli* and *S.typhimurium*. There is remarkable within-species codon usage variation among, for example, human genes (Table 4.1), but this is not obviously related to expression level. Rather, it can be most parsimoniously summarized as variation in the G+C composition of silent sites (Ikemura, 1985a). tRNA abundancies are known to vary among mammalian tissues, and there have been some attempts to relate codon usage patterns in different mammalian genes to adaptation for expression in different tissues. For example, Newgard *et al.* (1986) have recorded a tendency for genes expressed in muscle to have a higher G+C content than genes expressed in the liver. Tissue-specific codon usage does not appear to be a general phenomenon however; for example, the G+C content at third codon positions in human alpha-like globin genes (92%) is much higher than in beta-like globin genes (65%), even though both are expressed in reticulocytes, and also higher than in the myoglobin gene (79%) which is expressed in muscle. In fact, it would be surprising if translational selection was demonstrated to be a general influence on codon usage in mammals. Since the long-term effective population sizes of mammals appear to be rather small, of the order of 10^4 (Nei and Graur, 1985), the selective differences among

Table 4.1
Codon usage differences within and among six species

		E.coli high	E.coli low	B.subtilis high	B.subtilis low	S.cerevisiae high	S.cerevisiae low	S.pombe high	S.pombe low	Drosophila high	Drosophila low	Human G+C	Human A+T	
Phe	UUU	0.34	1.33	0.70	1.48	0.19	1.38	0.44	1.28	0.12	0.86	0.27	1.20	UUU
	UUC	1.66	0.67	1.30	0.52	1.81	0.62	1.56	0.72	1.88	1.14	1.73	0.80	UUC
Leu	UUA	0.06	1.24	2.71	0.66	0.49	1.49	0.28	1.79	0.03	0.62	0.05	0.99	UUA
	UUG	0.07	0.87	0.00	1.03	5.34	1.48	2.16	0.80	0.69	1.05	0.31	1.01	UUG
Leu	CUU	0.13	0.72	2.13	1.24	0.02	0.73	2.44	1.55	0.25	0.80	0.20	1.26	CUU
	CUC	0.17	0.65	0.00	0.93	0.00	0.51	1.13	0.31	0.72	0.90	1.42	0.80	CUC
	CUA	0.04	0.31	1.16	0.34	0.15	0.95	0.00	0.87	0.06	0.60	0.15	0.57	CUA
	CUG	5.54	2.20	0.00	1.80	0.02	0.84	0.00	0.68	4.25	2.04	3.88	1.38	CUG
Ile	AUU	0.48	1.38	0.91	1.38	1.26	1.29	1.53	1.77	0.74	1.27	0.45	1.60	AUU
	AUC	2.51	1.12	1.96	1.14	1.74	0.66	1.47	0.59	2.26	0.95	2.43	0.76	AUC
	AUA	0.01	0.50	0.13	0.48	0.00	1.05	0.00	0.64	0.00	0.78	0.12	0.64	AUA
Met	AUG	1.00	1.00	1.00	1.00	1.00	1.00	1.00	1.00	1.00	1.00	1.00	1.00	AUG
Val	GUU	2.41	1.09	1.88	0.83	2.07	1.13	1.61	2.04	0.56	0.74	0.09	1.32	GUU
	GUC	0.08	0.99	0.25	1.49	1.91	0.76	2.39	0.65	1.59	0.93	1.03	0.69	GUC
	GUA	1.12	0.63	1.38	0.76	0.00	1.18	0.00	1.06	0.06	0.53	0.11	0.80	GUA
	GUG	0.40	1.29	0.50	0.92	0.02	0.93	0.00	0.24	1.79	1.80	2.78	1.19	GUG
Ser	UCU	2.81	0.78	3.45	0.77	3.26	1.56	3.14	1.33	0.87	0.55	0.45	1.63	UCU
	UCC	2.07	0.60	0.00	0.81	2.42	0.81	2.57	0.52	2.74	1.41	2.09	0.80	UCC
	UCA	0.06	0.95	1.50	1.29	0.08	1.30	0.00	1.56	0.04	0.84	0.26	1.23	UCA
	UCG	0.00	1.04	0.00	0.94	0.02	0.66	0.00	0.67	1.17	1.30	0.68	0.13	UCG
Pro	CCU	0.15	0.75	2.29	0.99	0.21	1.17	2.00	1.21	0.42	0.43	0.58	1.50	CCU
	CCC	0.02	0.68	0.00	0.27	0.02	0.75	2.00	0.83	2.73	1.02	2.02	0.83	CCC
	CCA	0.42	1.03	1.14	1.08	3.77	1.38	0.00	1.51	0.62	1.04	0.36	1.57	CCA
	CCG	3.41	1.54	0.57	1.66	0.00	0.70	0.00	0.45	0.23	1.51	1.04	0.10	CCG
Thr	ACU	1.87	0.76	2.21	0.39	1.83	1.23	1.89	1.52	0.65	0.70	0.36	1.45	ACU
	ACC	1.91	1.29	0.00	0.98	2.15	0.78	2.11	1.04	3.04	1.58	2.37	0.92	ACC
	ACA	0.10	0.68	1.38	1.64	0.00	1.38	0.00	1.04	0.10	0.77	0.36	1.45	ACA
	ACG	0.12	1.28	0.41	0.98	0.01	0.60	0.00	0.40	0.21	0.95	0.92	0.18	ACG
Ala	GCU	2.02	0.61	2.94	0.78	3.09	1.07	2.30	1.79	0.95	0.91	0.45	1.59	GCU
	GCC	0.18	1.18	0.08	1.14	0.89	0.76	1.49	0.50	2.82	1.93	2.38	0.92	GCC
	GCA	1.09	0.79	0.60	1.19	0.03	1.49	0.21	1.14	0.09	0.59	0.36	1.38	GCA
	GCG	0.71	1.42	0.38	0.89	0.00	0.68	0.00	0.57	0.14	0.57	0.82	0.11	GCG

Notes: Codon usage is expressed as the observed number of codons divided by the number expected if synonymous codons are used equally.

For each species there is considerable heterogeneity among genes, and in each case multivariate analysis of codon usage has revealed a single major trend across genes. To illustrate these trends, genes have been ranked according to their position along the trend (as assessed by an index of codon bias; see below), and codon usage is presented for groups of genes comprising the 10% of genes at each end of the trend (data from Sharp *et al.*, 1988). In *E.coli*, *B.subtilis*, *S.cerevisiae* and *S.pombe* codon bias is measured by the Codon Adaptation Index

Table 4.1
Continued

		E.coli high	low	B.subtilis high	low	S.cerevisiae high	low	S.pombe high	low	Drosophila high	low	Human G+C	A+T	
Tyr	UAU	0.38	1.28	0.50	1.29	0.06	1.13	0.48	1.24	0.23	0.96	0.34	1.17	UAU
	UAC	1.63	0.72	1.50	0.71	1.94	0.87	1.52	0.76	1.77	1.04	1.66	0.83	UAC
ter	UAA	–	–	–	–	–	–	–	–	–	–	–	–	UAA
	UAG	–	–	–	–	–	–	–	–	–	–	–	–	UAG
His	CAU	0.45	1.21	2.00	1.28	0.32	1.16	0.56	1.44	0.29	0.86	0.30	1.28	CAU
	CAC	1.55	0.79	0.00	0.72	1.68	0.84	1.44	0.56	1.71	1.14	1.70	0.72	CAC
Gln	CAA	0.12	0.76	1.71	0.88	1.98	1.10	1.85	1.67	0.03	0.88	0.21	0.98	CAA
	CAG	1.88	1.24	0.29	1.13	0.02	0.90	0.15	0.33	1.97	1.12	1.79	1.02	CAG
Asn	AAU	0.02	1.12	0.47	1.21	0.06	1.28	0.30	1.41	0.13	1.13	0.33	1.20	AAU
	AAC	1.98	0.88	1.53	0.79	1.94	0.72	1.70	0.59	1.87	0.87	1.67	0.80	AAC
Lys	AAA	1.63	1.50	1.83	1.47	0.16	1.24	0.10	1.27	0.06	0.81	0.34	1.17	AAA
	AAG	0.37	0.50	0.17	0.53	1.84	0.76	1.90	0.73	1.94	1.19	1.66	0.83	AAG
Asp	GAU	0.51	1.43	0.53	1.16	0.70	1.38	0.78	1.56	0.90	1.10	0.36	1.29	GAU
	GAC	1.49	0.57	1.47	0.84	1.30	0.62	1.22	0.44	1.10	0.90	1.64	0.71	GAC
Glu	GAA	1.64	1.28	1.40	1.27	1.98	1.29	0.69	1.20	0.19	0.73	0.26	1.33	GAA
	GAG	0.36	0.72	0.60	0.73	0.02	0.71	1.31	0.80	1.81	1.27	1.74	0.67	GAG
Cys	UGU	0.60	0.94	0.00	0.94	1.80	1.10	0.14	1.56	0.07	0.71	0.42	1.09	UGU
	UGC	1.40	1.06	2.00	1.06	0.20	0.90	1.86	0.44	1.93	1.29	1.58	0.91	UGC
ter	UGA	–	–	–	–	–	–	–	–	–	–	–	–	UGA
Trp	UGG	1.00	1.00	1.00	1.00	1.00	1.00	1.00	1.00	1.00	1.00	1.00	1.00	UGG
Arg	CGU	4.47	1.71	3.11	0.54	0.63	0.64	5.17	1.89	2.65	0.69	0.38	0.64	CGU
	CGC	1.53	2.41	1.78	1.21	0.00	0.39	0.83	0.26	3.07	1.55	2.72	0.36	CGC
	CGA	0.00	0.52	0.00	0.74	0.00	0.65	0.00	0.86	0.07	1.12	0.31	0.81	CGA
	CGG	0.00	0.80	0.00	0.81	0.00	0.34	0.00	0.43	0.00	1.12	1.53	0.51	CGG
Ser	AGU	0.13	1.01	0.45	0.56	0.06	0.97	0.14	1.48	0.04	0.89	0.31	1.26	AGU
	AGC	0.93	1.62	0.60	1.63	0.16	0.70	0.14	0.44	1.13	1.01	2.22	0.94	AGC
Arg	AGA	0.00	0.37	1.11	2.02	5.37	2.51	0.00	1.71	0.00	0.56	0.22	2.40	AGA
	AGG	0.00	0.19	0.00	0.67	0.00	1.47	0.00	0.86	0.21	0.95	0.84	1.28	AGG
Gly	GGU	2.27	1.29	1.38	0.54	3.92	1.32	3.36	1.87	1.34	0.91	0.34	0.84	GGU
	GGC	1.68	1.31	0.97	1.30	0.06	0.92	0.59	0.27	1.66	1.65	2.32	0.76	GGC
	GGA	0.00	0.64	1.66	1.24	0.00	1.22	0.05	1.60	0.99	0.98	0.29	1.79	GGA
	GGG	0.04	0.76	0.00	0.92	0.02	0.55	0.00	0.27	0.00	0.46	1.05	0.61	GGG

(Sharp and Li, 1987a), which quantifies the extent of preference for particular (species-specific) codons which are favoured in highly expressed genes. In *D.melanogaster* codon bias is assessed by the deviation from equal usage of synonymous codons (Shields *et al.*, 1988). In *H.sapiens* the scale of codon bias is simply G+C content at synonymous sites. In the two bacteria and the two yeasts (and probably also drosophila), the preferred codons in the highly biased group appear to be those which optimise translation, and highly biased genes are those which are highly expressed.

codons would need to be considerably larger than in, say, *E.coli* or yeast (Li, 1987). This amount of selection would require a large reproductive excess (Nei, 1971), and it seems doubtful whether mammals have this excess.

In this context, it is interesting to note that the G+C content at silent sites in any mammalian gene is highly correlated with the G+C content of introns and flanking sequences of the same gene (Aota and Ikemura, 1986). Also, the results of density gradient centrifugation suggest that mammalian genomes comprise a mosaic of long (200–1000 kb) regions of differing base composition, the so-called 'isochores' (Bernardi *et al.*, 1985). Since the noncoding sequences are not known to be under any selective constraint, these observations are most easily explained if silent sites are effectively neutral in mammals, with the variation between genes in base composition reflecting different mutational biases in different regions of the genome.

Lastly, I consider insects. It has been known for some time that tRNA populations in the silk gland of the silkworm, *Bombyx mori*, are adapted to the unusual amino acid composition of the major components of silk, fibroin and sericin (Garel, 1974), but it has not been clear to what extent this may represent a rather extreme and unusual situation. There is very little other sequence data available for the silkworm, but molecular biologists have recently turned their attention to the geneticists' organism, *Drosophila melanogaster*. The accumulated sequences reveal large between-gene differences in codon usage, with a single major trend across genes (Shields *et al.*, 1988). At one end of this trend are genes with a high codon usage bias, including many which are known, or expected, to be highly expressed (e.g., those encoding ADH, GAPDH, actin and ribosomal proteins). The preferred codons are mainly C-ending (Table 4.1); this results in a gradient of silent site G+C content across genes, but (unlike the situation in mammals) this is not correlated with G+C content variation among introns. Thus, the variation in codon usage pattern among *D.melanogaster* genes may reflect different degrees of codon selection, in a fashion analagous to *E.coli* and yeast. The effective population size of drosophila is about 10^6 to 10^7 (Nei and Graur, 1985), larger than in mammals, and (apparently) sufficient to make codon selection effective.

Thus there is evidence of selective differences among synonymous codons in some eukaryotes, but not in others. Previously, it has seemed appropriate to discuss codon usage in either 'unicellular organisms' or 'multicellular organisms' (Ikemura, 1985a; Sharp and Li, 1986b), but since *D.melanogaster* appears to be more similar to *E.coli* and yeast than to mammals, it may be more appropriate to consider species according to whether codon selection is effective – the single most important determinant of this, at least in eukaryotes, is likely to be population size.

Rates of silent substitution

Most discussions of molecular evolutionary rates now start from the powerful conceptual framework of Kimura's neutral theory (see Kimura, 1983). Under this hypothesis only a small proportion of nucleotide substitutions result from positive selection, so that the rate of evolution at a DNA site depends on the frequency of occurrence of effectively neutral mutations. This frequency depends on the overall mutation rate, but is also inversely related to the strength of the selective constraint on a site and the effective population size. Then, silent sites in all genes in all organisms could be expected to evolve at the same rate only if either: (i) the rate of mutation is the same, and silent sites are equally constrained (presumably neutral), or (ii) the above factors vary in a coordinated way (presumably by coincidence) such that, for example, silent sites in genes with a high mutation rate are subject to more stringent constraint. In fact, there is considerable variation in synonymous substitution rates among genes.

Rate variation within species

The two species of enterobacteria featured in the discussion of codon usage also provide some of the

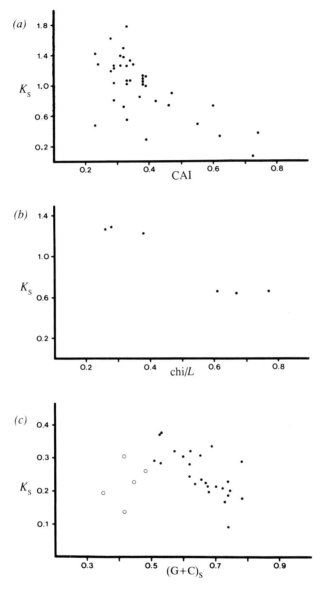

Figure 4.1. Silent substitution rates are related to codon usage bias, in (a) Enterobacteria, (b) Drosophila, (c) Rodents

Notes: Silent substitution rates are expressed as the number of synonymous substitutions per site (K_S) estimated by the method of Li *et al.*, (1985).

(a) 38 genes compared between *E. coli* and *S. typhimurium*; for each gene K_S is plotted against the Codon Adaptation Index (CAI; see Sharp and Li, 1987a). A large CAI value indicates high codon usage bias.

(b) 6 genes compared between *D. melanogaster* and *D. pseudoobscura*; for each gene K_S is plotted against chi/L (a chi[2] for deviation from equal usage of synonymous codons, divided by gene length; see Shields *et al.*, 1988). A large chi/L indicates high codon usage bias.

(c) 30 genes (each at least 200 codons long) compared between mouse and rat; for each gene K is plotted against G+C content at 4-fold degenerate sites. Genes with $(G+C)_S$ less than 50% are indicated by open circles.

best data for examining the rate of synonymous nucleotide substitution. Approximately 40 genes can be compared between *E.coli* and *S.typhimurium*, and the degree of divergence at silent sites varies greatly among them. In Figure 4.1(*a*) the genes with a highly biased codon usage (high CAI) can be seen to have diverged more slowly, as expected if codon choice exerts a constraint on silent sites (Sharp and Li, 1987b). Interestingly, the high negative correlation of codon bias and synonymous rate, and the paucity of low codon bias genes with anomalously low rates, suggest that horizontal gene transfer is not an important source of confusion here. There are as yet no other pairs of bacterial species for which such detailed comparisons can be made, while the two yeast species discussed above are so distantly related to each other that silent sites have become saturated with multiple substitutions.

Among species of drosophila, a few genes can be compared between *D.melanogaster* and *D.pseudoobscura*. By chance, these genes fall into two discrete groups with respect to codon usage bias. In Figure 4.1(*b*) it can be seen that silent sites in the genes with a low codon bias have diverged at twice the rate of those in highly biased genes (Shields *et al.*, 1988). This again implies a stronger selective constraint on silent sites in the genes with high codon bias, and thus corroborates the inference that these codons are under selection.

Among species of mammals, the two rodents, mouse and rat, provide the most suitable data set. Again, there is considerable variation among genes in the degree of synonymous divergence. By considering only longer genes (for which the estimates of substitution rate are less prone to stochastic variation) a clear pattern can be discerned: genes with an intermediate silent site G+C content, $(G+C)_S$, have diverged more rapidly (Figure 1(*c*)). Among genes with $(G+C)_S$ of 50% or higher, there is a strong negative relationship (correlation coefficient = 0.75) between G+C content and substitution rate, while among the few genes with $(G+C)_S$ lower than 50%, this relationship is positive. Comparisons can also be made with pseudogenes: silent sites in homologous functional genes have not evolved significantly more slowly, suggesting little if any selective constraint on synonymous codons. Is there a selective constraint to use particular (different) codons in GC-rich and GC-poor genes? Almost certainly not, since (as noted above) similar G+C content variation is seen in introns and flanking sequences. It seems far more likely that the observation in Figure 1(*c*) reflects differences in the rates at which mutations occur in different regions of the genome (i.e. isochores). There is evidence for different dNTP pool levels at different stages of replication (Holmquist *et al.*, 1982), and the observed pattern of variation in mutation rates, with a peak at an intermediate value of G+C content, would be predicted if G+C-rich sequences are replicated under conditions where dGTP and dCTP are abundant, whereas dATP and dTTP predominate during the replication of A+T-rich sequences. Such conditions would also serve to maintain the particular G+C content of each isochore.

Silent rates in different taxa

Estimation of the absolute rate of silent substitution requires some knowledge of the time since the common ancestor of the species being compared. The best fossil records exist for the mammals. For example, the rodent–primate divergence is thought to have occurred about 75 Myr ago, yielding an average synonymous substitution rate between these taxa of 4.7 substitutions per site per 10^9 yr (Li *et al.*, 1985). However, this average figure hides not only the heterogeneity among genes discussed above, but also apparently different rates in the rodent and primate lineages (Li *et al.*, 1987). For example, the best available estimates for the age of the common ancestors of mouse and rat (15 Myr ago) and of humans and Old World (OW) monkeys (25 Myr ago) yield average rates in the recent rodent and primate lineages of 7.9 and 2.5 substitutions per site per 10^9 years, respectively (Table 4.2). While these time estimates may be open to dispute, relative rate tests, which make use of an outgroup and are thus independent of fossil dating, confirm that molecular evolution has occurred far more rapidly in the rodent lineage than in that leading to humans (Li *et al.*, 1987).

The last common ancestor of *D.melanogaster* and *D.pseudoobscura* seems to have lived 40 Myr ago, at most (Throckmorton, 1975). This implies that the average silent substitution rate in drosophila is about three times the mammalian average (Table 4.2). Thus primates, rodents and flies each have different synonymous substitution rates. The rank order (although perhaps not the relative magnitude) of these rates is as would be expected if they reflect variation in the mutation rate per unit time, with the latter being inversely proportional to generation interval.

In contrast, the synonymous substitution rate in enterobacteria is surprisingly slow (Table 4.2). Bearing in mind the potentially short generation time of *E.coli*, the mutation rate (per unit time) in bacteria might be expected to be orders of magnitude higher than that in animals. The divergence time estimate for these species (from Ochman and Wilson, 1987) comes from 16S rRNA comparisons, and is itself subject to a molecular clock assumption. However, any error there (the date is unlikely to be wrong by an order of magnitude) cannot compensate for the slowness of the estimated synonymous rate. It may be that codon selection constrains the rate of 'silent' substitution in *E.coli*, even in genes where codon bias is comparatively weak.

Table 4.2
'Silent' DNA substitution rates in different taxa

Species compared		T	Rate	(Range)	N	L
Enterobacteria:	*E.coli* vs *S.typhimurium*	140	3.5	(1.0–5.3)	37	12,387
Drosophila:	*D.melanogaster* vs *D.pseudoobscura*	40	13.5	(8.0–16.1)	6	1884
Rodents:	mouse vs rat	15	7.9	(2.7–13.1)	72	13,771
Primates:	human vs OW monkey	25	2.5	(1.8–3.7)	10	1189

Notes:
T = time of the common ancestor (Myr ago)
rate = synonymous substitutions per site per 10^9 yr, averaged over genes
range = synonymous substitution rates for the slowest and fastest genes, in each case excluding the single most extreme case
N = number of genes
L = total number of silent sites

Conclusions

Degenerate sites in codons were heralded as candidates for neutral evolution, even before DNA sequences were known (King and Jukes, 1969) and on many occasions it has been suggested that silent sites should provide an excellent molecular clock, particularly since data might be pooled from different genes. Furthermore, it has often been assumed that the rate is the same in very distant taxa. Others had anticipated possible causes of selective differences among synonymous codons (Clarke, 1970). I conclude that there is as yet no firm evidence to contradict the neutrality of silent sites in mammalian genes. However, in other species, including not only bacteria and yeast but also drosophila, there are clear selective constraints on codon usage in at least some genes. As a consequence, silent sites in different genes from the same genome often evolve at different rates. What is more, rates of silent substitution vary among organisms, and even among regions of the mammalian genome; in both instances this probably reflects variation in the rate of mutation per unit time. These findings do not necessarily contradict the neutral theory (Jukes and Kimura, 1984), but rather it is clear that silent sites are not universally neutral. Consequently, silent molecular clocks must be used with caution.

Acknowledgements

My interest in molecular evolution was instigated by the teaching of Alan Robertson; it was subsequently greatly stimulated by my colleagues Wen-Hsiung Li and David McConnell. I am also grateful to Denis Shields, Ken Wolfe and Frank Wright for much discussion and collaboration on the work reported here (particularly that as yet unpublished), and to Masatoshi Nei and Manolo Gouy for their comments on the manuscript.

Chapter 5
DNA Sequence Polymorphisms

R.C. LEWONTIN

Museum of Comparative Zoology, Harvard University, Cambridge, MA 02138, USA

When gel electrophoresis of proteins was introduced more than twenty years ago as a method for studying the genetic variation within species, it was thought that the data generated would provide unambiguous evidence about genetic variation and the forces moulding it. It soon became apparent, however, that while some ambiguities had been cleared away, others remained. By the middle of the 1970s, it became clear that observations of protein variation were, intrinsically, incapable of distinguishing the roles of selection, mutation, gene exchange, population structure and random genetic drift, in determining the variation that electrophoresis had revealed. (See Lewontin, 1974, for a review of the situation.) Nothing that has happened since has changed the pessimism of that view. Studies of protein variation, by gel electrophoresis and other methods, have essentially disappeared from the literature of population genetics, although they continue to be important in species comparisons and in answering specific questions about population structure in particular cases.

The ambiguity of protein variation data arises from the fact that it is impossible to distinguish two alleles that are identical by descent from those that are identical by kind. If two populations have a very similar distribution of allele frequencies at a locus, is that a result of shared recent common ancestry of the genes in the different populations (identity by descent) or because selection has driven the frequency distributions to a similar state, although the apparently identical alleles have not shared any common ancestry in the recent past (identity by kind)? If a population is monomorphic for an allele is that because genetic drift has resulted in homozygosity in the roughly $4N$ generations since the allele first appeared in the population (homozygosity by descent) or because selection drives the allele to fixation irrespective of the origin of the alleles (homozygosity by kind)? There is no way to know by looking only at the amino acid sequence of a protein, or worse, by observing the electromorphic phenotype which is an imperfect detector of amino acid variation.

The development of technology for studying DNA sequence variation has changed the picture radically. By luck, DNA sequences turn out not to be in one-to-one correspondence with amino acid sequences. Were they in a one-to-one correspondence, nothing could be learned from them since they would contain no information not already contained in the protein sequences. However, there is added information in DNA sequences at three levels. First, the genetic code is degenerate so that at the codon level the same amino acid may be coded by 2, 3, 4 or 6 different codons. Variation at silent positions in the codon will not be under the same selective constraints as variation at amino-acid replacement sites, so a distinction between selective and nonselective forces is possible. Second, genes are composed of both introns and exons and, again, variation in nucleotides or introns will not be under the same constraints as the replacement sites in exons. Finally flanking sequences, especially 3′ nontranscribed regions, are again not subject to the same constraints as the positions that are significant to amino acid identity, or to 5′ nontranslated regions that are involved in transcription control. It is not necessary that there be *no* selective constraints on the redundant positions or introns, but only that such constraints be different from those acting on amino acid sequences. Thus, codon biases with very different frequencies of usage of equivalent codons are maintained across species differences despite wide nucleotide divergence, indicating some selection rules on codon usage. For example the codon usage tables for the alcohol dehydrogenase gene and the xanthine dehydrogenase genes are very similar between *D. melanogaster* and *D. pseudoobscura*, despite almost complete randomization of silent positions between these species (Schaeffer and Aquadro, 1987; S. Schaeffer, personal communication; Lewontin, 1989; Riley, 1988).

Questions

The differentiation of a very short region of DNA into gene and flanking sequence, exon and intron, and silent and replacement codon sites makes it possible to obtain unambiguous answers to a variety of questions about selection as opposed to history in the evolution of genic variation. Moreover, where unambiguous answers are not possible, no conceivable set of observations on genetic variation alone can provide that information because DNA sequence variation is the ultimate level to which observation of variation can be carried. Any remaining ambiguities can only be cleared up by direct ecological and physiological observation of the forces of migration, population structure and selection.

Questions that can be asked and answered with observations on DNA sequence variation, that are not accessible by other techniques include, among others:

(1) A comparison of the level of polymorphism between silent and replacement sites. If silent sites are highly polymorphic while there is little or no amino acid polymorhism, then consistent directional selection against protein variants is demonstrated. If silent and replacement sites are equally polymorphic then amino acid variation is selectively neutral or nearly so. If amino acid polymorphism is high, but there is little silent position variation within protein variants, then the protein variation has been recently established from single ancestral alleles. Whether there has been strong selection to create the polymorphism or a recent accident of drift may be determined by looking at several populations. Finally, if there is very low polymorphism at all sites then homozygosity by descent is detected. This homozygosity has two possible sources, drift or recent strong selection of an initially rare allele. These can be distinguished as discussed below.

(2) What kinds of amino acid polymorphism exist? DNA sequence variation is a direct characterization of amino acid polymorphism, so it is possible to ask whether some regions of a protein are variable, while others are more constrained, and whether different classes of amino acid substitutions are more or less likely.

(3) Are the apparently similar allelic frequency distributions in different populations really genetically the same? From the translation of DNA sequences it is possible to get complete and definitive answers about the ability of gel electrophoresis to discriminate amino acid substitutions. More important, even if electrophoresis is completely discriminating, will the polymorphisms in different populations have the same *silent* site variation? If not, then the very similar polymorphism cannot be ascribed to gene exchange, but must be a result of independent selection of polymorphisms in each population.

(4) Do the same silent variation haplotypes appear over and over in different populations? Patterns of gene exchange among populations can be reconstructed by finding repeated occurrences of the same haplotype on a background of considerable polymorphism.

(5) How important is intracistronic recombination (including gene conversion) in generating allelic diversity? Recombinant haplotypes can be observed directly in population surveys.

(6) Are there regions *within* a gene of excess or deficient polymorphism as compared with other regions? The observation of stretches of DNA with uncharacteristic silent site polymorphism is an indicator of selection. This effect is a consequence of genetic hitch-hiking of neutral sites that

are very closely linked to selected sites. If there is a heterotically maintained polymorphism at a nucleotide site coding for an amino acid change, then silent sites within several hundred base pairs around the selected site will have a higher than average heterozygosity (Strobeck, 1983). If, on the other hand, a site has recently been selected to homozygosity, there will be a deficiency of silent polymorphism in the surrounding region. In principle, it should even be possible to detect meiotic drive events that have driven an entire chromosome to homozygosity.

(7) Are there any truly neutral or unconstrained regions of DNA? The simplest view of DNA function would predict that silent sites within codons, introns and nontranscribed 3' flanking regions of genes ought to have the same amount of polymorphism since they are all 'functionless' or at least do not depend on their exact nucleotide composition. If, however, it is found that there are significant differences in polymorphisms of these nucleotide types, or differences between genes in the same genome in the intron or silent site polymorphic rates, then some functional constraints on this DNA must be looked for.

What characterizes this brief and nonexhaustive list of questions is that it covers most of the problems that population genetics has struggled with from its inception as a discipline: distinguishing selective from accidental historical changes, detecting gene exchange at low levels, distinguishing sources of new variation, reconstructing population structure, and detecting past selection events that are no longer occurring.

Data

There are essentially two levels at which observations of DNA sequence variation can be made. The first is complete sequencing of the gene and its flanking regions, or at least complete sequence information on representative stretches of DNA including flanking, intron and exon regions. The second is represented by restriction enzyme analysis that samples cleavage sites distributed along the DNA, but includes techniques that detect single site mismatches by either cutting of the DNA at mismatched sites or differential melting of mismatched DNA duplexes. There is an essential difference in the quality of information that can be recorded. Complete sequencing makes it possible to answer questions (1), (2) and (7) above, while restriction enzyme analysis does not, because the latter (and equivalent techniques) does not distinguish silent from replacement sites. Moreover, unless the gene region being investigated has been sufficiently well sequenced to mark the boundaries between introns, exons and flanking sequences, restriction analysis will not be able to distinguish polymorphism levels among these regions.

Restriction analysis can be made more powerful by using it only for DNA regions that have already been completely sequenced once, by replacing six-cutters, which only cut about every 1200 bp, with four-cutters, which cut about every 250 pairs, and by running the cut fragments on long sequencing gels so that fragment lengths can be determined to within 2–3 base pairs (Kreitman and Aguade, 1986a). Even with this technique, however, it is usually not possible to distinguish silent from replacement site polymorphisms.

Given the immense power to distinguish hypotheses in population genetics that resides in complete sequence polymorphism studies, it is noteworthy that not a single study using this technique has been published in the five years since Kreitman's (1983) pioneering study of sequence polymorphism of the *Adh* gene in *D. melanogaster*. This hiatus is a consequence of the major investment in equipment, supplies and retraining in molecular biology that has had to take place in population genetics laboratories and of the very high degree of experimental skill needed up to now for successful cloning and sequencing. This retooling period is now coming to an end and at least three further complete sequence polymorphism studies are near or at completion by various writers: esterase-6 in *D. melanogaster* (P. Cooke), alcohol dehydrogenase in *D. pseudoobscura* (S. Schaeffer) and xanthine dehydrogenase in *D. pseudoobscura* (M. Riley and co-workers).

Kreitman (1983) sequenced a 2.7 kb region, including the structural gene for alcohol dehydrogenase, in ten lines from natural populations and one laboratory line. The lines from nature were from five different populations spread over the world, each population being represented by one copy of the slow electrophoretic allele and one copy of the common fast allele. His results are listed below.

(1) Forty-three sites were polymorphic out of 2721 bp (1.6%) over all.
(2) 14/765 sites were polymorphic in the translated sequence (1.8%), but *all* of these were silent changes except for the single amino acid coding change between F and S already built into the sample. This gives an estimated silent site polymorphism of 13/192 = 6.8%. The complete lack of amino acid polymorphism despite the 6.8% silent polymorphism shows conclusively that consistent purifying selection operates against mutational change for this gene. The monomorphism of amino acids cannot be accounted for by genetic drift.
(3) A highly heterogeneous distribution of polymorphic sites among regions was revealed. Exon 4 (containing the single polymorphic replacement site) had 14.3% silent site polymorphisms while the other exons had only 4% polymorphism, and the introns 2.3%. Nontranslated and nontranscribed flanking regions were very conserved with only 0.7% polymorphism.
(4) There was clear evidence of intracistronic recombination exchanging left and right halves of slow and fast alleles, and some likely candidates for gene conversions (or double cross-overs in short regions).
(5) There was much greater variation within slow than within fast alleles pointing to the more recent derivation of fast from slow.
(6) There was a lack of any geographical differentiation, indicating high effective gene exchange or a world-wide basis in *D. melanogaster*.

Four-cutter analysis

Using the high resolution four-cutter enzymes with a panel of between 7 and 10 enzymes, Kreitman and Aguade (1986a,b) and Simmons *et al.* (1988) have studied very large samples of *Adh* sequences from several geographic populations (75–97 lines per population). Their findings amplify the original results of Kreitman:

(1) The *Adh* structural gene shows more silent polymorphism than does about 4 kb of 5' adjacent region.
(2) A recent founder event was inferred in a northern population which has much lower haplotype diversity than the southern populations. The most frequent haplotype (25% of the sample) was identical with a haplotype originally found in low frequency in a distant population, and it was found in low frequency in the other populations.
(3) Most haplotypes in all populations are separated by only a few mutational and recombinational events.
(4) There is considerable homogenizing migration among most populations. Aguade (1988) has confirmed the much more recent origin of fast from slow alleles, but has also shown that a common European inversion must have arisen very long before the derivation of fast from slow alleles.
Hudson and Kaplan (1988) have explained the excess of silent polymorphism within the *Adh* gene as a consequence of the single balanced amino acid polymorphism coded in exon 4 holding adjacent neutral variation in high heterozygosity as compared with more weakly linked distant sites. This is in agreement with considerable evidence that the F/S polymorphism is heterotic, but fails to explain the excess silent polymorphism *within* the slow allelic class.

Restriction site data

Since the first reports of restriction site polymorphism in humans (Jeffreys, 1979), peromyscus (Avise *et al.*, 1979) and drosophila (Shah and Langley, 1979), there has been a burgeoning literature on DNA sequence variation as detected with six-cutting enzymes. A great deal of this simply reports restriction fragment length polymorphisms (RFLP) based on one or several enzymes, but these have mainly resulted in the creation of a bank of genetic markers or in taxonomic studies. Intensive studies of population variation for haplotypes have been fewer. Six-cutter studies of long stretches of unsequenced DNA are restricted in the information they can produce, but they can measure average nucleotide diversity (without distinguishing silent from nonsilent variation on different functional regions of the genome), population differentiation and migration patterns and linkage disequilibria. They have the advantage that long sequences are sampled (10–60 kb) so that local effects average out.

The results of these studies can be easily summarized:

(1) Nucleotide heterozygosity ranges from a low of 0.6% and 0.7% in the *Adh* and *Notch* regions of *D. melanogaster* (Langley *et al.*, 1982; Schaeffer *et al.*, 1988) to above 2% for Adh in *D. pseudoobscura* (Schaeffer *et al.*, 1987).

(2) There may be considerable insertion/deletion variation in addition to single site variation (Cross and Birley, 1986; Aquadro *et al.*, 1986; Langley and Aquadro, 1987) even including up to 15 successive insertions of the same small sequence in barley (Saghai-Maroof *et al.*, 1984).

(3) The amount of polymorphism may vary strongly from one population to another for the same region. The *Adh* region of *D. melanogaster* has both the lowest measured heterozygosity (0.006) (Langley *et al.*, 1982) and the highest (0.025) (Cross and Birley, 1986).

(4) Haplotype diversity varies widely and this is related to the amount of linkage disequilibrium among polymorphic sites. Schaeffer *et al.* (1987) and Langley and Aquadro (1987) found high haplotype diversity and low linkage disequilibrium, while Cross and Birley (1986) and Lado *et al.* (1988) found high linkage disequilibrium among sites. Aquadro *et al.* (1986) found strong linkage association between the amino acid substitution at the *Adh* locus and the restriction sites. Linkage disequilibrium may appear at the species level because a single population is divergent in its allelic frequencies (Schaeffer *et al.*, 1988).

(5) Unique historical events can be seen. Jiang *et al.* (1987) found all copies of a null *Adh* allele from natural populations to have an identical haplotype, while active alleles had considerable haplotype diversity including the haplotype found in the nulls.

(6) As expected, mitochondrial DNA studies show very strong linkage disequilibrium and very strong differentiation among populations. Hale and Singh (1987) found 24 haplotypes in 92 lines from 18 populations of *D. melanogaster* around the world. Of these 24, 18 were unique to a single population, but a number of unique types were in high frequency within their population. An extreme case is in mtDNA of bats (Pumo *et al.*, 1988) where a four-cutter and six-cutter analysis showed essentially two haplotypes, despite a 10% site divergence between them.

In general it seems clear that studies of DNA sequence variation will reinforce the appearance uniqueness of each gene region, population and species, precisely because of the power of these studies to detect historical events and population structure, and to show the differential operation of constraints in different regions of the genome.

Chapter 6
Restriction Map Analysis in Population Genetics

Andrew J. LEIGH BROWN

Department of Genetics, University of Edinburgh, West Mains Road, Edinburgh EH9 3JN, Scotland

Use of restriction enzymes in evolutionary biology

Restriction enzymes can be purified from a large number of different prokaryotes and currently a very large number are commercially available. They usually recognize short (4, 6 or occasionally 8 base pairs) palindromic sequences and cleave at a defined position within the sequence leaving 3' or 5' overhangs of one or two bases, or flush ends. A single nucleotide substitution within the sequence will prevent recognition and cleavage.

In recent years extensive use has been made of restriction enzymes in evolutionary biology. There are four main reasons why this is so. First, the detection of restriction fragments is independent of variability. The importance of this feature in population genetic studies was discussed extensively by Lewontin (1974). Second, the ability to detect restriction fragments is completely independent of the phenotypic importance of the DNA segment. As the bulk of DNA in eukaryotes does not encode a protein product or indeed, have any detectable function at all, most restriction sites studied turn out to be in introns or intergenic DNA. However, the choice of the probes used is up to the investigator: thus it may be decided to concentrate on a protein-coding sequence (Kreitman and Aguade, 1986a) or probes may be used which detect restriction sites separated by greater distances on the molecular maps (e.g. Antonarakis et al., 1982; Beech and Leigh Brown, 1989). The latter approach is used for studies of linkage disequilibrium (see below). There is a third important advantage presented by restriction enzymes for surveys of population variation: the data generated can be directly related to the basic unit of the genome – the nucleotide – in a way which data from surveys of allozyme variation could not. Finally, restriction enzyme studies can detect phenomena that are exclusive properties of eukaryotic genomic DNA itself such as the insertion and excision of transposable genetic elements, and small-scale rearrangements.

Restriction enzymes have been employed in studies of nucleotide polymorphism and evolutionary divergence in three distinct areas: in mitochondrial genomes, in the analysis of repetitive nuclear DNA sequences (including for example ribosomal RNA genes (rDNA) and satellite sequences) and of unique or low copy number protein-coding nuclear genes. Within each area a very large amount of information has accumulated since the use of restriction enzymes were first adopted by evolutionary biologists. As the evolution of repetitive DNA has been reviewed elsewhere from different perspectives (Arnheim, 1983; Miklos, 1985; Ohta, 1988a) and the evolution of mitochondrial DNA has also been reviewed extensively (Palmer, 1985; Wilson et al., 1985; Avise, 1986; Avise et al., 1987; Moritz et al., 1987), I shall not deal further with either topic.

Evolution of single copy nuclear DNA sequences

Early studies of the restriction maps of protein coding genes and flanking regions showed a considerable level of variation within species and of divergence between related species (Jeffreys, 1979). Following the development of simple statistical procedures for estimating sequence divergence from restriction map data (Nei and Li, 1979; Engels, 1981; Ewens *et al.* 1981) some general features became apparent. Nucleotide divergence between *Drosophila melanogaster* and its sibling species *D.mauritiana* and *D.simulans* was estimated to about 3% in three regions of the genome (Leigh Brown and Ish-Horowicz, 1981; Langley *et al.* 1982). In primates, however, nucleotide divergence in the ß-globin gene cluster was much less. The estimated divergence between human and baboon was 4% while that between human and gorilla was only 0.9% (Barrie *et al.*, 1981). These results showed that, at randomly chosen restriction sites (with respect to variability), the morphologically almost identical drosophila species were almost as strongly diverged as human and baboon and much more so than human and gorilla.

Despite the differences in the results on restriction site divergence, the results of these three studies were in agreement on a second point. In each case the overall organization of the region of DNA under study, with respect to coding sequences and to the distance between restriction sites, remained the same, with the loss of occasional restriction sites in all these comparisons. Only in comparisons with much greater divergence in restriction sites was significant change generally seen in the organization of the gene clusters or in the distance between flanking restriction sites. A notable exception to this generalization is the *D.melanogaster* 87C heat shock locus (Leigh Brown and Ish-Horowicz, 1981) where centromeric repetitive DNA elements have inserted as a long tandem array between two coding sequences. Mills *et al.* (1986) have recently reported another example of localized variation in restriction map length between closely related species. There are two possible explanations for the conservation of restriction map length: either restriction map length is less liable to mutation than the nucleotide sequence of noncoding DNA or there are stronger selective constraints on the length of DNA between genes than on its actual base sequence. As we will see later, the latter is far from being the case so the indication is that changes in the length of non-coding DNA are not completely invisible to natural selection, although how they are expressed phenotypically is quite unclear.

Results from studies of intraspecific variation in restriction map do not show such concordance between mammals and drosophila. I shall review this area in greater detail, and examine two issues in particular: first, some of the results that have been obtained on linkage disequilibrium by analysis of restriction map variation, and second, the frequency and nature of restriction map variation due to DNA insertion and deletion.

Linkage disequilibrium in restriction map data

Although a newly arising mutation is found in only one combination of linked polymorphic variants, recombination rapidly causes these associations to decay with time. However, selection, natural or artificial, can generate nonrandom associations of existing alleles at different loci (gametic disequilibrium) when there is epistasis. In some cases this can lead to tight linkage of the loci involved into so-called 'supergenes' (Hedrick *et al.*, 1978). Such associations can also be generated in small populations by sampling effects and Alan Robertson made several important contributions on the complex interaction of selection and linkage (Hill and Robertson, 1966, 1968a; Robertson, 1970e, 1973c, 1977a; Robertson and Hill, 1983b). The demonstration of abundant polymorphism at the protein level using biochemical techniques led to a huge increase in the number of studies on disequilibrium in nature. However, allozyme data proved not to be ideal for the purpose (a large body of work has been reviewed succinctly by Langley, 1977). Polymorphic restriction sites are generally much more closely linked, can be unambiguously scored and represent a precise genetic locus, so in recent years restriction site data have come to be used extensively in studies of disequilibrium in nature.

In genetical terms, restriction maps represent a very fine scale analysis and restriction sites near the same gene or gene cluster will usually be very closely linked. It was therefore not surprising that early studies detected linkage disequilibrium among polymorphic restriction sites, and between certain restriction sites and genetic lesions such as ß-thalassaemia (Weatherall, 1985). Such disequilibria were found in both drosophila and mammalian genes (Langley *et al.*, 1982; Antonarakis *et al.*, 1982; Leigh Brown, 1983). Extensive work on the human ß-globin gene cluster has revealed a striking pattern of disequilibrium (Antonarakis *et al.*, 1982; Kazazian *et al.*, 1983; Chakravarti *et al.*, 1984). Eight restriction sites spread over 65 kb which span the entire cluster are polymorphic in all human races. Pairwise disequilibrium values between many of these polymorphic sites have been shown to be significant in several populations. However, in this gene cluster the probability of a disequilibrium value being significant is not simply a function of the molecular distance between the restriction sites. The restriction sites fell into two groups, one spanning 35 kb between the e and sß$_1$ genes and another spanning 19 kb to the 3' side of the ß-globin gene. Within each group, significant disequilibria are found frequently and in more than one population regardless of the distance between the restriction sites. Between the two groups, disequilibria are rarely significant even though the distance between the margins of the groups is only about ten kilobases (Antonarakis *et al.*, 1982; Chakravarti *et al.*, 1984).

This pattern has been interpreted as evidence for a recombination 'hot spot' around the δ-globin gene. Several features of the DNA sequence in that region have been put forward as possible mediators involved in recombination (Chakravarti *et al.*, 1984). However, interpretation of disequilibrium data in order to deduce recombination intervals is not entirely straightforward (Weir and Hill, 1986; Chakravarti *et al.*, 1986), and it should be noted that among American Blacks the 3' 'block' contains only one significant pairwise value in 15 estimates (Chakravarti *et al.*, 1984). So, while the recent observation of recombination events within families also indicates that there is an increased probability of recombination close to the d- and ß-globin genes (Gerhard *et al.*, 1984), it is not likely that all recombination in the cluster is restricted to this interval in all races.

Extreme localization of recombination events to 'hot spots' is known to be a feature of the major histocompatibility complex (MHC) of the mouse, the H-2 gene cluster on chromosome 17. There are 5 genes in this cluster which are highly polymorphic and, infrequently, recombinants between them have been detected. Two of these five genes, *H-2K* and *H-2D*, encode transplantation antigens and are known as Class I genes and three are Class II genes (*I-A$_\alpha$*, *I-A$_\beta$* and *I-E$_\beta$*) involved in the function of helper T cells. There are many other genes (over 50 coding sequences) which show lower levels of variability within the *H-2* complex. At the DNA level the entire gene cluster may span some 2500 kb or more and a genetic distance of about 2 cM (see Klein, 1986, for review). Extensive restriction site variation has been found between inbred strains of mice and these variable sites have been used to map the position of cross-over events detected immunologically (Steinmetz *et al.*, 1982). Remarkably, a high proportion of the cross-overs have occurred within very short regions of DNA (Steinmetz, 1986).

Such hot spots will clearly be important in determining the type of recombinants which arise at greatest frequencies. As hot spots appear to be sequence specific (Steinmetz, 1986) , they will differ in their location in related species or between populations. This will have complex implications for our expectations of linkage disequilibrium. Some particularly interesting data on the impact of hotspots on heterozygosity have been obtained from comparison of the restriction maps of the *H-2* gene complex. Cosmid clones covering almost the entire I region have been obtained from three unrelated strains of mice, Balb/c (*H-2^d* haplotype), AKR (*H-2^k* haplotype) and B.10.WR7 (*H-2^{WR7}*) (Steinmetz *et al.*, 1984). Restriction maps for 8 enzymes have been obtained giving up to 100 sites in each strain over the entire I gene region. A high level of restriction site polymorphism is present but within the region there are segments which differ significantly in their variability. The heterozygosity of the *I-A* region of the cluster is about three times that of the *I-E* region. The difference in variability is significant (Hudson, 1982) and the boundary between the variable and invariant regions coincides with the recombination hotspot in the *I-E$_\beta$* gene.

There is a remarkable parallel with the immunologically defined variability of the genes embedded in these two regions – the $I\text{-}A_\alpha$ genes and the $I\text{-}E_\beta$ genes in the variable segment being highly polymorphic while the $I\text{-}E_{\beta2}$ and $I\text{-}E_\alpha$ genes in the invariant segment are less variable. It has been considered for many years that the very high level of polymorphism found in the mouse and human MHC genes is the result of balancing selection. One of the reasons for this view is the allele frequency distributions observed in these gene clusters. Instead of the U-shaped frequency distribution expected of neutral alleles (Ewens, 1972; Watterson, 1978), those of the human HLA genes are flat, with many alleles found at similar frequencies (Hedrick and Thomson, 1983; Klitz et al., 1986). If we accept that such forces are acting on MHC genes we can go on to consider how they might influence the level of restriction site polymorphism.

It is well known that the action of selection at one locus can alter the gene frequencies and levels of heterozygosity of closely linked neutral alleles. For directional selection this has been termed 'genetic hitch-hiking' (Maynard Smith and Haigh, 1974). In the case of balancing selection, the effect, which can be very strong, is known as 'associative overdominance' or 'pseudoselection' (Hill and Robertson, 1968b; Sved, 1968; Ohta and Kimura, 1969b, 1970b; Thomson, 1977), as the heterozygosity of these linked loci will be raised. Because the intensity of the pseudoselection depends strongly on the recombination fraction between the selected and neutral loci, the alteration in heterozygosity due to this effect would be expected to fall dramatically at recombination 'hot spots'. This is precisely what is observed in the mouse H-2 genes (Steinmetz et al., 1984).

An earlier analysis compared enzyme variability on chromosome 17 with that for the rest of the genome but found little if any significant effect (Nadeau et al., 1982). Considering the short distances over which the increase in restriction site variation is detectable this would not be surprising but that study used the hitch-hiking model as the null hypothesis rather than pseudoselection, and it has been pointed out by Sved (1983) that that does not seem to be appropriate. We now appear to have an explanation – pseudoselection or associative over-dominance – for the unusual observations on restriction site polymorphism in the H-2 complex that depends solely on selection at the variable coding sequences. If this is correct, the effect is so strong that it should permit the detection of selective forces at other regions of the genome, given prior knowledge of the distribution of recombination events. With rapidly increasing knowledge at the molecular genetic level it could turn out that restriction site analysis succeeds where allozymes failed. It should be noted that Steinmetz (1986) interprets these data on the basis of gene conversion events which terminate at the recombination hot spots. This would require wide regions of sequence homology between the 'target' region and the DNA around the 'donor' coding sequences (which appear to be the class I genes in the Qa/TLa cluster (Mellor et al., 1983)). If it is found that the donor flanking DNA is not homologous, then the 'pseudoselection' explanation outlined above would be favoured.

What is known of the distribution of recombination in drosophila and its impact on restriction site variation? In drosophila, the genes which have been studied in greatest detail in this respect are not members of large complexes like the MHC. In addition, the analysis of the location of recombination events at the level of the restriction map is only just beginning (Clark et al., 1988). However, several genes have now been surveyed in natural populations for restriction map variation and for some of these there is a good background of genetic knowledge. Significant linkage disequilibria have been found in two autosomal and one X-linked locus (Langley et al., 1982; Aquadro et al., 1986; Leigh Brown, 1983; Cross and Birley, 1986; Beech and Leigh Brown, 1989). Two other X-linked loci showed no significant disequilibria, even between restriction sites only 2 kb apart (Langley and Aquadro, 1987; Schaeffer et al., 1988).

One recent study in the related species D. simulans, has raised an interesting contrast. In 20 kb on the third chromosome including the genes rosy (Xanthine dehydrogenase) and snake (a maternal effect pattern mutant), the level of heterozygosity was 4.0% per nucleotide, about 8 times that found in *D.melanogaster*. In the adjacent 20 kb studied, the heterozygosity was 0.5% (Aquadro *et al.*, 1988). There is a tantalising hint of a parallel with the situation at the mouse H-2 complex here, but a much more detailed genetic knowledge is required before any firm conclusions can be drawn.

Variation in restriction map due to DNA insertion

As indicated earlier, most restriction sites surveyed in population studies lie outside coding regions. As there appears to be only a relatively small region of DNA outside of coding regions whose sequence must be conserved for correct expression, the loss or gain of such restriction sites would not in general be expected to influence gene expression. However, restriction site analysis can also detect variation due to DNA insertion/deletion events and these may have a greater effect on the expression of neighbouring genes. The nature of the most common types of such insertion/deletion events appears to differ between mammals and drosophila. For example, in the human insulin gene region (Bell *et al.*, 1982) insertion/deletion variation has been shown to be due to changes in the number of copies of a dispersed tandemly repeated 'mini-satellite' sequence (Jeffreys *et al.*, 1985).

In contrast, in *D. melanogaster* there is a substantial amount of genetic variation due to the insertion of large (>500 bp) DNA segments (Langley *et al.*, 1982; Leigh Brown, 1983; Aquadro *et al.*, 1986; Langley and Aquadro, 1987; Beech and Leigh Brown, 1989; Schaeffer *et al.*, 1988). In several cases these have been shown to be complete or partial copies of transposable genetic elements (Leigh Brown, 1983; Aquadro *et al.*, 1986; Beech and Leigh Brown, 1989). The identification of DNA insertions 7–10 kb in length close to protein-coding genes in chromosomes which had been obtained from nature was a great surprise. The way in which such experiments were carried out necessitated that the regions under study should not carry mutations which would render them inviable or sterile. Nevertheless, it was clear that this type of variation was not subject to the same forces as restriction site variants. While many insertion events have been found in relatively small samples, in no case has any particular insertion reached a high frequency in any of the eight chromosomal regions studied to date. Once again, the observation of an allele frequency distribution that does not fit the U-shaped curve expected from the neutral, infinite allele model (Ewens, 1972; Watterson, 1978) has led to suggestions that selection is involved in determining the frequency of these restriction map variants (Langley *et al.*, 1982; Leigh Brown, 1983) although in this case selection appears to be acting to reduce variation by eliminating chromosomes bearing insertions.

The frequency distribution observed for DNA insertions could, however, also be produced if the rate of excision was high. Golding *et al.* (1986) adopted a phylogenetic model to investigate this question on the data of Aquadro *et al.* (1986) on the *Adh* locus and they found that a high excision rate (copy number independent) was not sufficient to provide a good fit to the observed frequencies of chromosomes bearing insertions. The most parsimonious tree which relates the observed *Adh* haplotypes results in haplotypes bearing transposable element insertions appearing at the tips of the tree. From no such chromosome has another arisen, apparently. This conclusion mirrors a point made earlier in this review that, in general, insertion/deletion changes in restriction maps have not been fixed during evolution.

The involvement, or otherwise, of selection in determining the frequencies of transposable element insertions can also be tested experimentally. While the excision rate per element will not be expected to be systematically different between loci located on different chromosomes, selection pressures may differ. Any deleterious mutation which is at least partly recessive will be eliminated more rapidly if it is X-linked, because of the hemizygosity of males. Several recent studies have set out to examine whether this is the case for transposable element insertions (see also Charlesworth – this volume). Beech and Leigh Brown (1989) compared the results from these studies with the autosomal data from the *Adh* and 87A heat-shock regions. There were significantly fewer insertions, on a per chromosome, per kilobase basis, in the X-linked AS-C and *notch* regions than in the autosomal loci, as would be predicted if insertions were mildly deleterious and partly recessive in their effects. However the *white* locus, also X-linked, was not significantly different from the autosomal loci in the frequency of large insertions. In addition, Montgomery *et al.* (1987) found that only one out of three elements studied had significantly fewer copies on the X chromosome than on the autosomes. Clearly there is variation between loci that is not explained by their chromosomal location and there appears to be variation between element families. The factors responsible for these differences are as yet unidentified.

The suggestions from these results that transposable elements have detectable, if mild, deleterious effects, raises the question of why they should be so abundant in *D. melanogaster* by comparison even with mammals or its sibling species *D. simulans* (Aquadro *et al.*, 1988; Leigh Brown and Moss, unpublished data), which even shares many of the same families of elements (Young and Schartz, 1981). The answer to that question will only be obtained by studies which combine population genetics and evolutionary biology with molecular biology and genetics. The foresight shown by Alan Robertson in recognizing the need to combine these fields and his encouragement of those who attempted to do so were important contributions to this area of research.

Chapter 7
Analysis of Disequilibrium Coefficients

B. S. WEIR and C. Clark COCKERHAM

Department of Statistics, North Carolina State University,
Raleigh, NC 27695-8203, USA

In a population genetic setting, *disequilibrium* refers to the departure of the joint frequency of a set of genes from the product of their single frequencies. Although the term is used mainly for genes at different loci on a single gamete, it need not be so restricted. The oldest work on disequilibrium refers to pairs of genes at the same locus, and we could make the case that much of population genetic theory is concerned with treating situations where this Hardy–Weinberg equilibrium does not hold. Much of population genetic data analysis is concerned with detecting departures from Hardy–Weinberg or other equilibria. As we shall note, Alan Robertson has contributed to both these theoretical and data-analytic activities.

Coefficients of disequilibrium

In large random mating populations, when there are no forces such as selection or mutation, the Hardy–Weinberg law tells us that genotypic frequencies are completely specified by gene frequencies. In terms of genes at a single locus **A**, the frequency of homozygotes formed by the union of two gametes both carrying allele A_i is

$$P_{A_i}^{A_i} = p_{A_i}^2,$$

while the frequency of heterozygotes from the union of A_i and A_k gametes is

$$P_{A_k}^{A_i} + P_{A_i}^{A_k} = 2p_{A_i}p_{A_k}, \qquad i \neq k.$$

We do not usually wish to distinguish between $P_{A_k}^{A_i}$ and $P_{A_i}^{A_k}$, and use twice either quantity to denote the total frequency of heterozygotes. These genotypic frequencies are reached after a single generation of random mating.

There is an obvious extension to more than one locus. For loci **A** and **B**, for example, double homozygotes formed by the union of two A_iB_j gametes have frequency

$$P_{A_iB_j}^{A_iB_j} = p_{A_i}^2 p_{B_j}^2,$$

while the single heterozygotes formed by the union of A_iB_j and A_iB_l or of A_iB_j and A_kB_j gametes have frequencies

$$2P_{A_iB_l}^{A_iB_j} = 2p_{A_i}^2 p_{Bj}p_{B_l}, \qquad j \neq l$$

$$2P_{A_kB_j}^{A_iB_j} = 2p_{A_i}p_{A_k}p_{B_j}^2, \qquad i \neq k,$$

and the double heterozygote frequencies are

$$4P^{A_iB_j}_{A_kB_l} = 4p_{A_i}p_{A_k}p_{B_j}p_{B_l}, \qquad i \neq k, j \neq l.$$

These two-locus Hardy–Weinberg frequencies are reached only asymptotically in random mating populations, at a rate that depends on the degree of linkage between the loci. Eventually, though, genotypic frequencies are completely specified by gene frequencies.

At a single locus, there are several ways of characterizing departures from Hardy–Weinberg equilibrium, and Robertson and Hill (1984) have recently given a discussion in terms of inbreeding coefficients. A coefficient F can be introduced that allows genotypic frequencies to be parameterized as

$$P^{A_i}_{A_i} = p^2_{A_i} + Fp_{A_i}(1 - p_{A_i})$$

for homozygotes, and

$$2P^{A_i}_{A_k} = 2p_{A_i}p_{A_k} - 2Fp_{A_i}p_{A_k}, \qquad i \neq k.$$

for heterozygotes. The parameter F is a convenient means of measuring departures from Hardy–Weinberg equilibrium, and values can be specified in the neutral case when there are no selective forces acting. For data analysis we prefer to adopt other measures of disequilibrium, one for each genotype:

$$P^{A_i}_{A_i} = p^2_{A_i} + D^{A_i}_{A_i},$$
$$2P^{A_i}_{A_k} = 2p_{A_i}p_{A_k} + 2D^{A_i}_{A_k}, \qquad i \neq k.$$

Since genotypic frequencies must add to gene frequencies

$$p_{A_i} = P^{A_i}_{A_i} + \frac{1}{2} \sum_{k \neq i} 2P^{A_i}_{A_k} = \sum_k P^{A_i}_{A_k},$$

there are relations among the disequilibrium coefficients,

$$D^{A_i}_{A_i} + \sum_{k \neq i} D^{A_i}_{A_k} = 0,$$

and it is sufficient to work with one coefficient for every heterozygote. To make things concrete, for a two-allele case, the genotypic frequencies can be expressed as

$$P^{A_1}_{A_1} = p^2_{A_1} + D^1_2,$$
$$2P^{A_1}_{A_2} = 2p_{A_1}p_{A_2} - 2D^1_2,$$
$$P^{A_2}_{A_2} = p^2_{A_2} + D^1_2,$$

where the signs have been reversed from those given in the definitions.

The measures D are seen to refer to the failure of pairs of allelic genes to have joint frequencies that are products of single frequencies. An entirely parallel set of parameters refer to pairs of non-allelic genes. We can define the coefficient of *linkage disequilibrium* for alleles A_i and B_j at loci **A** and **B** as

$$D_{A_iB_j} = P_{A_iB_j} - p_{A_i}p_{B_j}$$

where $P_{A_iB_j}$ is a gametic frequency. Strictly, then, we should refer to *gametic linkage disequilibrium*. Several authors have pointed out that even the word *linkage* is misleading since linkage is not necessary for there to be disequilibrium. At this point though, it seems that the original nomenclature is here to stay. There is an analogous *nongametic disequilibrium coefficient*

$$D_{\cdot B_j}^{A_i} = P_{\cdot B_j}^{A_i} - p_{A_i} p_{B_j},$$

where the dots indicate summation over alleles in those positions

$$P_{\cdot B_j}^{A_i} = \sum_k \sum_l P_{A_k B_l}^{A_i B_l}.$$

It is simpler to write this frequency as P_{A_i/B_j}. We can find gametic frequencies in the same way, and could also use dots to show the summations that have occurred, although this is probably not necessary

$$P_{\cdot \cdot}^{A_i B_j} = P_{A_i B_j} = \sum_k \sum_l P_{A_k B_l}^{A_i B_l}$$

It is convenient to maintain the parallel treatment with Hardy–Weinberg disequilibrium. When loci **A** and **B** are completely linked, the Hardy–Weinberg and nongametic linkage disequilibrium coefficients are identical. It will simplify developments if we focus attention on particular alleles A and B at loci **A** and **B**, and modify the notation for these two-allele disequilibrium coefficients to

$$D_A = P_A^A - p_A^2, \qquad D_B = P_B^B - p_B^2,$$

$$D_{AB} = P_{AB} - p_A p_B, \qquad D_{A/B} = P_{A/B} - p_A p_B.$$

The collection of alleles other than A and B can be written as $\bar{A}$ and $\bar{B}$.

To illustrate the notation, suppose that the 10 two-locus genotypes for these two loci have the following frequencies – arranged to show the one-locus marginal frequencies

$P_{AB}^{AB} = 0.20$	$2P_{AB}^{A\bar{B}} = 0.18$	$P_{AB}^{\bar{B}} = 0.02$	$P_A^A = 0.40$
$2P_{\bar{A}B}^{AB} = 0.26$	$2P_{\bar{A}B}^{A\bar{B}} = 0.08$ $2P_{A\bar{B}}^{A\bar{B}} = 0.04$	$2P_{\bar{A}\bar{B}}^{A\bar{B}} = 0.02$	$2P_A^{\bar{A}} = 0.40$
$P_{\bar{A}B}^{\bar{A}B} = 0.04$	$2P_{\bar{A}B}^{\bar{A}\bar{B}} = 0.10$	$P_{\bar{A}\bar{B}}^{\bar{A}\bar{B}} = 0.06$	$P_A^{\bar{A}} = 0.20$
$P_B^B = 0.50$	$2P_B^{\bar{B}} = 0.40$	$P_B^{\bar{B}} = 0.10$	1.00

$$p_A = 0.60$$
$$p_B = 0.70$$

The two Hardy–Weinberg coefficients are then seen to be

$$D_A = P_A^A - p_A^2 = 0.04,$$

$$D_B = P_B^B - p_B^2 = 0.01.$$

We next arrange the genotype frequencies to show gametic and nongametic marginal frequencies

$P_{AB}^{AB} = 0.20$	$P_{AB}^{A\bar{B}} = 0.09$	$P_{AB}^{\bar{A}B} = 0.13$	$P_{AB}^{\bar{A}\bar{B}} = 0.04$	$P_{AB} = 0.46$
$P_{A\bar{B}}^{AB} = 0.09$	$P_{A\bar{B}}^{A\bar{B}} = 0.02$	$P_{A\bar{B}}^{\bar{A}B} = 0.02$	$P_{A\bar{B}}^{\bar{A}\bar{B}} = 0.01$	$P_{A\bar{B}} = 0.14$
$P_{\bar{A}B}^{AB} = 0.13$	$P_{\bar{A}B}^{A\bar{B}} = 0.04$	$P_{\bar{A}B}^{\bar{A}B} = 0.04$	$P_{\bar{A}B}^{\bar{A}\bar{B}} = 0.05$	$P_{\bar{A}/B} = 0.26$
$P_{\bar{A}\bar{B}}^{AB} = 0.02$	$P_{\bar{A}\bar{B}}^{A\bar{B}} = 0.01$	$P_{\bar{A}\bar{B}}^{\bar{A}B} = 0.05$	$P_{\bar{A}\bar{B}}^{\bar{A}\bar{B}} = 0.06$	$P_{\bar{A}/\bar{B}} = 0.14$
$P_{A/B} = 0.44$	$P_{A/\bar{B}} = 0.16$	$P_{\bar{A}B} = 0.24$	$P_{\bar{A}\bar{B}} = 0.16$	1.00.

The two linkage disequilibrium coefficients can be calculated as

$$D_{AB} = P_{AB} - p_A p_B = 0.04,$$

$$D_{A/B} = P_{A/B} - p_A p_B = 0.02.$$

Inferences about disequilibria

When data are available from a single population, we may wish to estimate the various coefficients and to test hypotheses about them. Estimates and tests are constructed by assuming that, over all possible samples from the population sampled, genotypic frequencies are multinomially distributed.

Looking just at locus **A**, suppose that a sample of n individuals has n_{AA} homozygotes AA, $2n_{A\bar{A}}$ heterozygotes $A\bar{A}$ and $n_{\bar{A}\bar{A}}$ homozygotes $\bar{A}\bar{A}$. The number n_A of A genes in the sample is therefore

$$n_A = 2n_{AA} + n_{A\bar{A}}.$$

Sample gene and genotypic frequencies are denoted by tildes

$$\tilde{p}_A = \frac{n_A}{2n}, \qquad \tilde{P}_A^A = \frac{n_{AA}}{n},$$

and these are also maximum likelihood estimates of the population frequencies. The Hardy–Weinberg disequilibrium coefficient therefore has a maximum likelihood estimate, denoted by a caret, of

$$\hat{D}_A = \tilde{P}_A^A - \tilde{p}_A^2.$$

The estimate has a small bias, since its expected value over all samples, found from properties of the multinomial distribution, is

$$\mathscr{E}\hat{D}_A = D_A - \frac{1}{2n}(p_A p_{\bar{A}} + D_A).$$

There is a rich literature on testing procedures, and we have found that a satisfactory procedure for testing the hypothesis that $D_A = 0$ is to assume the estimate is approximately normally distributed, and to regard

$$X_A^2 = \frac{\hat{D}_A^2}{\text{var}(\hat{D}_A)}$$

as a chi-square variable with one degree of freedom. Under H_0, the variance of $\hat{D}_A$ is approximately $p_A^2 p_{\bar{A}}^2 / n$ so that the usual test statistic for Hardy–Weinberg is recovered

$$X_A^2 = \frac{n\hat{D}_A^2}{\tilde{p}_A^2 \tilde{p}_{\bar{A}}^2} = n\left(\frac{n_{AA}n_{\bar{A}\bar{A}} - n_{A\bar{A}}^2}{(n_{AA} + n_{A\bar{A}})(n_{\bar{A}\bar{A}} + n_{A\bar{A}})}\right)^2.$$

This statistic can also be found as a goodness-of-fit chi-square on the three genotypic classes, by comparing observed numbers to those expected on the basis of the estimated gene frequencies.

If we wish to take simultaneous account of m multiple alleles, instead of testing for Hardy–Weinberg for each allele separately we can use a goodness-of-fit test on the $m(m+1)/2$ genotypic classes. This $m(m-1)/2$ degree-of-freedom chi-square can be regarded as providing a test statistic for the joint hypothesis

$$H_0: \quad D_{A_k}^{A_i} = 0, \qquad \text{for all } i \neq k.$$

If this hypothesis is rejected, likelihood ratio tests can be established to test hypotheses about subsets of the coefficients. Hernández and Weir (1989a) have shown that it is much easier, and quite satisfactory, to use variance-based tests as for two alleles. The appropriate test statistic for $H_0: D_{A_k}^{A_i} = 0$ is

$$X_{A_iA_k}^2 = \frac{2n\hat{D}_{A_iA_k}^2}{\tilde{p}_{A_i}\tilde{p}_{A_k}[(1-\tilde{p}_{A_i})(1-\tilde{p}_{A_k})+\tilde{p}_{A_i}\tilde{p}_{A_k}]+\sum_{l\neq i,k}(\tilde{p}_i^2\hat{D}_{A_l}^{A_k}+\tilde{p}_k^2\hat{D}_{A_l}^{A_i})}.$$

If these one-locus tests indicate the absence of Hardy–Weinberg disequilibrium, we may wish to assume that two-locus genotypic frequencies are simply the products of appropriate gametic frequencies. Even if coupling and repulsion double heterozygotes cannot be distinguished, W. G. Hill (1974b) showed that maximum likelihood estimates $\hat{P}_{AB}$ of the gametic frequencies can then be found. It is a simple matter to estimate the gametic linkage disequilibrium from the gametic frequency (the non-gametic coefficient having been assumed to be zero):

$$\hat{D}_{AB} = \hat{P}_{AB} - \tilde{p}_A\tilde{p}_B.$$

A variance-based chi-square can be constructed as for the Hardy–Weinberg case, and it is usually assumed that the sampling variance for the estimate is

$$\text{var}(\hat{D}_{AB}) = \frac{1}{2n}[p_A(1-p_A)p_B(1-p_B)+(1-2p_A)(1-2p_B)D_{AB}-D_{AB}^2].$$

Under the hypothesis $H_0: D_{AB}=0$ the test statistic becomes the familiar

$$X_{AB}^2 = \frac{2n\hat{D}_{AB}^2}{\tilde{p}_A(1-\tilde{p}_A)\tilde{p}_B(1-\tilde{p}_B)}$$

which also follows from a goodness-of-fit on the four gametic classes AB, $A\bar{B}$, $\bar{A}B$ and $\bar{A}\bar{B}$.

Progress can also be made without the Hardy–Weinberg assumption. It is sometimes possible, through family studies for example, to determine the phase of double heterozygotes and so obtain directly the gametic frequencies in a sample. It is the genotypic, rather than gametic, frequencies that have a multinomial distribution, however. We have shown (Weir, 1979; Weir and Cockerham, 1989) that the variance of the estimated gametic linkage disequilibrium coefficient requires the introduction of a four-gene disequilibrium coefficient. The variance of $\hat{D}_{AB}$ also involves D_A, D_B and $D_{A/B}$. The test statistic must reflect the fact that the data are collected on genotypes and that between-gamete associations need to be invoked in sampling theory.

Without direct knowledge of gamete frequencies, and without assuming Hardy–Weinberg equilibrium, we advocate the use of a composite measure of linkage disequilibrium, Δ_{AB}, defined as the sum

$$\Delta_{AB} = D_{AB} + D_{A/B}.$$

Maximum likelihood estimation of Δ_{AB} is straightforward

$$\hat{\Delta}_{AB} = \tilde{P}^{AB} + \tilde{P}_{A/B} - 2\tilde{p}_A\tilde{p}_B,$$

where the sum of gametic and nongametic frequencies follows directly from the data as

$$\tilde{P}^{AB} + \tilde{P}_{A/B} = \frac{1}{n}[2n^{AB}_{AB} + (2n^{AB}_{A\bar{B}} + 2n^{AB}_{\bar{A}B}) + \tfrac{1}{2}(2n^{AB}_{A\bar{B}} + 2n^{A\bar{B}}_{\bar{A}B})]$$

and the total count of double heterozygotes $2n^{AB}_{A\bar{B}} + 2n^{A\bar{B}}_{\bar{A}B}$ does not require knowledge of gametic phase.

The sampling variance of $\hat{\Delta}_{AB}$ depends on three- and four-gene coefficients of disequilibrium, and these should be tested for first (Weir and Cockerham, 1989). If they can be shown to be not significantly different from zero, then a reduced variance formula is used to construct a test statistic for H_0: $\Delta_{AB} = 0$. This statistic takes explicit note of possible Hardy–Weinberg disequilibrium at each locus:

$$X^2_{AB} = \frac{n\hat{\Delta}^2_{AB}}{[\tilde{p}_A(1-\tilde{p}_A) + \hat{D}_A][\tilde{p}_B(1-\tilde{p}_B) + \hat{D}_B]}.$$

We have recently confirmed (Hernández and Weir, 1989b) that the variance-based tests for linkage disequilibrium and higher-order disequilibria perform well when compared with likelihood-ratio tests.

Example

Hill (1975) analysed some isozyme data on the marine fish *Fundulus heteroclitus* supplied by J. B. Mitton. For loci **A** and **B**, the counts are

$n^{AB}_{AB} = 19$	$2n^{AB}_{A\bar{B}} = 9$	$n^{A\bar{B}}_{A\bar{B}} = 2$	$n^A_A = 30$
$2n^{AB}_{\bar{A}B} = 75$	$2n^{AB}_{A\bar{B}} + 2n^{A\bar{B}}_{\bar{A}B} = 34$	$2n^{A\bar{B}}_{\bar{A}B} = 11$	$2n^A_{\bar{A}} = 120$
$n^{\bar{A}B}_{\bar{A}B} = 144$	$2n^{\bar{A}B}_{\bar{A}\bar{B}} = 45$	$n^{\bar{A}\bar{B}}_{\bar{A}\bar{B}} = 10$	$n^{\bar{A}}_{\bar{A}} = 169$
$n^B_B = 208$	$2n^B_{\bar{B}} = 88$	$n^{\bar{B}}_{\bar{B}} = 23$	$n = 319.$

The estimates and test statistics for Hardy–Weinberg coefficients indicate a clear departure from equilibrium at locus **B**. As we cannot, therefore, assume random-mating to estimate gametic frequencies, and as double heterozygotes cannot be distinguished, we proceed with the composite coefficient:

$$\hat{D}_A = 0.0144, \qquad X^2_A = 1.62;$$
$$\hat{D}_B = 0.0280, \qquad X^2_B = 9.08;$$
$$\hat{\Delta}_{AB} = -0.0100, \qquad X^2_{AB} = 0.76.$$

Application of the methods for higher-order coefficients (Weir and Cockerham, 1989) shows that they also are not significantly different from zero.

Discussion

We have given a brief treatment of genetic disequilibrium, stressing the parallels between Hardy–Weinberg and linkage disequilibrium. If mating truly is at random, then it is possible to work with gametic rather than genotypic frequencies, and we can concentrate on the gametic linkage disequilibrium. In the more general situation, the genotypic nature of the data must be reflected in the analyses. In either case, we cannot treat the two sets of disequilibria (Hardy–Weinberg and linkage) as being independent. The requirement for genotypic frequencies to be positive and to be bounded by gametic (and nongametic) frequencies imposes constraints on the various coefficients (Weir and Brooks, 1986). As an example, we see that gene frequencies of $p_A = 0.5$, $p_B = 0.8$ might suggest that D_{AB} has an upper bound of 0.10, but this bound is reduced to 0.08 if $D_A = -0.25$ and $D_B = 0$.

We have made brief reference to analyses that go beyond those for one allele at one or two loci. Hardy–Weinberg coefficients for multiple alleles may be tested easily. At two loci, the numbers of coefficients increased too rapidly with the number of alleles to allow reliable analyses. We have supplemented the two basic quantities D_{AB} and $D_{A/B}$ by coefficients for three and four of the four genes present per individual at two loci. Extensions to three or four loci are manageable when gametic data are available (e.g. Weir and Brooks, 1986), but not for genotypic data.

Although we have concentrated on methods of inference for disequilibrium, we point out that a great deal is known about the expected behaviour of the coefficients over time once population and mating structures are specified. Hill and Robertson (1966, 1968a) showed how linkage disequilibria changed over time. They also (Hill and Robertson, 1968a) gave transition equations for squared linkage disequilibria, which are of interest because they give rise to quantities that tend to nonzero limits over time even when the mean disequilibrium becomes zero. The difficulty of drawing inferences about squared disequilibria was illustrated by Hill and Weir (1988) who showed that these quantities have very large sampling variances.

Acknowledgements

Paper number 11577 of the Journal Series of the North Carolina Agricultural Research Service, Raleigh, NC 27695-7601. This investigation was supported in part by NIH Grant GM 11546.

Chapter 8
The Population Biology of Transposable Elements

Brian CHARLESWORTH

Department of Ecology and Evolution, The University of Chicago, Chicago, IL 60637, USA

Introduction

The genomes of most species contain families of repeated DNA that have the property of self-replication and movement to novel locations within the genome (Shapiro, 1983). This finding has generated a considerable amount of debate concerning their significance in evolution and the means by which they are maintained within their host populations (Doolittle and Sapienza, 1980; Orgel and Crick, 1980; Campbell, 1983; Syvanen, 1984). While much of this debate was initially conducted in the absence of data on the distribution of transposable elements between host individuals in natural populations and of clearly-formulated models of their population biology, these deficiencies have to some extent been remedied in recent years (Charlesworth, 1985, 1988; Brookfield, 1986; Hartl *et al.*, 1986). Alan Robertson was quick to take an interest in these issues, and encouraged research on the population biology of transposable elements (Yamaguchi *et al.*, 1987).

Both theoretical and empirical studies will be reviewed here. The focus will be on models and data relating to randomly mating, diploid populations with elements dispersed over multiple genomic sites and with relatively high frequencies of recombination between different sites, since these provide the most clear-cut means of discriminating between alternative hypotheses. An excellent review of results pertaining to bacterial populations, where there are strong correlations between element frequencies at different sites due to the limited opportunities for transfer between host individuals, is provided by Hartl *et al.* (1986).

Population statics

Distribution of element numbers between individuals

Consider a particular family of elements in a given generation. It is assumed that there is a number m of chromosomal sites in a haploid genome, into which members of the family are capable of inserting. It seems clear that m is usually a large number, probably of the order of several thousand in a multicellular organism such as maize or drosophila. The number of elements belonging to a given family in the diploid genome of a given individual is denoted by n. Studies of populations of yeast, drosophila and bacteria have shown wide variation between individuals in the sites at which elements are located (see below), indicating that n is very much less than the total number of sites available for occupation in a genome. Such studies also indicate some variation between individuals of a population with respect to the value of n, which has mean and variance over individuals denoted by $\bar{n}$ and V_n.

Simple quantitative genetic considerations enable some predictions to be made about the properties of the distribution of the number of elements between host individuals. The number of elements per individual can be regarded as a strictly additive, polygenic trait, since it is merely the sum of the number of elements at each site. If there is no linkage disequilibrium and no variance in element frequencies between different sites, it follows from standard theory that there is a binomial distribution of n across individuals. If $\bar{n} < 2m$, then this reduces to a Poisson distribution, such that $V_n \simeq \bar{n}$. Variation between sites in element frequencies reduces the variance below binomial expectation; if there is a tendency for the frequency of occupation of one site to be higher if a neighbouring site is occupied (positive linkage disequilibrium), then the variance will be greater than for the binomial (Charlesworth and Charlesworth, 1983).

Population data to test these predictions seem to be available only for *Drosophila melanogaster* and *E. coli*. Data on the former are mostly from the technique of *in situ* hybridization of labelled element probes to the polytene salivary chromosomes, enabling counts of the numbers of elements in the euchromatin of each chromosome arm to be made (e.g. Montgomery *et al.*, 1987). Overall, agreement of the data with Poisson expectation is generally fairly good. For example, Table 8.1 shows the results of a survey of the distribution of ten families of elements over 14 X chromosomes isolated from a Maryland population (Charlesworth, unpublished data). Data on bacterial *IS* sequences, obtained by using Southern blot estimates of copy numbers for individual isolates of *E. coli*, show wide deviations from the binomial distribution, as would be expected from the linkage disequilibrium generated by the sporadic nature of genetic transmission in this species. In contrast to the results for drosophila, a considerable fraction of strains lack copies of one or more families of elements (Hartl *et al.*, 1986).

Table 8.1
Means ($\bar{n}$) and variance (V_n) of number of elements on a sample of 14 X chromosomes from a Maryland population for ten families of *D. melanogaster*

Element	$\bar{n}$	V_n	Element	$\bar{n}$	V_n
412	2.29	2.53	2217	1.79	1.41
copia	1.21	0.80	2161	4.07	5.46
2181	1.21	0.49	2210	0.64	0.55
2158	0.21	0.18	roo	11.4	4.40
297	4.43	4.88	2156	0.50	0.58

Note: The copy numbers shown omit values for the base of the X chromosome (18D1-20A4) because many elements show a differential accumulation of elements in this region. Details on the elements are given by Brookfield *et al.* (1984).

At present, there seems little reason to doubt that the nature of the distribution of elements between individuals in natural populations simply reflects the consequences of Mendelian transmission, although more population data for a wider variety of species are needed.

Frequencies of elements at individual chromosomal sites

Identification of the sites of labelling of probes to sets of isolated salivary chromosomes also permits analysis of the frequencies of elements at individual chromosomal sites. The resolution of this technique is somewhat coarse, since the chromosomal locations of elements can only be determined down to the level of a salivary chromosome band. Nonetheless, a useful picture of the general features of the distribution of element frequencies can be obtained (Montgomery and Langley, 1983; Montgomery *et al.*, 1987; Leigh Brown and Moss, 1987; Charlesworth, 1988). The results of population surveys show that usually a site is occupied only once in the sample for each element in the family. Because of the lack of resolution of the method, it is possible that at least some instances of apparent multiple occupancy of the same site are the result of insertions into different sites within the same salivary chromosome band (Kaplan and Brookfield, 1983b). An alternative method is to employ restriction fragment mapping of a defined region of the genome to determine with much greater accuracy the locations of insertions of elements in a sample of genomes from a population. In *D. melanogaster*, this method has been applied to the *Adh* region (Langley *et al.*, 1982; Aquadro *et al.*, 1986), to the heat-shock region (Leigh Brown, 1983), and to the *white* locus (Langley and Aquadro, 1987). Each of these studies indicates that element insertions appear to occur at unique locations within a sample of a few tens of genomes, indicating low frequencies at these locations in the population from which the samples were drawn. Southern restriction mapping studies of complete yeast (Cameron *et al.*, 1979) and *E. coli* (Hartl *et al.*, 1986) genomes similarly indicate low element frequencies at individual sites. Further analysis and interpretation of these data require consideration of possible dynamic processes affecting element frequencies and abundances, which will now be discussed.

Population dynamics

Infinite population models

Models of the transmission of a family of elements from generation to generation incorporate the following features.

Change in copy number due to transposition

The probability per generation that a given element in the germ-line of an individual with n elements of the family in question produces a new copy that is inserted elsewhere in the genome is denoted by u_n. The subscript n is used to take into account the possibility of functional dependence of the probability of transposition of an element on the number of elements in the same cell; there is evidence for self-regulation of transposition probability (such that u_n is a decreasing function of n) in a variety of species (Charlesworth and Langley, 1986). The value of u for most eukaryote elements seems to be of the order of 10^{-4} or less per generation (Charlesworth and Charlesworth, 1983; Eggleston *et al.*, 1988).

Excision of elements

The probability per generation that an element is excised from its location in the genome, and hence not transmitted to the next generation, is denoted by v. (v could obviously also be written as a function of n, but there is little evidence that this is often so.) There is experimental evidence for such excision events in bacterial transposable elements (Kleckner, 1981) and drosophila *P* elements (Engels, 1986), but no firm evidence for excision of the abundant retroviral-like elements of drosophila (Finnegan and Fawcett, 1986). If transposition is regulated, and if the rate of transposition at low copy number (u_0) exceeds v, there will be an increase in copy number until $n \simeq \hat{n}$, where $u_{\hat{n}} = v$, or all available sites have filled up with elements (Charlesworth and Charlesworth, 1983). In the former case, there will be a stable equilibrium at $\hat{n}$, with the frequency of elements at each site equal to $\hat{n}/(2m)$. If there is no regulation, transposition leads to fixation of elements at each site if $u_0 > v$. Unless regulation of transposition is so strong that there is no transposition at all above a certain threshold copy number, maintenance of an equilibrium with element frequency polymorphism requires a nonzero excision rate.

Natural selection

There are several ways in which natural selection can affect the abundance of elements within populations, which will be outlined below. All of them have the common feature that the fitness of a host individual with copy number n for a given family is assumed to be a decreasing function, w_n, of n. If this is the case, and if similar assumptions to those used for equation (8.1) are used, the following equation is obtained for the change in mean copy number per generation (Charlesworth and Charlesworth, 1983; Charlesworth, 1985):

$$\Delta \bar{n} \simeq \bar{n}(1-\bar{n}/[2m])(\delta \ln \bar{w}/\delta \bar{n}) + \bar{n}(u_{\hat{n}}-v) \qquad (8.1)$$

where $\bar{w}$ is the mean fitness of the population ($\bar{w} \simeq w_{\bar{n}}$).

The approximate equilibrium value of $\bar{n}$, $\hat{n}$, is given by setting $\Delta \bar{n}$ to zero. It is relatively easy to find functional forms that result in an equilibrium with low element frequencies at each site; for example, with $w_n = \exp(-tn^2/2)$, we have $\hat{n} = (u_{\hat{n}}-v)/t$. In the absence of regulated transposition, the stability of this equilibrium requires that the logarithm of fitness declines more steeply than linearly with increasing n (Charlesworth and Charlesworth, 1983; Charlesworth, 1985). Given that low frequencies of transposition seem to be the norm, these results imply that even a weak selection pressure (as measured by t, which is the slope of the relation between the logarithm of fitness and copy number at a copy number of 1) is capable of maintaining a balance with transpositional increase in copy number. For example, a mean copy number of 50 elements per diploid genome would be maintained if t is one-fiftieth of the excess of rate of transposition over excision. With transposition rates of the order of 10^{-4} or less, this implies that extremely small selection coefficients would be needed to maintain a balance. The loss in mean fitness to the population would be negligible, since the mean fitness of the population is given by $\bar{w} = \exp[-(u_{\hat{n}}-v)]$, regardless of the mode of selection (Charlesworth, 1985).

One possible mode of selection on transposable elements is through the deleterious mutational effects of the insertion of elements into or near genes (Shapiro, 1983). Recent studies of P element mutagenesis in $D.$ $melanogaster$ have shown that transposition is frequently accompanied by detrimental mutations (Yukuhiro et al., 1985; Mackay, 1985b, 1986; Fitzpatrick and Sved, 1986).

Theoretical considerations (Charlesworth, 1988) lead to the conclusion that, although selection against insertional mutations may be a factor in stabilizing element frequencies in natural populations, it may not be the only force involved, since sites where such mutations are nearly neutral would accumulate elements at high frequency. One possibility is that regulation of transposition rates may act in addition to selection. This makes it much easier to achieve stabilization of element abundances at reasonable levels. Another possibility is that crossing over between homologous elements located at different chromosomal sites will lead to the production of deleterious chromosome rearrangements (Montgomery et al., 1987), and models of this process have been developed (Langley et al., 1988). These models predict higher equilibrium abundances of elements in regions where exchange is infrequent.

Genetic drift

Since the deterministic forces acting on transposable elements seem to be so weak, the full interpretation of statistics on element frequencies in natural populations requires that effects of genetic drift be taken into account. Stochastic models of element frequencies have been constructed by Ohta and Kimura (1981), Ohta (1981,1983a), Charlesworth and Charlesworth (1983), Langley et al. (1983), Kaplan and Brookfield (1983a), Charlesworth (1985) and Kaplan et al. (1985). The form of the probability distribution of element frequencies for a given family that is attained when the forces of transposition, excision and selection considered above come into statistical equilibrium with random changes in element frequencies – the stationary distribution of element frequencies – has been derived (Langley et al., 1983; Charlesworth and Charlesworth, 1983).

The following parameters provide a complete approximate description of the form of $\bar{\phi}(x)$, the stationary probability density for element frequency x: N_e is the effective size of a local population; $\bar{n}$ is the expected value of the mean copy number, which can be approximated by the equilibrium value $\hat{n}$ for a large population; $\alpha = 4N_e u\bar{n}/(2m - \bar{n})$ measures the effect of drift, and the effect of transposition in causing insertions into a given site; $\beta = 4N_e(s + v)$, where s is the value of $-\delta \ln \bar{w}/\delta\bar{n}$ at $\bar{n} = \hat{n}$, measures the joint effects of drift, excision and selection.

ϕ is approximated by a β distribution:

$$\phi(x) \simeq [\Gamma(\alpha + \beta)/\Gamma(\alpha)\Gamma(\beta)]x^{\alpha-1}(1-x)^{\beta-1} \qquad (8.2)$$

(Charlesworth and Charlesworth, 1983). This is the formula for a closed population; if $\bar{n} < 2m$, the effect of migration can be included by adding $4N_e M$ to β, where M is the frequency of immigrants into a local population. If m is sufficiently large compared with $\bar{n}$, α can be neglected in equation (8.2). This yields the formula derived by Langley et al. (1983) for the expected number of sites per haploid genome with element frequency x, $\Phi(x) = m\phi(x)$.

Tests of the models

Given the small probable magnitudes of the forces involved in determining element frequencies, apart from exceptional situations such as hybrid dysgenesis (Engels, 1986), it seems that attempts to discriminate between the various possibilities for the maintenance of stable element frequencies in natural populations by direct measurement of the quantities concerned are doomed to failure. Instead, tests of the predictions of the models for the statistical properties of populations are needed.

Tests of the role of insertional mutations

One such test has been devised and applied by Montgomery et al. (1987) and Langley et al. (1988). They noted that the hypothesis that frequencies are stabilized by the deleterious effects of insertional mutations predicts that element frequencies at equilibrium will be lower for X chromosomal sites than for autosomal sites, because X chromosomal mutational effects are expressed in the hemizygous state in males whereas rare autosomal genes are predominantly in the heterozygous state.

The results of scoring the numbers of copies of the retroviral elements *copia*, *roo* and *412* on sets of 20 X, second and third chromosomes from a natural population of *D. melanogaster* showed no significant difference between the observed abundance of elements on the X and the expectation on the null hypothesis of no selection for *copia* and *roo*, whereas *412* shows a significant ($P<0.01$) deviation from the null expectation, but agreed with the expectation on the insertional mutation hypothesis. It is unclear why these differences between families should exist; further data on a wider variety of families are clearly desirable. On the whole, these results suggest that, while there may be some effect of insertional mutations on copy number, it is not necessarily the chief mechanism involved in controlling copy number.

Tests for effects of unequal exchange

A test for the role of selection against chromosome rearrangements produced by unequal exchange between elements is provided by asking whether or not the abundance of elements tends to be higher in regions where exchange is reduced in frequency. *D. melanogaster* again provides useful material for carrying out such a test, since there is pronounced suppression of meiotic crossing over in the telomeric and centromeric regions of the euchromatin of the chromosome arms (Lindsley and Sandler, 1977). It is straightforward to use *in situ* hybridization to determine the distribution of the numbers of elements over the salivary chromosome maps in chromosomes sampled from natural populations, and to compare this with the distribution expected on a null hypothesis of no variation in element abundance with respect to the rate of crossing over. The results of a study of *roo* by Langley *et al.* (1988) showed clear evidence for an excess of elements at the base of the X chromosome, of the order of three-fold over the number of elements that would be expected if elements were inserted in proportion to the physical size of the region. The picture is less clear for the autosomes. A survey of a sample of 14 X chromosomes from a Maryland population for ten families of elements showed an excess of elements near the base in 7 families; there was no evidence for an excess at the tip (Charlesworth, unpublished). The study of Miklos *et al.* (1984, 1988) of DNA cloned from subdivisions 19E/F and 20A, which are close to the base of the X, also demonstrates an unusual abundance of elements in this region. The data thus suggest that elements do tend to accumulate in regions where exchange is limited, although the effect seems to be much stronger for the base than the tip for reasons which are unclear. Ajioka and Eanes (1989) have found a strong tendency for *P* elements to be associated with the extreme tip of the X chromosome in natural populations, but this may be due to site specificity of insertion.

Data on the distribution of element frequencies

Equation (8.2) can be applied to the analysis of *in situ* hybridization data on element frequencies, using the sampling theory developed by Charlesworth and Charlesworth (1983), Langley *et al.* (1983), and Kaplan and Brookfield (1983b). The model of Langley *et al.* assumes that $\alpha = 0$, and uses an extension of Ewens' (1972) sampling theory to estimate ß (their $\bar{\theta}$). A more general method is to estimate α and ß jointly by a minimum χ^2 technique (Charlesworth and Charlesworth, 1983). This method has the advantage that the estimated value of ß is not affected by the frequency estimation bias introduced by the *in situ* technique; the confounding of separate sites within the same salivary chromosome band merely has the effect of reducing the value of m in the expression for α, and hence inflating the value of α as far as the true insertion sites are concerned.

Charlesworth (1988) has reviewed estimates of α and ß obtained with this method for a number of elements scored on chromosomes sampled from natural populations of *D. melanogaster* (*copia*, *412*, *297*, *roo* and *I*). Values of α range from 0 to 1, and ß from 13 to 40. It should be stressed that the fits obtained with this procedure are usually not significantly better than those obtained by assuming that $\alpha = 0$. Nevertheless, the generally large values of ß obtained give highly significantly better fits than small values, reflecting the low frequencies of multiple occupation. This effectively rules out the possibility that element frequencies are controlled by regulated transposition in the absence of excision, such that transposition rates are zero in an equilibrium population (see above). An excellent fit to the observed occupancy profiles is obtained with the minimum χ^2 estimates of the parameters. If the expected mean number of elements per individual is unchanging over time, then from equation (8.1) $ß \simeq N_e u$. Estimates of N_e for East Coast *D. melanogaster* populations from the frequencies of allelism between recessive lethals suggest values of the order of 2×10^4 (Mukai and Yamaguchi, 1974). With a ß value of 20, this yields an estimate of 2.5×10^{-4} for u, which is of the same order as direct estimates (Charlesworth and Charlesworth, 1983; Eggleston *et al.*, 1988). The estimates of N_e are subject to considerable uncertainty, so that this value should not be taken too seriously.

Conclusions

The main conclusion from the models and data presented above is that there is nothing that we know at present that is strikingly inconsistent with the view that transposable elements are maintained in populations as a result of transpositional increase in copy number. This view also has the merit of providing detailed explanations for a range of phenomena associated with transposable elements. In particular, it convincingly explains the fact that elements in bacteria yeast, and drosophila are usually present at low frequencies at individual chromomsomal sites into which they can insert; this fact is almost impossible to accommodate on the hypothesis that elements persist as a result of favorable mutations associated with their transpositional activities (Syvanen, 1984). Indeed, this observation suggests that elements in these organisms rarely or never have a positive role in directly promoting evolutionary change in association with an increase in their own frequency. It is possible, however, that certain classes of mutation, such as chromosome rearrangements (Engels and Preston, 1984; Mikus and Petes, 1982; Roeder, 1983; Goldberg *et al.*, 1983; Davies *et al.*, 1987), could be generated as a result of excision of elements or exchange between elements at different locations, and could be fixed in populations by drift or selection without leading to an increase in element frequency at the sites in question. Thus, transposable elements could act indirectly as a source of mutational variation for evolutionary change, but this would not have much influence on their distribution within populations.

It should be emphasized that low frequencies of elements at individual sites are not necessarily predicted by the models; it is relatively easy to generate situations in which elements rise to high frequencies or even fixation at many chromosomal sites. There are, indeed examples of such elements, the classic one being the very abundant mammalian *Alu* sequences (Schmid and Shen, 1985), although the status of *Alu* as a genuine transposable element has been questioned (Rogers, 1985).

The evidence on the distribution of elements within populations suggests that transpositional increase in copy number is generally balanced by some force or forces; it is at present unclear whether this involves self-regulation of transposition, selection against mutational effects of insertions, selection against chromosome rearrangements produced by exchange between elements at non-homologous sites, or some combination of these.

Acknowledgements

This work was supported by grants NSF grant BSR-8-16629 and PHS grant GM 36405-01, and a grant from the Louis Block fund of The University of Chicago. I am grateful to Chuck Langley for several years of discussions and collaboration on this topic. I thank Angela Lapid and Darlene Canada for technical assistance.

Chapter 9
Gene Families

Tomoko OHTA

National Institute of Genetics, Mishima 411, Japan

The remarkable development of molecular biology has produced a number of new findings on the genetic organization of higher organisms, and the manner in which they differ from lower organisms. One such discovery is the prevalence of gene families. Multigene families are defined as groups of genes with sequence homology and related or overlapping functions that exist in multiple copies per individual. Supergene families are broader groups of genes containing one or more domains of common origin. Their widespread occurrence has significant implications for evolution and variation in higher organisms.

The mechanisms of evolution are closely related to those which cause the differentiation among highly selected domestic animals and plant species. Robertson and his associates (Clayton and Robertson, 1955, 1957) have recognized the importance of new variation in contributing to the response to artificial selection in their selection experiments. Experimental populations are much smaller than natural species both in space and in time, so one can infer from these results that the occurrence of new variation for quantitative traits must be very common in nature. Recent observations on gene families suggest that a substantial part of such new variation arises from organizational changes in gene families, such as variation in copy number or other rearrangements (Frankham et al., 1978; Ohta, 1980).

In general, gene families evolve through various mechanisms of molecular interaction among homologous DNA segments, such as gene conversion, unequal crossing-over, replication slippage and duplicative transposition. As a result of the continued occurrence of such interactions, 'concerted evolution' of gene families often takes place (Ohta, 1980, 1983b; Hood et al., 1975; Dover, 1982; Arnheim, 1983), so that genes belonging to a family within a species, even if their chromosomal locations differ, are more similar than those in different species. The various gene families range from those with extremely diverse functions, such as the protein kinase family (Hunter, 1987), to those with very variable members, like immunoglobulin genes (Hood et al., 1975), and to those consisting of uniform members, like ribosomal RNA genes (Dover, 1982; Arnheim, 1983). All existing gene families are the evolutionary products of millions of years of evolution, and the organization and diversity of each family reflect the effects of natural selection. Therefore it appears that the variability contained in each gene family is the product of natural selection for optimum diversity. When the rate of molecular interaction is high, the turnover of gene members becomes rapid and the gene family is expected to be uniform, and vice versa (Ohta, 1983b). Adjustment of the turnover rate of each gene family by natural selection can occur by a mechanism which I called indirect selection (Ohta, 1986). On the other hand, if selection works in such a way that more and more differentiated functions of redundant gene members are beneficial, the gene family evolves into a supergene family. Then molecular interactions, like gene conversion and unequal crossing-over, no longer occur.

At the time I started to investigate the complicated problems of the population genetics of multigene families more than ten years ago, Robertson (1975d) remarked in his review of the influential book of Lewontin (1974) that the evolution of repeated gene families was totally outside its scope. This encouraged me to work on the problem intensely.

The mutant dynamics of multigene families apply at two different levels: in the cell, by unequal crossing-over, gene conversion and transposition; and in the population, by random genetic drift and natural selection. In addition, gene members are scrambled among individual cell lineages by sexual reproduction. Various simple models of the evolution of multigene families have been considered which incorporate unequal crossing-over, gene conversion and duplicative transposition (Nagylaki, 1984; Ohta, 1985). The results of such analyses have significant implications regarding the mechanisms for maintaining genetic variability within species; for example, the enormous diversity of antibodies and histocompatibility polymorphisms are generated by a combination of gene multiplicity, molecular interactions and selection.

It is important to appreciate how gene families may have originated in order to understand organismal evolution. There is a remarkable array of supergene families: for example, the genes of many protein kinases, oncogenes, growth factors, hormones and their receptors (Ohta, 1988a). Many of these are essential for cell differentiation and organismal development, and so they may affect morphological characters. It would be impossible to formulate exactly how they originated, but it is highly desirable to know, even if incompletely, how random drift and natural selection may have contributed to their origin.

In theoretical models of the steady state gene family, only purifying natural selection is needed to maintain the *status quo*, but more positive natural selection is required to model their origin. I have attempted to clarify the interaction among natural selection, genetic drift and unequal crossing-over by carrying out extensive Monte Carlo simulations (Ohta, 1987). For simplicity, gene conversion and duplicative transposition are not taken into account in these simulations.

In one type of Monte Carlo experiment, it is assumed that there is initially only one gene on a chromosome, and unequal crossing-over occurs infrequently to produce either deletion of the gene or two tandem genes. This gene is assumed to be indispensable, so the deletion is lethal; but, by continued unequal crossing-over, the copy number may increase by chance. At a low constant rate, a gene mutates to become nonfunctional, i.e. a pseudogene, and, at a still lower rate, a gene mutates to a new allelic type that may be beneficial to the organism. The population is finite and consists of the chromosomes that independently undergo unequal crossing-over and mutation, and no interchromosomal recombination is considered (i.e. a haploid population).

Natural selection is assumed to favour the chromosomes containing more beneficial mutant genes, but pseudogenes have no effect on survival. By using steady state theory together with molecular data, realistic rates of mutation and unequal crossing-over were chosen relative to the population size. Each simulation experiment was continued for a number of generations equal to 50 times the population size, and the evolution of the gene family was examined.

Figure 9.1 shows a typical sample path of the simulation experiment which shows how number of copies, number of beneficial genes (both starting from unity) and number of pseudogenes change over generations (starting from zero), all measured per chromosome. It can be seen from the figure that the copy number gradually increases and both beneficial and deteriorated genes accumulate. In this particular sample path, the rate of mutation to pseudogenes was assumed to be ten times higher than that to useful mutations; therefore pseudogenes accumulated more rapidly than beneficial genes, even though the latter were favoured by selection.

Several significant results were obtained from a large number of such simulation experiments:
(i)   The spreading of a beneficial gene has to be more rapid than that of a pseudogene, otherwise more pseudogenes (so-called 'junk' DNA) accumulate than beneficial genes in clusters of tandem copies, which is contrary to observation in nature.
(ii) There is a large fluctuation in the outcome even if the parameters take similar values, because the whole process depends greatly on chance.
(iii) If the unequal crossing-over rate increases, the gene system evolves more quickly.

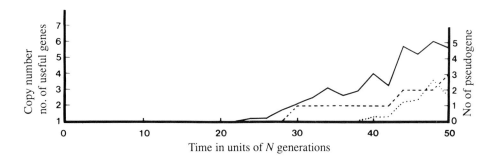

Figure 9.1. A sample path from a simulation of the evolution of a multigene family, with positive selection (from Ohta, 1987). The copy number (full line), the pseudogene number (dotted line) and the number of different beneficial alleles (broken line) increased. Parameters are: effective population size = 50; unequal crossing-over rate = 0.0025; detrimental mutation rate = 0.001; beneficial mutation rate = 0.0001; selection coefficient as in Ohta (1987) = 0.2.

The first result is particularly important. It implies that positive natural selection is needed for the evolution of gene families because the rate of production of beneficial mutations is usually much lower than that of deteriorating mutations, and only through the help of natural selection can the former have a higher chance of spreading than the latter.

If a gene is treated as a linear sequence of integers, it is possible to examine, in addition to gene organization, sequence divergence among duplicated gene copies (Ohta, 1988b). Figure 9.2 shows phylogenetic trees of duplicated genes on a chromosome randomly chosen from the simulated population at the 50Nth generation. Two replications with the same parameters are given. It can be seen that the beneficial mutants often spread just after duplication.

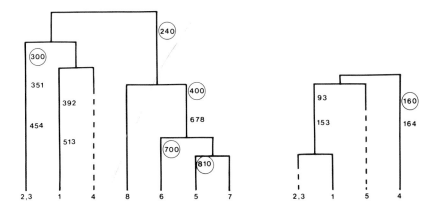

Figure 9.2. Phylogenetic trees of duplicated genes on single chromosomes (from Ohta, 1988b). The two chromosomes were samples from different runs at the 50Nth generation. Full lines represent beneficial genes and broken lines deteriorated genes. Integers beside the branches are numbers of mutants, and those which are beneficial (multiples of 10) are circled. Integers at the bottom of the tree indicate the sequence of the genes on the chromosome. Parameters are the same as in Figure 9.1, but the neutral mutants were also incorporated.

How does this finding relate to actual data of molecular evolution? Goodman (1976) and his associates, by examining phylogenetic trees constructed from amino acid sequences of globin genes, have noted that molecular evolution is accelerated at the time of duplication. By comparing the rate of amino acid replacing with that of synonymous substitutions in an analysis of DNA sequences which included the genes for haemoglobins, somatostatin, cytochrome c, and growth hormone, Li (1985) unambiguously showed an acceleration of amino acid substitution after gene duplication. He concluded that this result could be explained under the neutral theory (Kimura, 1983), in that selective constraints are relaxed by gene redundancy, whereas Goodman (1976) proposed that Darwinian selection is responsible for the acceleration. My analyses suggest that both random drift and positive selection contribute to the origin of gene families.

A final topic is the effect of sexual recombination on evolution by gene duplication. By extending the above analysis to a diploid population with sexual recombination (interchromosomal unequal crossing-over at meiosis), I have shown that evolution of gene families is more rapid in diploid than in haploid populations (Ohta, 1988b). Two reasons are:

(i) Two beneficial genes may exist in a diploid cell as two alleles at the same locus without duplication after the mutants' frequencies have increased.

(ii) Two segregating beneficial genes may be combined onto one chromosome by interchromosomal unequal crossing-over.

Table 9.1 presents some properties of simulated populations which clearly show the effect of diploidy

Table 9.1
Properties of gene families at the 50Nth generation in the simulated populations (average of 15 runs)

Designation of beneficial mutants[1]		Copy no	No of pseudogenes	No of different beneficial alleles[3]
Multiple of 5	haploid	3.26	1.02	1.87
	diploid[2]			
	1:0	6.07	2.31	3.72
	3:1	6.34	2.33	3.39
	1:1	8.54	3.20	5.68
	1:3	7.83	2.73	4.92
	0:1	7.77	2.89	5.18
Multiple of 10	haploid	3.73	1.75	1.58
	diploid[2]			
	1:0	4.17	1.94	2.03
	3:1	4.68	2.02	2.25
	1:1	5.96	2.70	2.80
	1:3	6.16	2.42	3.25
	0:1	5.75	2.50	2.86

[1] A gene is a sequence of ten integers and each integer may mutate. Beneficial and neutral mutations are stored as positive integers, and at each occurrence, the integer is increased by one. In one set, the beneficial mutants correspond to integers that are multiples of 5, and in the other set, multiples of 10.

[2] Relative rates of intra- versus inter-chromosomal unequal crossing-over.

[3] The number is per genome in the haploid model, and per diploid cell in the diploid case. Other numbers are counted per genome.

Other parameters are: total rate of unequal crossing-over per gene copy = 0.002; effective population size = 50; detrimental mutation rate = 0.001; rate of neutral and beneficial mutations = 0.001; selection coefficient as in my previous model (Ohta, 1988b) = 0.2.

and meiotic recombination. This finding should be considered seriously when the advantage of sexual reproduction is discussed, in view of the many gene families that have been found in recent years. If the type of selection discussed above operates for a long period of time, gene families would evolve into supergene families. Information from various supergene families suggests that 'exon shuffling' plays an important role (Gilbert, 1978). Functional differentiation of redundant genes might have occurred by exchange of 'domains' among different genes, or by domain duplication in addition to nucleotide substitutions. Theoretical examination of such processes is an excellent subject for future studies by evolutionary geneticists.

Acknowledgement

I thank Dr Kenichi Aoki for his many helpful comments on the manuscript.

Chapter 10
Phenotypic Models of Evolution

J. MAYNARD SMITH

School of Biological Sciences, University of Sussex, Falmer, Brighton, Sussex, BN1 9QG

Usually, when we think about the evolution of a population, we think of changes in the frequencies of genes and of genotypes, brought about by natural selection and drift. In this review, I discuss two alternative approaches – optimization and game theory – that concentrate on phenotypes and largely ignore genetics. I aim to show that these approaches are a necessary supplement to population genetics, although they cannot wholly replace it.

The nature of optimisation models

To illustrate the nature of optimization models, consider the evolution of tetrapod limbs and gaits. This topic was discussed by Maynard Smith and Savage (1956); considerable progress has been made since then, but the nature of the models remains the same. One question we asked is the following: what gait is optimal, in the sense of giving the maximum speed for a given expenditure of energy (or, equivalently, the minimum expenditure of energy for a given speed)? Our model agreed qualitatively with observed vertebrate gaits, in predicting that J (the proportion of time spent with all feet off the ground) should increase with speed, and decrease with size. However, we predicted that J should increase with speed more rapidly than it actually does: the discrepancy arose because we ignored wind resistance, and failed to allow properly for the elastic properties of muscle.

The first general point illustrated by the model is that, in order to optimize, one must assume that certain things remain constant. Some of these constant features consist of laws of nature – in this example, the law of levers and the conservation of energy. Others, such as the mechanical strength of bone and tendon, the efficiency of muscle in converting chemical into mechanical energy, and the fact that vertebrates do not have wheels, are not universal laws of nature; but it is probably safe to treat them as constant when thinking about tetrapod evolution. Still others, such as the fact that ungulates tend to have stiff backbones whereas carnivores have flexible ones, or that archosaurs and terrestrial birds run whereas mammals gallop or hop, refer to features that have changed during tetrapod evolution, but which can be treated as constants when considering evolution within, say, a family.

Such constancies are referred to as 'constraints'. Except in the case of laws of nature, constraints are not absolute, and may themselves require an evolutionary explanation. The snag, of course, is that the explanation may involve historical events no longer readily accessible to investigation, or aspects of developmental mechanics about which little is known.

A second general point is that one must make some assumption about what property has been maximized by natural selection. The best assumption would be that what is maximized is fitness, or inclusive fitness if interactions between genetically similar individuals are relevant. This is sometimes possible. For example, Charnov et al. (1976), analyzing how a hermaphrodite should allocate resources between male and female functions, took, as the trait to be maximized, the number of genes passed on to the next generation. More often, as, for example, in foraging theory, the best one can do is to maximize some trait that contributes to fitness. For example, Savage and I maximised speed for a given energy expenditure. This would be reasonable if the only thing legs are used for is running, but in fact they may also be used for swimming, digging, scratching, kicking or capturing prey. One reason for the difference in gait and limb proportions between ungulates and carnivores is that the latter do more things with their legs.

Complications arise because there is often a trade-off between success in performing different functions. For example, it follows from the law of levers that any change in limb shape that increases speed will decrease strength, and *vice-versa*. Other constraints are less absolute, and require experimental demonstration. For example, in *Drosophila subobscura* any treatment that prevents a female from laying eggs (including a dose of radiation) lengthens her life (Maynard Smith, 1958). Rose and Charlesworth (1980) have demonstrated (in a different species, *D. melanogaster*) that there is a negative genetic correlation between longevity and fecundity. In optimization theory, such trade-offs can often best be analyzed using 'fitness sets' (Levins, 1962), as illustrated in Figure 10.1. This figure can be used to make a point about the nature of genetic variance. The optimal phenotype will lie on the boundary of the fitness set. When optimizing, one considers a series of points on this boundary, seeking for that point which maximizes fitness. However, not all variation will move the phenotype along the boundary of the set (direction A): an unconditionally deleterious mutation will move the phenotype in direction B. For example, in the case of longevity/fecundity, selection for increased longevity will reduce fecundity, and the phenotype will move along the boundary of the set. But inbred females are both shorter-lived and less fecund (Maynard Smith *et al.* 1955): homozygosity moves the phenotype in direction B.

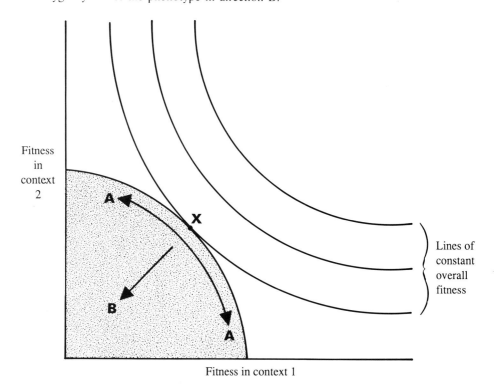

Figure 10.1. Fitness set and adaptive function (Levins, 1962). The shaded area defines the set of possible phenotypes. Point X is the optimum phenotype.

A final point, which should perhaps have been discussed first, concerns what it is that optimization models are trying to explain. Such models have sometimes been criticized (e.g. Gould and Lewontin, 1979) as if the hypothesis being tested were the optimization assumption itself – that all traits are such as to maximize fitness. I have argued (Maynard Smith, 1978) that this is a misunderstanding. In any particular model, what is being tested is not the optimization assumption itself, but the assumptions about constraints and about selective forces. If a model fails adequately to predict observed behaviour or structure, the most likely explanations are either that the range of possible phenotypes is different from that assumed, or that the selective forces responsible for the evolution of the trait have been wrongly identified. Of course, it may also be true that some phenotypic traits are far from a selective optimum, but this seems unlikely in the case of costly traits such as the tails of peacocks, the fighting behaviour of spiders, or the ovaries of fruit flies.

Optimisation and game theory

The optimal gait for a mammal does not depend on what other members of the species are doing (or, at least, Savage and I assumed that it does not, but if predators take only the slowest member of a group, the optimal gait may be frequency-dependent). In such cases, optimization is the appropriate technique. But if fitnesses are frequency-dependent the appropriate method is evolutionary game theory. Instead of seeking an optimum, one seeks an evolutionarily stable strategy, or ESS. That is, one seeks a phenotype, I, such that, if almost all members of the population are of type I, no alternative mutant phenotype, M, can invade the population. I have discussed methods of finding ESS's at length elsewhere (Maynard Smith, 1982).

The importance of context to optimization is nicely illustrated by an example discussed by Mirmirani and Oster (1978). They discuss the allocation of resources by an annual plant between growth and reproduction, such that seed production is maximized, allowing for the fact that rate of growth and resource acquisition depend on size, and hence on past growth. Consider three cases:

(i) A plant grows by itself, without competition. The optimum is to allocate all resources to growth up to some time, T, and then to switch all resources to reproduction: this is 'bang-bang control'.

(ii) A plant grows in pure stand with other unrelated members of its species. The growth rate of a plant at any instant depends not only on its own size, but also on the size of its neighbours, which compete with it for light and nutrients. We now seek an ESS, because the best growth strategy depends on what others are doing.

The ESS is again to choose a time, T', at which all resources are switched from growth to reproduction, but it turns out that T' is later than T. In competitive situations, a plant must allocate a larger proportion of its resources to growth. For an example agreeing with this prediction, see Gadgil and Solbrig (1972) on dandelions.

(iii) A plant grows in pure stand, but its neighbours are also its relatives. If neighbours were genetically identical, the solution would be that phenotype which maximizes the seed production of the population as a whole. The optimal switching time would be earlier than in case (ii), because, for the population, it is wasteful to allocate resources to leaves whose effect is to shade out competitors. It is interesting that plant breeders, concerned to maximize the population yield, have recently tended to produce dwarf varieties with rather few leaves.

The general case, when there is an average relatedness between neighbours of r, is more difficult. A natural approach, illustrated in Table 10.1, is to replace 'personal' by 'inclusive' fitness, and then to seek an ESS. It is not clear, however, that such an approach is always justified (see Grafen, 1979; Hines and Maynard Smith 1979). For the simple case of two pure strategies, it seems that the evolutionarily stable state of the population is always an ESS of the inclusive fitness matrix, but that there can be ESSs of that matrix that can be invaded. The problem of 'games against relatives' needs further work. But the distinction between an optimum in the constant-fitness case, and an ESS in the frequency-dependent case, is clear and important.

Table 10.1
The inclusive fitness matrix

Personal pay-offs				Inclusive pay-offs		
	A	B			A	B
A	a	b		A	$a(1+r)$	$b+rc$
B	c	d		B	$c+rb$	$d(1+r)$

Numerical example

	A	B			A	B
A	1	6	$r=0.5$ A		1.5	7
B	2	5	B		5	7.5

Note: In each matrix, the entries are the pay-offs, or changes of fitness, to an individual adopting the strategy on the left, if its partner adopts the strategy above. The personal pay-off is the change in the expected number of offspring to an individual; the inclusive pay-off is the personal pay-off plus r times the pay-off to the partner, where r is the coefficient of relatedness. The numerical example shows that the ESS of the inclusive fitness matrix is more cooperative than that of the personal fitness matrix: always do B, instead of ($\frac{1}{2}$A,$\frac{1}{2}$B).

The limitations of game theory

Evolutionary game theory was first developed (Maynard Smith and Price, 1973) to analyze the evolution of fighting behaviour. Changes in fitness (pay-offs) were supposed to arise from random pairwise interactions in an infinite population. Each individual had a fixed phenotype, that might be 'pure' (always do A) or 'mixed' (do A with probability p, do B with probability $1–p$). After interaction, individuals reproduced their kind (pure or mixed) exactly, the number of offspring being proportional to pay-off. ESSs then correspond to the stable fixed points of the dynamics (Taylor and Jonker, 1978). The model was therefore of an asexual population with pairwise interactions, and heredity was exact.

Some ways in which the model has since been extended are as follows:

(i)   To 'games against the field'. The fitness of an individual is determined, not by interaction with a single opponent, but by the frequency of certain traits in the population. For example, the effectiveness of producing offspring in a given sex ratio depends on the sex ratio in the population as a whole (Fisher, 1930), or in some neighbourhood (Hamilton, 1967).

(ii) To phenotypes that vary continuously. Suppose that x is some phenotypic variable, and x^* is a (pure) ESS. Let $W(x,x^*)$ be the fitness of a rare mutant of phenotype x in an x^* population. Then, since x^* is an ESS, $W(x,x^*)<W(x^*,x^*)$. That is, $W(x,x^*)$ reaches a maximum when $x = x^*$. The ESS can therefore be found by solving the equation

$$[\partial W(x,x^*)/ \partial x]_{x = x^*} = 0,$$

and checking that $\partial^2 W(x,x^*)/ \partial x^2$ is negative. There may be no pure ESS, but some distribution $p(x)$ may be uninvadable, as in the 'war of attrition' (Maynard Smith, 1982).

If there are more than two possible pure strategies, or if the possible phenotypes are continuously distributed, there may be no ESS. In the discrete case, the simplest type of game with no ESS is the Rock–Scissors–Paper game (R beats S, S beats P, P beats R). A simple continuous game with no ESS is the 'size game' (Maynard Smith and Brown, 1986), in which the breeding success of an individual increases with the proportion of the population smaller than itself. This leads to continuing oscillations in size, or, if asexual reproduction is replaced by polygenic inheritance, to an indefinite increase in size, with a corresponding decrease in the proportion of the population reaching breeding size.

Suppose that the ESS is the mixed strategy: do A, B, C. . . with probabilities p_A, p_B, p_C . . ., where $\sum p = 1$. Let this strategy be M^*. In a population of M^* individuals, the expected pay-offs to the alternatives, A, B, C . . . are equal (Bishop and Cannings, 1978). Given that this is so, an asexual population could achieve evolutionary stability in one of two ways:

(i) A genetically uniform population of M^* individuals.

(ii) A genetically polymorphic population, in which genetically A, B, C . . . individuals are present in the proportions specified by the ESS.

In fact, there is a line of neutral equilibria: any population consisting of a proportion of p of M^* individuals, and $1-p$ of pure strategists in the ESS proportions, is also stable. One can take the matter a little further. For any population, we can specify the frequencies with which the acts A, B, C . . . are actually encountered. Hines (1980) has shown that, if these frequencies are not those specified by the ESS, then the mean fitness of M^* individuals is greater than the mean fitness of the population. Hence M^* will increase in frequency so long as the act frequencies depart from the ESS. In a finite population, frequency fluctuations will result in the mixed strategists having a higher geometric mean fitness than any pure strategist (Maynard Smith, 1988). This may be one reason why, in almost all cases, a 1:1 sex ratio is achieved by a randomizing device in one or other parent, and not by a device ensuring that half the parents of one sex produce only sons, and half produce only daughters. Taylor and Sauer (1980) analyzed a related problem.

In particular cases, there may by physiological or developmental reasons why a genotype producing the required mixed strategy is unlikely or impossible. For example, it is hard to imagine a genotype in *Papilio dardanus* that caused an individual to develop as a mimic of one model species with probability p, and of another with probability $1-p$. In such cases, we expect a stable genetic polymorphism rather than a mixed ESS.

This raises the question of how far the conclusions of a game theoretic analysis will be modified by the introduction of diploid genetic inheritance. If a single true-breeding genotype can give rise to the required ESS phenotype (pure or mixed), then the introduction of sex and diploidy makes little difference: a diploid population of M^* individuals would be uninvadable. Difficulties arise when there is a mixed ESS, but no homozygous genotype has the required range of phenotypes. Will a diploid sexual population evolve to a stable polymorphism, with the phenotypes in the ESS proportions? In the simple game with two pure strategies, the answer is that it will if the genetic system allows. That is, if there exists a set of gene frequencies that generate the phenotypes in the ESS proportions, then those frequencies correspond to a stable polymorphism (Maynard Smith, 1981).

There are, unfortunately, cases in which no set of gene frequencies can generate the appropriate phenotypic frequencies. For example, the ESS of the War of Attrition game is the phenotypic distribution, $p(x) = (1/V)\exp(-x/V)$. No polygenic system could produce a negative exponential phenotypic distribution. Only if a single genotype can generate the appropriate distribution can the ESS be achieved.

I conclude this section by describing a method of analysis that is a kind of hybrid between population genetics and game theory. Suppose that a series of alleles s_1, s_2, s_3 . . . determine some quantitative trait with frequency-dependent effects on fitness. The usual approach of population genetics would be to ask what will happen to a population with two (or more) alleles present: will one allele be fixed, or is there a stable polymorphism? In the spirit of game theory, however, we can seek an allele s^* which cannot be invaded by any alternative allele s. This is now a standard way of analyzing sex ratio problems.

In general, the introduction of diploid genetics does rather little to modify the conclusions drawn from optimisation and game theory. What it does is to make it possible to discuss some important questions that game theory does not address. A representative but not exhaustive list of such questions is:

(i) What is the cause of genetic variation, and how is it maintained in natural populations? Optimization theory tacitly assumes the existence of some additive genetic variance for the traits being optimized. Game theory goes a little further, by emphasizing the many situations in which frequency-dependent fitnesses can give rise to polymorphism. Neither method has anything to say about such topics as mutation-selection balance or heterosis.

(ii) Rates of change. Both theories are concerned with equilibria (and, in the case of game theory, with the absence of stable equilibria in some situation), and say nothing about the rate of approach to equilibria.

(iii) Sexual reproduction. Although a number of important topics that arise because of the existence of sex (sex ratio theory, resource allocation in hermaphrodites, parental care, the origin of anisogamy) have been treated by game theory, I do not at present see how the theory could be used to analyse the evolution of sex and recombination themselves.

(iv) Linkage disequilibrium and its consequences. So far, game theory has not contributed to an understanding of Fisher's (1930) runaway process of sexual selection (although Maynard Smith and Brown, (1986) did show that male–male competition can lead to runaway). I think that this is because Fisher's process depends on linkage disequilibrium between genes for choice and for the chosen trait.

The limitations of population genetics

There is a sense in which population genetics is a more precise theory than optimization. If the predictions of the two theories differ (for example, because the optimal phenotype is a heterozygote), it is population genetics that should be accepted. However, it does not follow that, given an adequate theory of population genetics, one can do without optimization and game theory. Population genetics avoids most of the difficult problems of evolution theory by ignoring them. A model in population genetics usually takes the form, 'If the following genotypes have the following fitnesses, then the population will evolve in the following way'. Usually nothing is said about how the genotypes come to have the assumed fitnesses. There are exceptions. In sex ratio theory, the fitnesses emerge naturally from the phenotypes – that is, from the progeny sex ratios. In quantitative genetics, first developed as a theoretical underpinning to animal and plant breeding, fitnesses are assigned to phenotypes by artificial selection. But in general, population geneticists avoid the difficulties involved in specifying phenotype sets, and their corresponding fitnesses, by simply not discussing the matter.

A second weakness of population genetics is its reluctance to discuss frequency-dependent selection. It is symptomatic that Crow and Kimura (1970), in their classic textbook, discuss frequency-dependent selection in less than half a page, and refer to only one possible example (self-sterility alleles). The reason, of course, is that theorems are hard to come by if fitnesses are not constant. But that should not lead us to think that constant fitnesses are a common feature of the real world.

It has recently become fashionable to argue that, if one wants to understand the response of a population to selection, one must measure the variance–covariance matrix for the relevant phenotypic traits. Apart from the practical difficulties of making such measurements, I suspect that they may be of rather limited utility. A genetic covariance between two traits can arise from a number of causes: depending on the cause, it may be constant in evolutionary time, or it may change rapidly and differ betwen local populations. Referring back to my earlier discussion of trade-offs in optimization models, a negative covariance between speed and strength is an inevitable consequence of the law of levers. The negative covariance between longevity and fecundity observed in drosophila by Rose and Charlesworth (1980) arose from physiological constraints whose existence was well established experimentally, and which would probably be hard to alter by selection, although they are not understood at a biochemical level. However, the sign of the covariance between the two traits would depend on the nature of the genotypes being compared. If the comparison was between populations well adapted to different selective regimes, the covariance would probably be negative. But if the comparison was of inbred and outbred individuals the covariance would be positive.

Genetic covariance can arise from pleiotropism. In the case of longevity and fecundity, the pleitropism probably arises from fairly fundamental features of the physiology of drosophila. Consequently, genes at many different loci will have similar pleiotropic effects, and different populations will tend to have similar covariances. But other pleiotropisms may be peculiar to particular loci. In a population of mice segregating for the grey lethal mutation, there would be a correlation between abnormal skeletons and grey colouration, but one would not expect to find a similar correlation in other populations.

Finally, genetic covariance can arise from linkage disequilibrium. This, in turn, can arise from stochastic events in finite populations, and will differ from population to population. Alternatively, it may arise from past nataural selection: if so, it will last only while the selection lasts.

The conclusion is as follows. The response of a population does indeed depend on the genetic covariance matrix. But the covariance matrix itself may not be a fundamental or permanent feature of the population. In so far as it is stable, its stability is a consequence of the kind of engineering and physiological constraints that are considered in optimization theory. Perhaps the most useful outcome of measuring genetic covariance will be an insight into the nature of these constraints.

Chapter 11
Population Structure

J.S.F. BARKER

Department of Animal Science, University of New England, Armidale, NSW 2351, Australia

Population structure is generally considered as a concept clearly within the domain of population and evolutionary genetics. The null hypothesis of population genetics is the ideal panmictic population – genetically homogeneous and random mating. When a population does not fit this ideal, it is said to be structured. Such structure may be exhibited as a nonuniform distribution of individuals in an essentially continuous population through varying degrees of subdivision to localized subpopulations with limited migration among them. The effects of structure on the genetic composition of populations, together with those of natural selection, are then central to the study of evolution.

This is the usual view of population structure for evolutionary geneticists, one which traces back to Wright (1931). However, Wright's consideration of structure in natural populations was influenced by his earlier analyses of inbreeding and crossbreeding in guinea pigs (Wright, 1922), and of structure in breeds of livestock (McPhee and Wright, 1925, 1926). Studies of structure in livestock breeds and effects of structure in animal improvement have since been largely independent of studies in natural populations, but each of these three areas will be considered here.

Alan Robertson has contributed to all three areas, exemplifying one of the features of his work, i.e. his ability to move back and forth between theoretical and empirical studies, and between population genetics, quantitative genetics and animal breeding.

Structure of livestock breeds

Pure breeds of livestock are closed (or essentially closed) populations, with all individuals identified and pedigreed. Thus the current and past breeding structure (inbreeding, migration of animals between herds) can be readily assessed. Robertson's first paper on genetics (Robertson and Asker, 1949) was the first study of migration between herds within a breed. This movement was so great as to preclude the possibility of much local genetic differentiation, a finding that had important implications for the development of the artificial insemination and bull progeny testing programmes that were to become the primary basis of genetic improvement of dairy cattle (King, this volume). Robertson and Asker (1951a) showed that movement of animals (particularly bulls) between herds was not at random, but was a directional process because of the existing breed structure. Pedigree herds of the British Friesian breed were structured in three levels, with a top group of some 20 to 30 herds determining the genetic composition of the whole breed. A similar hierarchical structure was demonstrated (Robertson, 1953a) for a number of British breeds (dairy cattle, beef cattle, pigs, goats), and in all cases, the number of herds having any permanent genetic effect on the breed was less than 30. The genetic consequences of this structure were clearly recognized – only a small fraction of the differences between herds in average performance could be genetic (Robertson and McArthur, 1955). As Alan once put it to me: 'To the extent that differences among herd averages are genetic, they are due to the genes of the human managers, not of the cows'.

Studies of a number of breeds of all livestock species were done during the 1960s and 1970s in Great Britain, Australia, Canada and USA. All showed a hierarchical breed structure, with the number of herds or flocks in the top group ranging from as low as 5 to about 60. Partly due to the breed structure, but also to linebreeding to particular males and the high variance in the number of offspring per sire, the effective number of males used per generation averaged about 30. As some of these breeds numbered tens of thousands of individuals, this figure may seem surprisingly small. Natural populations are not the same as breeds of livestock, but this small effective size of livestock populations, relative to the often very large census size, should sound a note of caution in thinking about the effective size of natural populations.

Structure in artificial selection and animal improvement

Wright (1922) first discussed the effects of population structure on response to artificial selection. He concluded that improvement would be more rapid if a population were subdivided into small lines, with regular selection between and crossing among lines, than if the same total population were maintained as a single line. Subsequently, these ideas applied to evolution became the basis of his shifting balance theory.

The idea that population structure might be manipulated in this way in artificial selection programmes to increase rates of improvement continues to attract attention, but a clear advantage for structured populations has not yet been demonstrated. One problem is that the consequences of a particular population structure will differ for different types of gene action. Wright stressed always that his conclusions were dependent on selection being for epistatic combinations, i.e. epistasis with respect to fitness. No advantage would be expected, therefore, when selecting for a quantitative trait where the genetic variation was additive, unless epistatic variance arose during selection due to the interaction of artificial and natural selection.

A number of studies have compared artificial directional selection in subdivided and undivided populations (Bowman and Falconer, 1960; Hill, 1963; Enfield, 1970; Goodwill, 1974; Madalena and Robertson, 1975; Katz and Young, 1975; Katz and Enfield, 1977; Rathie and Nicholas, 1980; and Banks, 1986).

Except for Katz and Young (1975) and Banks (1986), these studies showed no real advantage for population division (see also López-Fanjul, this volume). These negative results are best interpreted as indicating essentially additive gene action for the selected traits (drosophila bristle number and tribolium pupa weight). On the other hand, Katz and Young (1975) and Banks (1986) selected for adult body weight in *D. melanogaster*, a trait that is likely to exhibit more epistasis (F.W. Robertson, 1954). Unfortunately, it is not clear whether the greater advances in structured populations in these two experiments were due to epistatic gene action, or to other factors. In contrast to other experiments where sublines were mixed by crossing at intervals of some generations, these experiments allowed regular migration among the sublines in each (or every second) generation. One of Banks' treatments included a more realistic modelling of diffusion among sublines than any previous experiment, and the response in one of two replicates of this treatment suggested the presence of a major nonadditive effect.

As these empirical results are inconclusive, larger experiments are necessary (Barker, 1988) or we must rely on computer simulation. Enfield and Anklesaria (1986) extended the computer simulations of Madalena and Hill (1972) to include epistatic models. Using the same structure of selection within sublines, and regular cycles of selection between sublines and crossing, their results confirm Wright's expectation that when multiple peak epistasis is important, such a structured population is superior to a single large population. For a simple two-locus epistatic model, Rutledge (1970) showed that the favoured genotype can have a much higher probability of fixation in a subdivided population. Effects of regular migration between sublines have not been studied theoretically, so that, for the present, increased genetic gains in subdivided populations appear to depend on there being substantial epistasis for the selected trait.

Recognition of the hierarchical breed structure led animal breeders to develop specific designs (nucleus breeding schemes) which concentrate selection of males in a nucleus herd or flock, but with an optimum structure to maximize rates of genetic improvement and the most rapid spread of this improvement through the total population (Mueller and James, 1984).

In addition to within-breed structure, livestock species are structured into the distinct subpopulations recognized as separate breeds. Thus the animal breeder has available not only the genetic variation within breeds, but also that between breeds. As these breeds have been genetically isolated for various lengths of time while subjected to varying selection pressures and genetic drift due to small effective sizes, they are differentiated genetically to some, but generally an unknown, degree. This structure at the species level, with concomitant between-breed genetic variation, can be used in three ways (Dickerson, 1969):

(i)   to 'grade-up' one breed to another by continuous backcrossing, i.e. simple breed replacement,
(ii) to form a new synthetic (or composite) population from a mixture of two or more existing breeds, and then to select for further genetic improvement within the new population.
(iii) in systematic crossbreeding programmes to take advantage of complementarity (differential expression of particular traits) and heterosis in the crossbred individuals.

In development of synthetics, the aim generally has been to produce a single population combining characteristics from the parental breeds, e.g. crosses of *Bos taurus* and *B. indicus* for tropical beef cattle production. In addition, an expected increase in genetic variability in the synthetic population, as compared with that in each of its constituent breeds, should allow greater selection gains in both the short and long-term. Such increased genetic variability was demonstrated by Howe and James (1973) for synthetic populations derived from long-term selection lines of *D. melanogaster*, but not by Baffour-Awuah and Barker (1988) for populations derived from unselected isofemale lines. Many new synthetic breeds, particularly of beef cattle, have been developed in recent years, but further experimental study is clearly indicated.

Except for Merino sheep in Australia and Friesian (or Holstein–Friesian) dairy cattle in many countries, some form of crossbreeding is used in producing most of the commercial livestock in the developed world, the main reason being to take advantage of heterosis. The optimum utilization of crossbreeding was reviewed by Robertson (1971), where he noted that 'the expected gain by crossing in any specific instance will depend on the degree of relationship between the breeds being crossed'. This degree of relationship is largely unknown, and the choice of breeds has been a matter of experience and empirical data. Genetic distances based on electrophoretic loci, however, could be useful in predicting the expected heterosis (Goddard and Ahmed, 1982; Graml and Pirchner, 1984) and in formulating crossbreeding policies. Once breeds are chosen, they are commonly used in systematic designs such as F_1 crosses, three-way crosses or rotational crossbreeding. In defining an optimum programme, cost and logistical factors must be accounted for, and Kinghorn (1986) has developed an approach which does this and which also avoids the specification of a particular design. In this approach, a design emerges as a consequence of a breeding policy implemented at the level of the breeding pair or mating group.

Structure of natural populations

Natural populations of many species are spread over large areas, and a primary ecological question concerns the number and distribution of individuals. Only rarely, if ever, are individuals distributed at random, and the distribution may range from essentially continuous to highly clumped. For the population geneticist, it is not merely the distribution of individuals as such that matters, but the distribution of kinds of individuals (genotypes), i.e. the genetic composition of the population and possible heterogeneity of that composition within the population. This should be referred to as the genetic structure of the population, but where the context is clearly genetic, population structure often is used as equivalent to genetic structure. Here population structure will be used strictly for numbers and distribution of individuals, and genetic structure for numbers and distribution of genotypes or alleles.

Where there are disjunct subpopulations, because of ecological or behavioural factors, the genetic connections among them depend on the amount of genetically effective migration (gene flow) between subpopulations. With extensive gene flow, the population may breed as a single panmictic unit, while for no gene flow each subpopulation is an independent evolutionary unit. Where the population distribution is essentially continuous, it can still be genetically structured and show local genetic differentiation if the range of the population is large relative to the distances that individuals move during their lifetime. These two situations, which represent the extremes of possible structures in natural populations, were first analyzed theoretically by Wright (1951) as the island model and the model of isolation by distance. Theoretical treatment for these and other models is standard in population genetics texts (e.g. Hartl, 1980; Hedrick, 1983).

The present genetic structure of any population depends on the balance of evolutionary forces – genetic drift leading to genetic differentiation among subpopulations, selection that may lead to differentiation if selective forces differ among subpopulations or to homogeneity if selective forces are similiar in all, and gene flow among subpopulations leading to genetic homogeneity. The potential difficulties in analyzing and understanding genetic structure thus are apparent – it may be affected by all forces that can change genotype or allele frequencies. In this situation, laboratory studies would be expected to provide valuable insights, but surprisingly little empirical work has been done. One notable exception is the work of Wade and McCauley (1984) on effects of population size and island model migration in tribolium populations, where significant subpopulation differentiation was found for a wide range of population structures.

How then should the study of the genetic structure of natural populations be approached? In essence the relevant questions are:
(i)   What are the expected effects of different population structures on the genetic structure of a population?
(ii) What are the genetic structures of particular populations, and what factors have operated to determine them? That is, can we infer past evolutionary processes?
(iii) Given the genetic structure of a population, what factors operate to maintain it?

As noted previously, the theoretical analysis of structure (question (i)) was pioneered by Wright (for review, see Wright, 1969), with other models introduced by Kimura (see Weiss and Kimura, 1965, and for review, see Felsenstein, 1976). At the same time as he was working on the structure of breeds of dairy cattle, Robertson's first contribution to this area appeared as an acknowledgement by Wright (1951) for simplification of a formula relating to the isolation by distance model.

In developing the theory, Wright (1951) also introduced three parameters, F_{ST}, F_{IT} and F_{IS}, to summarize genetic structure, where:

F_{ST} is the correlation of genes between individuals within the same subpopulation,

F_{IT} is the correlation between genes within individuals,

F_{IS} is the correlation between genes within individuals within subpopulations, and

$$F_{ST} = (F_{IT} - F_{IS})/(1 - F_{IS}).$$

When allele and heterozygote frequencies are known for subpopulations, these F-statistics may be estimated and used to describe genetic structure. However, if generalizations concerning genetic structure are to emerge from empirical studies, estimated F-statistics must be comparable. Unfortunately, a variety of assumptions and methods have been used, but Weir and Cockerham (1984) have given estimation formulae which are more general and hence more useful in comparing results from different studies.

Although there were some early studies of genetic structure based on blood groups in man and allelism of lethal genes in drosophila, it was only with the introduction of protein electrophoresis to population genetics in 1966 that suitable extensive data could be obtained (reviewed by Selander and

Whittam, 1983; Taylor and Powell, 1983). These empirical studies (question (ii)) have focused largely on subpopulation differentiation, using F_{ST} estimated as the standardized variance of gene frequency. That is, for a particular locus, if $\bar{p}$ is the mean allele frequency over several subpopulations and $\dot{\sigma}^2$ the variance of allele frequency over the same subpopulations,

$$F_{ST} = \sigma^2/[\bar{p}(1-\bar{p})].$$

Not surprisingly perhaps, a wide range of degrees of differentiation has been found for different species (Wright 1978, Table 7.31) and many are strongly structured genetically.

However, identifying the factors that have operated to determine existing structures has been more problematical, and closely tied to the continuing debate over the relative importance of random drift and selection. For any number of loci in a given set of subpopulations, F_{ST} values will be expected to be homogeneous since all loci have been subject to the same structure, unless selection is acting at some loci. This expectation appears to provide a basis for testing whether selection is acting on any loci, and various statistical tests for the homogeneity of F_{ST} values were developed by Lewontin and Krakauer (1973). However, Robertson (1975a,b) has shown these to be invalid if the subpopulations show hierarchical relationships, i.e. to determine if the tests can be applied, one needs to know what generally cannot be known – the history of the subpopulations. Even where a test is valid (e.g. of temporal variation within a population), significant heterogeneity of F_{ST} values does not indicate which loci are affected by selection, nor can any inferences as to the type of selection be made (Nicholas and Robertson, 1976).

If selection is indeed affecting some loci, the problem of inferring genetic structure is further compounded, in that it will not be known if the mean F_{ST} value has been increased or decreased and genetic differentiation thus either overestimated or underestimated. Clearly one would like to know which of the loci (or alleles within loci) are effectively neutral, so that they might be used to infer the true degree of differentiation. But even for a set of neutral loci, only a rough estimate of the importance of drift can be obtained. Assuming the subpopulations are at equilibrium, F_{ST} is given approximately as $1/(1+4Nm)$, where N is the effective size of subpopulations (assumed equal) and m the proportion of migrants entering each subpopulation each generation. Thus given an estimate of F_{ST}, Nm can be estimated. If $Nm < 1$, genetic drift will lead to substantial subpopulation differentiation (Wright, 1931), but the degree of differentiation cannot be apportioned to the relative effects of migration rate and effective population size.

If an independent estimate of either N or m were available, the other could then be inferred, necessarily with caution, given the assumptions in estimating Nm itself. Both N and m are notoriously difficult to estimate (see Slatkin, 1985a, 1987b for reviews of gene flow in natural populations) although N is sometimes estimated (guessed ?) from census data. As it is the average effective size over previous generations that will have affected the genetic structure (again history is important), and remembering the very small effective sizes of livestock breeds noted earlier, many such 'estimates' of N may well be overestimates and consequently m may be seriously underestimated.

An alternative method for estimating Nm using data for rare alleles (Slatkin, 1985b) does not depend on the assumption of neutral loci. For the same data set, it is then possible to estimate Nm (and infer the magnitude of gene flow), and to estimate F_{ST} as a measure of genetic differentiation. Using this approach, Singh and Rhomberg (1987) found that D. melanogaster and D. pseudoobscura showed similar and extensive gene flow, but that the former had much more geographic differentiation. Absolute levels of gene flow were not estimated, but comparison of results for the two species provided much more information than would either alone. In this case, the differences in genetic structure were interpreted as due to migration in combination with habitat selection causing the greater geographic uniformity in D. pseudoobscura. Clearly this inference of selection affecting genetic structure suggests further empirical study to elucidate the evolutionary forces acting to maintain that structure (question (iii)).

There has been in recent years a rapidly growing interest in quantitative characters in natural populations (Barker, 1988), and Lewontin (1984) discussed the problem of comparing differentiation among populations in quantitative characters with that in gene frequencies. While the ensuing discussion (Rogers, 1986; Felsenstein, 1986; Lewontin, 1986) did not emphasize the point, it did serve to indicate that F_{ST} can be estimated from data on quantitative variation between and within populations (again with the assumption of a selectively neutral quantitative character). That is, F_{ST} values estimated from allele frequencies and from quantitative characters may be compared to provide further information on the factors affecting population differentiation. What Lewontin (1986) does emphasize is what is really needed, i.e. estimates of F_{ST} based on neutral genes, and he notes that this will be available in the future using third-position or intron polymorphism in DNA sequences.

However, we do not need to wait for this particular future before using molecular variation in analyzing the genetic structure of populations and making inferences about evolutionary mechanisms. In particular, analyses of mitochondrial DNA (mtDNA) provide a powerful tool, although still subject to the assumption that mtDNA variants are neutral (see reviews of Avise et al., 1987, Moritz et al., 1987). Within species, a high level of polymorphism is usual and the maternal inheritance of mtDNA allows the possibility of inferring historical information on subpopulation differentiation (i.e. within population phylogenies). As direct knowledge of history is usually lacking, this information from molecular variation, together with empirical study of effects of selection and magnitudes of gene flow, will be major components contributing to future understanding of the genetic structures of populations and the evolutionary mechanisms that have determined them and are acting to maintain them.

B GENETICS OF METRIC TRAITS

Chapter 12
Alan Robertson's Contributions to Quantitative Genetics

R. FRANKHAM

School of Biological Sciences, Macquarie University, Sydney, NSW 2109, Australia

Few fields of science have been as strongly influenced as quantitative genetics has been by the Edinburgh group in general and by the contributions of Alan Robertson and his students and collaborators in particular. Three collaborations deserve special mention, that with George Clayton on experimental studies, that with Bill Hill on theoretical problems (both former students of Alan's) and that with Douglas Falconer. Alan's influence has been paramount from the late 1950s till the present. The Institute of Animal Genetics at Edinburgh has been a mecca for collaborators throughout the world to visit and work, and Alan's morning coffee session has attained an aura similar to that of the Columbia University Fly Room.

Quantitative genetics research by Alan and his students has encompassed theoretical work (both analytical and using computer simulations) and experimental work with *Drosophila melanogaster*, in its practical applications in animal breeding (reviewed by King, this volume) and in its fundamental underpinnings in population genetics (reviewed by Felsenstein, this volume). It has involved a happy marriage of skills that is a model of how to make advances in fields involving complex systems. Details of the contributions of Alan and his students and collaborators to quantitative genetics are given below.

Effects of inbreeding on the genetic variation due to recessive genes

The first unambiguous contribution to quantitative genetics (as opposed to animal breeding or population genetics) by Alan Robertson was his theoretical investigation (Robertson, 1952) of the effects of inbreeding on the variation due to recessive genes. He showed that the variation within lines due to rare recessive genes increases under inbreeding up to an inbreeding coefficient of 0.5 and declines to zero when inbreeding is complete, while the total variance of the population of lines increases almost linearly with the inbreeding coefficient. These increases in variance are greatly reduced in the face of complete selection against the recessive allele. He also considered the impact of inbreeding on the variation due to overdominant loci. This work extended the classic early work of Wright to encompass the type of genetic variation likely to be found for reproductive fitness characters.

Predicting short-term response to selection

The most prominent early experimental work was the series of papers 'An experimental check on quantitative genetical theory' with George Clayton, John Morris and G.R. Knight. Even though the prediction equations arose from the work of Fisher (1918) and particularly Wright (1921) and were the basis of animal breeding practices from the 1930s due to the influence of J.L. Lush, the first of these papers (Clayton, Morris and Robertson, 1957) provided the first convincing evidence that the equations were adequate to predict short-term response to mass or family selection. Further, they evaluated the effects of relaxing selection and of inbreeding. Their results indicated that animal and plant breeders could confidently predict selection response in the short term. McBride and Robertson (1963) extended such evaluations and showed that assortative mating had little effect on selection response for characters with heritabilities in the range considered.

Drosophila as an experimental model for quantitative genetics research

Prior to the work of Clayton, Morris and Robertson (1957), drosophila had been used to demonstrate that selection was effective. However, their study began the long and continuing use of drosophila as a model system for *quantitatively* evaluating prediction from quantitative genetic theory (see Robertson, 1958b, 1967b; Hammond, 1974; Frankham, 1982). This study was notable in that they estimated genetic parameters in their base population, deliberately used a large noninbred base population, the experiments involved replicated selection lines and control lines were kept to check on environmental changes. The drosophila model for quantitative genetics has been highly effective. I know of no case where results from drosophila have been misleading when applied to equivalent characters in other species of laboratory or domestic animals.

Long term effects of selection

The first of many contributions on the long-term effects of selection began with the second paper in the 'An experimental check . . .' series (Clayton and Robertson, 1957; briefly reported in Clayton *et al.*, 1953). In contrast to short-term response, long-term response was unpredictable. This paper set the scene for the most original and important of Alan's contributions to quantitative genetics, namely his 1960 paper on 'A theory of limits in artificial selection'. He showed that the limit to selection in any population should be a function only of Ni, where N is the effective population size and i is the standardized selection differential. He predicted that for small values of Ni, the limit to selection would be $2N$ times the gain in the first generation. Further, he predicted that the furthest limit would be attained when half the population was selected each generation. This paper (Robertson, 1960c) provided a basis for predicting the long-term response to artificial selection. It represented the first clear indication that population size was of importance in plant and animal breeding and led to a large number of experimental and theoretical studies in Edinburgh and elsewhere. Several experimental evaluations of the theory of limits (Jones *et al.*, 1968; Eisen, 1975) demonstrated the qualitative validity of the theory, namely that long-term response increases with increases in Ni. As a consequence, population size has been recognised as an important variable in selection lines, in addition to selection intensity.

James (1962), Verghese (1974) and Nicholas and Robertson (1980) extended the theory to situations where the limits to selection were due to the opposing forces of natural selection rather than to the exhaustion of genetic variation, and came to similar qualitative conclusions regarding the importance of population size and selection intensity in determining response. Robertson himself (Robertson, 1970e, 1973b, 1977c), along with Hill (Hill and Robertson 1966; Robertson and Hill, 1983b) and Latter (1965) extended the theory to encompass more complex models involving linked genes. These theoretical studies indicated that linkage was likely to have only minor effects on the limits to selection. In turn these spawned experimental investigations in his laboratory (McPhee and Robertson, 1970) and elsewhere (Markow, 1975; Thompson, 1977) which demonstrated that suppression of recombination had a relatively minor effect on selection response, as predicted.

Allan and Robertson (1964) investigated the effects of an initial reverse selection on the subsequent limit to selection. They predicted that if the duration of reverse selection, t, is less than $N/2$, the loss in selection advance due to the reverse selection is roughly t/N.

An assumption of the theory of limits to selection is that all the genetic variation comes from the base population, i.e. that mutations arising in selection lines do not contribute significantly to response. This assumption is almost certainly untrue (see Frankham, 1982). Hill and co-workers (see Hill, 1982b; this volume) have subsequently extended the theory of limits to include the contribution of mutations to selection response.

The applications of Wright's shifting balance theory of evolution to animal breeding was evaluated by determining the effects of population structure on response to selection (Madalena and Robertson, 1975). They compared response in a large population with that in a series of smaller partly or totally isolated populations and found no clear effect of subdivision for the additive character used, as predicted from theory (Madalena and Hill, 1972).

Osman and Robertson (1968) carried out an empirical test of means for breaking selection plateaux, showing that limits could be broken and that they were an artifact of the selection regime.

Effects of population-size bottlenecks on response to selection

A second issue relating to the effects of population size on selection response is the impact of population size bottlenecks. Consideration of this effect was included in the theory of limits paper (Robertson, 1960c). This theory was extended in a seminal paper by James (1971), another worker who collaborated extensively with Alan. He showed that the effects of a population-size bottleneck of size n was expected to reduce the subsequent short-term response to selection by the proportion $(1 - 1/2N)$, while the long-term effects were less predictable, depending on the nature of the genetic variation for the particular character. Several experimental studies in Edinburgh and elsewhere have supported the predictions of this theory (Da Silva, 1961 – see Robertson, 1966a; D. E. Robertson, 1969 – see James, 1971; Hammond, 1974; Frankham, 1980b). It is galling that a recent commentary in *Science* (Lewin, 1987) on the Bryant *et al.* (1986) paper ignored both this prior theoretical and experimental work and emphasized the deviations from expectations in the Bryant *et al.* work.

Optimum selection intensities in artificial selection

The gap in our understanding of the selection intensities required to achieve maximum response to selection after a single generation (as intense as possible) and at the limit to selection (50%) was filled by Alan (Robertson, 1970c) and Smith (1969). They predicted that the selection intensity required to produce maximum response after any length of time (in selection units of a given total size) declined in a sigmoid fashion with generations. The practical validity of their predictions are however, open to question in the light of experimental work (Frankham, 1977). Contributions of mutations to selection response may need to be added to the theory to provide better predictions in the longer term.

Correlated response to selection

The third paper in the series, 'An experimental check . . .' was concerned with correlated responses to selection (Clayton *et al.*, 1957). Correlated responses were reasonably predictable in the short term but somewhat anomalous under family selection and asymmetrical in the long term. Replicate variation was large and genetic drift appeared to be an important factor, based on the empirical observations and a brief theoretical consideration. A later theoretical and simulation study (Bohren *et al.*, 1966) showed how asymmetry in correlated responses could develop rapidly from gene frequency changes. Baptist and Robertson (1976) documented asymmetrical correlated responses when selecting on a composite trait based on a few compound characters which interact in a nonlinear manner.

A major contribution to the asymmetry in the responses and correlated responses to selection in the long term selection lines of Clayton and Robertson (1957) was the anomalous sex-limited response in females in the low lines. They failed to determine the basis of these anomalies, but suggested that recombination events were involved and Robertson (in Bowman, 1958) showed that the X chromosome was the major contributor to response in one of the anomalous low lines. We later found a similar sex-limited selection response and were able to attribute it to unequal exchanges occurring between the rRNA multigene families on the X and Y chromosomes, generating *bobbed* mutations on the X chromosome with partial rDNA deficiencies (see Frankham *et al.*, 1978; Frankham, 1988). These events only contributed to selection response in low lines. When I suggested the above explanation to Alan, while on study leave with him in 1976, he commenced (unbeknown to me) a new low selection line from the same base population and a similar event occurred. We were able to confirm that a *bobbed* allele was responsible.

An experimental evaluation of two trait selection theory (Sen and Robertson, 1964) compared tandem, index and independent culling levels selection and found that they were approximately as predicted by Hazel and Lush (1943) and Young (1961).

Mutation and quantitative variation

Another of the early experimental contributions was the 'Mutation and quantitative variation' paper, again the result of collaboration with George Clayton (Clayton and Robertson, 1955). This and their 1964 paper (Clayton and Robertson, 1964) provided evidence for the generation of new spontaneous and irradiation-induced quantitative variation. However, the spontaneous rates were very low. This led to the view that mutations could largely be ignored as a source of genetic variation and as a corollary that essentially all genetic variation utilized to produce selection response came from pre-existing variation in the base population. In the short term this is almost certainly true, but the long-term implications were overplayed (see Hill, 1982b; this volume). Some of the evidence for this change of emphasis was supplied by Robertson (1978) himself. He showed that the delayed time of detection of a variety of visible mutations in selection lines could only be explained by their being mutations that arose *de novo* in the lines, rather than being present in the base population.

Use of computers in quantitative genetics

The use of computers to investigate theoretical problems in quantitative genetics was pioneered by Fraser (1957) and by Martin and Cockerham (1960). They used memory bits as genes and simulated all the steps in reproduction from sampling of gametes, recombination, fertilization, environmental effects, generation of phenotype and selection on phenotype. This involved much complexity and a large stochastic variation in outcome. Consequently, it resulted in only minor improvements in insight into quantitative genetics. Alan Robertson and his students greatly improved the insights gained from computer simulations by simplifying the use of the computer (reviewed by Robertson 1967b, 1970e, 1973b, 1980b). Firstly, they reduced the process to one of selection on gametes to simulate selection with additive gene action, thus reducing the computer time and the cost involved. Secondly, they were the first to use the computer to derive the gene frequency distribution, i.e. randomness is dealt with algebraically, rather than by Monte Carlo simulations (Allan and Robertson, 1964). To do this the computer was programmed to produce the matrix of transition probabilities relating the gene frequency distribution in one generation to that in the next (Narain and Robertson, 1969). This reduces the uncertainty of the outcomes and reduces the computing time and costs. Thirdly, an infinite locus model was simulated (Robertson, 1973b, 1977c).

The most novel use of the computer for theoretical investigations in quantitative genetics by Alan Robertson and his students and colleagues has been to use them to determine the nature of the theoretical solution as a forerunner to providing algebraic solutions to theoretical problems. This was done for the case of the effects of an initial period of reverse selection (Allan and Robertson, 1964), for the effects of linkage on the limits to selection (Robertson, 1970e) and for the conflict between natural and artificial selection (Nicholas and Robertson, 1980).

Reproductive fitness

Reproductive fitness has been a long term interest of Robertson's. The earliest of these was a theoretical paper (Robertson, 1956b) on stabilizing selection due to phenotypic deviation (the optimum model) or to heterozygosity. He showed that stabilizing selection led to a reduction in genetic variation in contrast to the maintenance of variation when selection favoured heterozygotes. He returned to this theme briefly a number of times (Robertson, 1965c, 1974a, 1980c) to criticize the validity of the optimum model.

The first successful attempt to select for sexual isolation between marked lines of *D. melanogaster* was carried out by Knight *et al.* (1956). This was the forerunner to many similar attempts by others.

Shortly afterwards Knight and Robertson (1957) devised the competitive index method for measuring fitness of lines and established that a series of inbred lines had reduced fitness. In its original form or in a modern version using a compound chromosome strain their competitive index is arguably the best available method to assess the fitness of lines (Haymer and Hartl, 1982, 1983). This study was a forerunner to the Latter's and Robertson's (1962) analysis of the effects of inbreeding and artificial selection on reproductive fitness. They showed that the logarithm of reproductive fitness in inbred lines declined linearly with the inbreeding coefficient. The mean fitness of artificial selection lines declined but the effects of direction of selection were often different and the behaviour of individual replicates was highly variable. This was preceded by the important theoretical paper (Robertson, 1961c) on the inbreeding effects of selection. Inbreeding in selection lines was shown to depend on the intensity of selection and heritability of the character. Equations were presented to predict the extent of these effects.

Nonlinear heritabilities

It has long been a tacit (but unspecified) assumption that offspring–parent regressions and heritabilities are linear, in spite of evidence that this is not necessarily true. Robertson (1977b) was prompted to investigate the linearity of heritabilities by the discovery in his laboratory of a non-linear heritability for sternopleural bristle number in drosophila. He pointed out that heritabilities could be nonlinear due to skewness in genetic and environmental distributions, to nonadditive scales of observation, to genotypic and environmental effects combining multiplicatively, to nonadditive genetic effects or to directional gene frequencies. He concluded that perhaps the most important possible cause of curvilinearity is a recessive at low or moderate frequencies contributing significantly to the genetic variance. Maki-Tanila (1982) has produced the most comprehensive investigation of this problem in his unpublished PhD thesis, done under Alan's supervision.

An important but unnoted implication of their theoretical results is that they suggest that heritabilities for reproductive fitness characters may be nonllinear. This prediction arises from the observations that genes affecting fitness characters show directional dominance and directional gene frequencies. Gowe (1983 and personal communication), another collaborator, has predicted that the heritabilities of reproductive fitness characters will be higher at the bottom of the distributions due to the segregation of rare deleterious recessive alleles and that culling on fitness will be more effective than generally recognized. We (Frankham *et al.*, 1988) have shown that the usual declines in reproductive performance in selection lines can be prevented by culling on fitness, as predicted by Gowe. Consequently, investigation of the linearity of heritabilities for reproductive fitness characters may lead to new insights into the organization of genetic variation for fitness characters and to improved means for manipulating them in quantitative genetics and animal breeding. The contribution of Robertson (1977b) and Maki-Tanila (1982), along with Robertson's (1952) work on the effects of inbreeding on the genetic variation due to recessive genes, may well have a significance well beyond that currently recognized.

Nature of quantitative genetic variation

A major emphasis of work in Alan's laboratory has been on the nature of quantitative genetic variation (Robertson, 1966a, 1967a, 1968a; Madalena and Robertson, 1975; Robertson et al., 1977; Shrimpton and Robertson, 1988a,b). This can be said to have begun with the discovery, in long-term selection lines, of high frequency lethals presumed to have effects on the selected character (Clayton and Robertson, 1957). Much of this work remains unpublished in PhD theses of his students (Louw, 1966; Piper, 1972) or was only presented verbally at meetings.

A notable feature of this work was its critical and thorough nature (a feature of all of Alan's work) typified by McMillan's and Robertson's (1974) study on the power of methods for locating polygenes. They showed that the magnitude of effects of located polygenes may be grossly overestimated due to the effects of undetected loci close to them on the chromosome. Other cautionary notes regarding the polygene location technique were passed on by personal communications, word of mouth or in the above mentioned theses. On one notable occasion, Alan presented a paper at a meeting in which he first presented the results of a polygene location experiment, demonstrating genes of large effect. After appropriate heads had wagged approvingly, he then put up the actual model he had assumed in his simulations. The actual model and the results of the simulations of a polygene location experiment that he had done on the computer bore little relation to each other. These issues urge caution in accepting the simple interpretations of related work by Thoday and colleagues.

Two major papers (Shrimpton and Robertson, 1988a, 1988b) provide an up-to-date coverage of all the accumulated evidence and include some of the earlier evidence. This work indicates that genes with a range of effects contribute to selection response, a few genes having relatively large effects and an increasing proportion with smaller and smaller effects, i.e. there appears to be an exponential distribution of gene effects. It does not seem to be a coincidence that it was one of Alan's above-mentioned students who was involved in identifying one of the few major genes of practical importance in commercial livestock, namely the Booroola high fertility gene (Piper and Bindon, 1982).

The range of allele frequencies for genes affecting quantitative traits remains unclear. On the basis of the effects of population size bottlenecks on selection response, Robertson (1967a, 1968a) argued that rare alleles were of minor importance.

The investigations on the nature of quantitative genetic variation have relevance to the likelihood that genetic engineering can be used to introduce useful genetic variation across species boundaries to speed animal and plant improvement, an issue in which Alan has taken a very active interest (see King, this volume).

Automated scoring of quantitative characters

The most serious impediment to quantitative genetics research in drosophila is the labour involved in scoring individuals for quantitative characters. Baptist and Robertson (1976) made a novel device to measure body size automatically in an apparatus with a series of chambers with openings of decreasing size. A notable feature of their successful selection experiment was the use of internal controls of a genetically marked strain that were both raised in the same culture and run together. They documented clearly different types of response on the two directions, those responding to selection for high score having lowered activity but little change in body size, while the low lines showed lowered body size and little change in activity. This apparatus has not come into general use, but is one of the forerunners to the current series of devices that are creating considerable and growing interest, especially the inebriometer and other devices made by Ken Weber and collaborators (for example see Cohan and Graf, 1985).

Perspective

Surprisingly Alan Robertson has not written a textbook and has edited only one volume (Robertson, 1980a), based on the highly successful Harrogate meeting that he organized on 'Selection Experiments in Laboratory and Domestic Animals'. His interactions with Douglas Falconer are only represented in one paper (Falconer and Robertson, 1956) on the effects of disruptive selection on body size in mice. However, their collaboration and interaction occurred on a daily basis and had much to do with the outstanding textbook in this field (Falconer, 1960a) that contains the following acknowledgment 'It is no exaggeration to say that without Dr Alan Robertson's help this book could not have been written'.

Alan Robertson has been the dominant focus of quantitative genetics over the last 30 years. He has led the field by providing a plethora of new theories to test, by carrying out experimental tests of many theoretical developments, by training many of the current crop of quantitative geneticists and by strongly influencing the many scientists who have flocked to work in the Institute of Animal Genetics in Edinburgh. His papers have been highly original, invariably reliable in both theory and interpretations, models of clarity and accessible to the mathematically less able. The experimental studies from his laboratory have shown refreshing rigour (mites not withstanding!). The contribution from his groups have been a testament to the essential interplay of theory and experimentation in quantitative genetics. At all times Alan's erudite contributions have been made with humility and a lack of rancour and with respect for the conventions of correct scientific behaviour.

Acknowledgements

I am grateful to Stuart Barker, Roger Frankham, Bill Hill and Trudy Mackay for comments on a draft manuscript.

Chapter 13
Fitness Variation in Natural Populations

James F. CROW

Genetics Department, University of Wisconsin, Madison, WI 53706, USA

R. A. Fisher (1930) in his 'Fundamental Theorem of Natural Selection' stated that 'the rate of increase in fitness of any organism at any time is equal to its genetic variance in fitness at that time'. By genetic variance, Fisher meant the additive or genic component and he measured fitness in a logarithmic unit, the Malthusian parameter. Many authors have been at pains to point out limitations of the principle, but its utility in compressing a great deal of information into one succinct statement is, I think, not in doubt. It also reminds us that the variance (standardized by dividing by the square of the mean fitness when generations are discrete) is the appropriate measure of fitness variability.

Alan Robertson was the first to carry the theorem a step further. His 'Secondary Theorem of Natural Selection' (Robertson 1966b, 1968a) states that the change produced by selection on any character is equal to the additive covariance between fitness and the character. He (Knight and Robertson, 1957) was the first to demonstrate that in drosophila genetic differences in fitness are much larger than those in viability alone. He was also (Robertson, 1955b) the first to articulate clearly the expectation that genic variance in fitness is small and that the correlation between fitness and one of its components becomes negative as the value of the component increases (see Falconer, 1981, p. 305). Finally, he (Clayton and Robertson, 1955) was the first to compare the genetic variance in an equilibrium population to that arising from a single generation of mutation.

This chapter is a brief discussion of further work in these areas pioneered by Alan Robertson. But first I shall consider total variance and covariance rather than their genetic components. This may seem like measuring the oyster rather than the pearl, but there are reasons for this approach. For one thing, genetic variance in fitness is notoriously difficult to measure reliably, even in the laboratory. For another, the total variance can reveal patterns of selection, especially when different parts of the life cycle are compared.

The index of opportunity for selection

Starting with a population of zygotes (or newborns) let p_i be the proportion of the class that leaves W_i descendants, counted at the same stage next generation. The mean fitness is $\bar{W} = \sum p_i W_i$, where $\sum_i p_i = 1$.

With perfect heritability the proportion of the ith class in the next generation is $p'_i = p_i W_i / \bar{W}$, and the mean fitness of the progeny is $\bar{W}' = \sum_i (p_i W_i) W_i / \bar{W} = \sum_i p_i W_i^2 / \bar{W}$. The fitness of the progeny relative to their parents is $\bar{W}' / \bar{W} = \sum_i p_i W_i^2 / \bar{W}^2 = \overline{W^2} / \bar{W}^2$ and the relative increase is

$$\Delta \bar{W} / \bar{W} = (\bar{W}' - \bar{W}) / \bar{W} = (\overline{W^2} - \bar{W}^2) / \bar{W}^2 = V_W / \bar{W}^2 = I. \tag{13.1}$$

I has been called the *index of total selection* or the *index of opportunity for selection* (Crow, 1958, 1972). It is also V_w, the variance of relative fitness, where $w_i = W_i / \bar{W}$. If fitness were completely heritable, I would measure the relative increase of fitness after one generation of selection.

Fitness and selection differential

Let Z_i be the measurement of a trait of interest in the ith individual. Then the mean values of the measure before and after selection are $\bar{Z} = \sum_i p_i Z_i$ and $\bar{Z}' = \sum_i p_i' Z_i$. Noting that $p_i' = p_i w_i$ and $\sum_i p_i w_i = 1$, the selection differential, is $S = \bar{Z}' - \bar{Z} = \sum_i p_i w_i Z_i - \sum_i p_i w_i \bar{Z} = \text{cov}(Z, w)$, a special case of the equation of Robertson (1966b, 1968a). Standardizing Z by measuring in units of the phenotypic standard deviation, σ_Z, (i.e. $z_i = Z_i / \sigma_Z$), the standardized selection differential is $i = \bar{z}' - \bar{z} = \text{cov}(z, w)$.

This can be related to the index I by noting that $S = r(Z, w) \sigma_Z \sigma_w$ where r is the correlation coefficient. Since $|r| \leq 1$ and $V_w = I$, it follows that $|S| \leq \sigma_Z I^{1/2}$ or, in standardized units,

$$|i| \leq I^{1/2}. \tag{13.2}$$

Thus the square root of I sets an upper limit on how many standard deviations the phenotypic mean can change in one generation of selection (Arnold and Wade, 1984a).

Subdivision of I

Suppose that, starting with zygotes, the life cycle is divided into $T+1$ stages, with fitnesses $W_0, W_1, \ldots, W_T$, where W_k is the number of descendants at stage $k+1$ of the ith individual at stage k. Thus the lifetime fitness of the ith zygote is $W_i = \prod_k W_{ki}$ and the population mean is $\bar{W} = \prod_k \bar{W}_k$. For a single stage of selection, $\bar{W}_{k+1} / \bar{W}_k = \overline{W_k^2} / \bar{W}_k^2$, from which the ratio of mean fitness after a full generation of selection to that at the beginning is $\bar{W}_T / \bar{W}_0 = (\bar{W}_1 / \bar{W}_0)(\bar{W}_2 / \bar{W}_1) \cdots (\bar{W}_T / \bar{W}_{T-1}) = \prod_k (\overline{W_k^2} / \bar{W}_k^2$. In the notation of equation (13.1),

$$1 + I = \prod_k (1 + I_k). \tag{13.3}$$

When selection operates entirely through differential mortality and the members of an age class survive independently, the index for mortality selection is particularly simple. If the fitness of survivors is 1 and of nonsurvivors is 0, and s is the proportion of survivors, the mean fitness before selection is s. After selection the mean fitness is 1, so the increase is $1 - s$. The relative increase because of mortality during the specified period is

$$I_m = (1 - s)/s. \tag{13.4}$$

For another derivation of equations (13.3 and 13.4), see Crow (1958, 1972).

In the derivation of equation (13.3) I have assumed that each generation starts with zygotes. In an asexual population each generation can begin at any stage. Starting at a stage other than the zygote can also provide a useful approximation even for a biparental population.

The analysis of fitness components would be simpler if fitness were measured in logarithmic units, which are additive rather than multiplicative. Indeed a more exact treatment for populations with overlapping generations would utilize the Malthusian parameter as a measure of fitness (Fisher, 1930). No such difficulty arises in subdividing the selection differential, for its components are additive.

The index of opportunity due to mortality becomes enormous in organisms, such as some insects and fish, that produce a large number of eggs or larvae. Undoubtedly, most of the deaths in such a species are random with respect to genetic fitness, although this reproductive pattern permits a rapid response to a new selective agent, such as an insecticide. In general, however, whether a species produces a large number of progeny most of which die or a small number and takes care of them is surely a reproductive, not an evolutionary, strategy.

Data

The opportunity for selection has been measured in several species, especially in humans. I shall consider three examples.

Red deer on Rhum Island (Clutton-Brock *et al.*, 1982)

This study has provided abundant, detailed demographic data for a wild mammal. The relevant data are shown in Table 13.1. The total number of offspring per calf is more than twice the replacement number of 2, indicating (if the data are complete) that the population more than doubled in this generation. There is considerably greater opportunity for selection among males than among females, especially in the fertility component for which the male index is 6 times that of the female.

Table 13.1
Index of opportunity for selection in the red deer population on Rhum Island

	Females	Males	Ratio
Proportion surviving to reproduce, s	0.64	0.55	
Index of mortality (and sterility), I_m	0.56	0.82	1.45
Mean number of offspring per calf	4.72	4.20	
Mean number of offspring per fertile adult	7.38	7.64	
Variance of offspring number among fertile adults	6.05	38.73	
Index of fertility, I_f	0.11	0.66	5.98
Total index of opportunity, I	0.74	2.03	2.75

Humans in Utah (Jorde and Durbize, 1986)

Table 13.2 shows data from an exceptionally large study involving more than a half-million Utah Mormons. Since 1840 there has been a steady reduction in the mortality index as economic and health standards have increased. (I have omitted the mortality index for the years prior to 1840 because of incomplete death records in the early years.) The high fertility index in males from 1800 to 1830 is largely due to polygyny. In recent years I_f is about the same for both sexes, as expected in a monogamous population. Furthermore, in the 1920 cohort the index, 0.23, is equal to its Poisson value ($V = \bar{x}_a$, $I = 1/\bar{x}_a$, $= 0.23$), suggesting that recent fertility differences are largely random.

Table 13.3 brings out the large effect of even a small fraction of polygynous males. While the indices for polygynous and monogamous males are both small, although larger than their Poisson expectations, that for the total population is substantially larger as the consequence of the two kinds of families. Various writers (e.g. Neel, 1970) have called attention to the large potential for male-fertility selection in polygynous societies. The index, I, has been measured for many human populations; for a summary, see Spuhler (1976).

Bullfrogs (Howard, 1979)

Arnold and Wade (1984a,b) have used Howard's data on the progeny of 38 male bullfrogs to measure I for three life-cycle stages. The results are given in Table 13.4. Howard's data gave the mean and variance for stage 1, for 1 and 2 combined, and for all three combined, but we can use equation (13.3) to calculate the separate components. (The values in Table 13.4 differ slightly from those of Arnold and Wade, who did a more complex analysis taking into account covariances between stages rather than the simpler equation (13.3.) Despite the large number of eggs laid and the small fraction hatching, the largest contribution to I comes from sexual selection.

We can also use the Robertson 'Secondary Theorem' to measure the selection differential for male body size. The calculations (Arnold and Wade, 1984b) are given in Table 13.4. As for fitness,

Table 13.2
Index of opportunity for selection, and its components, in Utah Mormons (data from Jorde and Durbize, 1986). I_m, mortality index; $\bar{x}_a$, mean number of children per adult; I_f, fertility index; I, total index

Birth cohort	Females				Males			
	I_m	$\bar{x}_a$	I_f	I	I_m	$\bar{x}_a$	I_f	I
1800		6.2	0.48			11.1	0.75	
1810		7.4	0.28			8.1	0.40	
1820		6.7	0.32			11.1	0.35	
1830		7.6	0.24			11.6	0.28	
1840	0.26	8.4	0.20	0.51	0.26	10.3	0.22	0.54
1850	0.24	8.1	0.20	0.50	0.24	9.3	0.18	0.46
1860	0.26	7.7	0.20	0.51	0.22	8.1	0.18	0.44
1870	0.25	7.0	0.22	0.52	0.26	7.0	0.24	0.51
1880	0.21	6.1	0.27	0.53	0.22	6.0	0.26	0.53
1890	0.19	5.1	0.32	0.57	0.20	5.0	0.31	0.57
1900	0.16	4.1	0.34	0.55	0.18	4.1	0.32	0.55
1910	0.11	4.0	0.29	0.44	0.14	4.0	0.27	0.44
1920	0.11	4.3	0.24	0.37	0.13	4.3	0.23	0.38

Table 13.3
Comparison of fertility index for polygynous and monogamous males, 1800–1860 birth cohorts

	Number	$\bar{x}_a$	σ_x	I_f	Poisson
Polygynous males	1,778	17.4	6.5	0.14	0.057
Monogamous males	33,488	7.7	3.6	0.23	0.130
Total	35,266	8.4	4.4	0.28	0.119

Table 13.4
Index of opportunity for selection and selection differential in bullfrogs. (Data from Howard, 1979, as analyzed by Arnold and Wade, 1984b.) S, absolute selection differential (log length in mm); i, relative selection differential, S/σ

Stage	Index of opportunity		Selection differential	
	Mean	I	S	i
Sexual selection (mates/male)	0.711	1.38	0.069	0.626
Female fertility (eggs/female)	15,691	0.25	0.010	0.088
Viability (hatchlings/egg)	0.501	0.12	0.013	0.120
Total (hatchlings/male)	5,589	2.33	0.092	0.832

the largest component comes from sexual selection. Selection on mating success is such that after selection the mean size has increased by 0.626 standard deviations. The total change when other stages are included is only 0.832. Taken at face value, the data also suggest a slight effect of male size on female egg laying and zygote survival, but more data are needed to establish this.

Genetic components

The foregoing section has shown that measurements of fitness variance and its life-cycle components are simple, at least in principle. Data for more than one generation can provide estimates of the extent to which fitness differences are transmitted. Although they do not separate genetic from environmental correlations, such data are of interest when social and biological inheritance are both of interest. Data on genetic variation in fitness come mainly from plants and from laboratory experiments. I shall discuss only those from drosophila.

An early estimate of genetic variance of viability (Hiraizumi and Crow, 1960) was based on chromosomes from natural populations. Using as markers the second-chromosome mutants *cn* (cinnabar eye colour) and *bw* (brown eye colour), they mated *cn bw/cn bw* females to + +/*cn* + males; the + + chromosome came from nature. The background genotypes were randomized, the *cn bw* and *cn* + chromosomes were each essentially isogenic, and there is no recombination in drosophila males. One-hundred-and-seventy-seven wild-type chromosomes, some of which were lethal or semi-lethal when homozygous, were tested in 20 replications each. The measured variable was p, the proportion of wild type progeny (approximately 0.5). The variance among wild chromosomes, after removal of variance among replicates, was 0.000288 (see Table 2 in Hiraizumi and Crow, 1960). $V/\bar{W}^2 = V/p^2 = 0.00115$. This is for a single chromosome making up about 1/5 of the diploid genome, so the standardized genotypic variance is $V/\bar{W}^2 = 0.00575$.

Mukai (1985) found that random pairs of drosophila second chromosomes have a genic variance in viability (survival from egg to adult of random wild-chromosome heterozygotes relative to a marked laboratory chromosome) of about 0.012. For the whole genome, this is $5/2 \times 0.012 = 0.030$, while the corresponding dominance component is 0.002, not significantly different from zero. Mukai's data varied greatly from one population to the other, but the dominance variance was low in every case. Female egg-laying shows a similar pattern – significant additive variance and no measurable dominance component (Charlesworth, 1987). This argues that the genetic variance for fitness components is largely mutation-maintained.

Why Mukai found more genetic variance than Hiraizumi is not clear. Hiraizumi's experiments were genetically less complicated, and may possibly be more reliable for this reason. The two studies agree for flies collected from colder latitudes, Wisconsin and northern Japan. Perhaps the variability in warmer regions of southern Japan has substantial contributions from sources other than mutation.

How much variance is generated by one generation of mutation? Mukai *et al.* (1972, 1974), measured the decrease in homozygous viability per generation of spontaneous mutation (excluding lethals) and compared this to the value in a natural population, presumably at equilibrium. The decrease per generation of mutation is $\Sigma \mu_i v_i$ where μ_i and v_i are the mutation rate and homozygous viability decrease of an individual locus and the summation is over all relevant second chromosome loci. In the natural population each locus is weighted by the persistence, P, the number of individuals affected by the mutant before its elimination. Thus the homozygous load is $\Sigma \mu_i v_i P_i = \bar{P} \Sigma \mu_i v_i$. The ratio of the two is $\bar{P}$, the weighted mean persistence. The observed values are 0.004 for the decrease per generation and 0.3 for the load in the natural population, so $\bar{P} \simeq 75$. This value is supported by the comparison of the variance in homozygous viability introduced by one generation of mutation to that of chromosomes from a natural population. The two values were, respectively 9.4×10^{-5} and 0.011; the ratio is $\bar{P} \simeq 117$ in good agreement with the value 75 given by the load ratio, considering the uncertainties of both measurements. This argues that the typical mutant affects 75 – 117 individuals before elimination; in other words it has a heterozygous fitness disadvantage of about 0.01. For bristle characters in drosophila the ratio of genetic variance to that produced by one generation of mutation is of the order 10^3 (Clayton and Robertson, 1955), arguing that the fitness effects of individual bristle mutants are an order of magnitude smaller than those of viability mutants. For isozyme loci, the ratio appears to be still larger, of order 10^4 (Mukai, 1985), as would be expected for nearly neutral loci.

Lynch (1988) has summarized the data from a number of organisms on the variance introduced by one generation of mutation. He standardized this value by the environmental variance. For bristle characters this procedure gives about the same value as that in the previous paragraph for which the genetic variance was used as the standard, a ratio of about 10^{-3}. For viability, however the ratios are strikingly different. Standardized by genetic variance, the input variance is about 10^{-2}; standardized by environmental variance it is about 10^{-5}. Although this seems like an enormous discrepancy it is actually quite reasonable since the environmental variance of survival is very large. In the drosophila experiments of Mukai *et al.* (1972) on which Lynch based his analysis, the expected ratio of wild-type to marked chromosomes is $p = 1/3$, so the binomial variance is $p(1–p) = 2/9$. Scaled to a mean value of 1, the variance is $3^2(2/9) = 2$. The input variance from a single generation of mutation is 9.4×10^{-5}, giving a ratio of 4.7×10^{-5}. Lynch's calculated value of 3×10^{-5} implies environmental variance 50% larger than binomial, a value typical of such experiments.

Several studies, starting with Robertson (Knight and Robertson, 1957) have shown that a chromosome that has a small deleterious affect on viability has a much larger effect on fitness (Sved and Ayala, 1970; Sved, 1975). Simmons *et al.* (1978) contrived a multigeneration test in which two competing chromosomes were always heterozygous and could be followed in a population cage for many generations. The selection was between two kinds of males, one of which had been mutagenized, competing for survival and for mating with a single kind of female. The males with heterozygous mutagenized chromosomes were significantly less fit; in fact heterozygous effects on male fitness were about the same as homozygous effects on viability. Similar results were obtained for heterozygous selection on X chromosomes in females (Mitchell, 1977). Finally, Wijsman (1984) arranged competition within and between drosophila species for the entire genome. This was done within *D. melanogaster* by using two strains, one of which had compound autosomes, so that all matings between strains produced inviable zygotes. The strains with mutagenized chromosomes were much less competitive. Again an increased mutation rate produced genetic variance for fitness.

There is no direct measurement in drosophila of the standing genetic variance or of the contribution of a single generation of mutations to fitness itself. Taking the standardized genetic variance for viability of 0.006 (Hiraizumi and Crow, 1960), the observations of Sved (1975), Sved and Ayala (1970), and Simmons *et al.* (1978) would suggest that variance for fitness must be an order of magnitude higher, perhaps 0.05. The data of Mukai suggest that this variance is mostly additive. Yet a population at equilibrium under selection theoretically has no additive or genic variance in fitness (for a recent discussion, see Charlesworth, 1987). The value 0.05, or an even higher value if the data of Mukai are used, seems too high.

I suspect, however, that the observed positive correlation between viability and fitness effects, and the greater magnitude of the latter are misleading. These observations came from flies of less than normal viability, either inbred or mutagenized. Hiraizumi (1960) confirmed Alan Robertson's (1955b) prediction that such correlations become negative at high performance levels. Although there was a positive correlation between a chromosome's effect on viability and longevity when viability was low, this correlation was close to –1 for high viability chromosomes. Rose (1982) has given similar data for other fitness components. Likewise, Simmons *et al.* (1980), measuring the heterozygous effects of chromosomes from an equilibrium population, found the regression of fitness (measured by competition in caged populations) on viability was essentially zero; taken at face value the regression was slightly negative. All these studies argue strongly for negative correlations between fitness components of 'good' chromosomes. The implication is that genetic, and especially genic, variance of fitness is much smaller than suggested in the previous paragraph, but no satisfactory estimate exists.

Conclusion

It is relatively easy to measure the realized variance in fitness, and this has been done for a number of species. All that is required is reliable census and life-table information. In contrast, genetic components are much more difficult. Laboratory experiments in drosophila have measured the genetic variance in viability and have shown that this is mainly additive. The implication that this is largely mutation-maintained is supported by direct experiments on induction of viability variance by mutation. A large additive component is supported by independent evidence for partial dominance in both viability and fitness of new mutations. The mean persistence of a mutant affecting viability is of the order of 100 generations, an order of magnitude less than that for bristle number, which in turn is an order less than that for isozymes, as expected from their presumed fitness effects.

Although genotypic and genic variance for components of fitness have been measured with varying degrees of certainty, the genetic variance for fitness itself is much more elusive. Negative correlations that are large, but not accurately measured, mean that one cannot compute this from its components.

Acknowledgement

Most of this article was written while I was a guest of the Genetics Department at the University of Edinburgh. I am thankful for both hospitality and many helpful discussions.

Chapter 14
The Measurement of Fitness in Drosophila

J.A. SVED

School of Biological Sciences A12, University of Sydney, Sydney, NSW 2006, Australia

The systematic study of fitness in drosophila can be said to date from the paper by Knight and Robertson (1957), which introduced the 'competitive index' (CI) as a single readily interpretable measure of fitness. The CI technique was applied in detail by Latter and Robertson (1962) to evaluate the fitness of strains which had been subjected to various levels of inbreeding and artificial selection.

The paper by Knight and Robertson was not the first occasion on which fitness components had been measured in drosophila. Prior to this a number of such studies had been made, perhaps most notably the attempt by Wallace (1948) to account for the population frequency of *Sex Ratio* in *D.pseudoobscura* using fitness components. However the paper by Knight and Robertson represents the first attempt to give a single readily estimable measure of the overall fitness of a particular strain which takes account of all the major fitness components. Perhaps equally importantly, the method was the first to ensure that all components were obtained under competitive conditions.

This chapter is intended as a review of the CI test and of the more closely related methods, rather than as a general review. However to give an overview of the field, and the place in it occupied by the CI test, it is convenient to start by attempting a classification of all the different ways in which fitness has been studied in Drosophila. Table 14.1 divides the field according to the level being studied, i.e. whole organism, chromosome or single locus, and then lists the most frequently used approaches within that level. Naturally there is a degree of arbitrariness involved in the table. In some studies, for example, it is a moot point as to whether the effects being studied are at the level of the single locus or whether they are being influenced by large chromosomal regions. Similarly some of the approaches overlap, and are relevant to more than one level. All of the approaches will be mentioned during the following discussion, together with their number from Table 14.1; only topics 1, 2 and 4 will be covered in any detail.

Table 14.1
Grouping of topics on fitness determination in drosophila by level of study

Organism	*Chromosome*	*Locus*
(1) Competitive index (CI)	(4) Balancer equilibration (BE)	(7) Population frequencies
(2) Interspecific competition (IS)	(5) Viability	(8) Fitness components
(3) Biomass and productivity	(6) Natural inversions	

Before embarking on details of the methods, some comment should be made on the question of measuring natural selection in natural populations. Of the different approaches, only the naturally occurring inversion (6) can be said to be primarily directed at natural populations. The study of electrophoretic alleles is another area in which much work has been done on natural populations, but this work cannot be said to have contributed a great deal to the study of fitness. It is unfortunate, given the dominant role played by drosophila in the study of natural selection in the laboratory, that it plays such a minor role in studies of 'natural selection in the wild' (Endler, 1986).

The competitive index (CI): (1)

The method involves mixing a known proportion of the strain to be tested (labelled 'strain' in Figure 14.1) with a genetically marked stock (tester). The tester must be one which allows hybrids to be recognized. If males and virgin females of the two types are mixed in known proportions the expectations in the offspring can readily be calculated, as in Figure 14.1. Male fertility, which in this test is essentially the probability of mating, is assumed to be in the ratio M:1 for the two strains. Similarly female fertility, or fecundity, is in the ratio F:1, and is assumed to be independent of which male is involved in the mating. Finally, there is a viability factor V. The expected frequencies of strain and tester offspring can thus be given in terms of the parameters F, M and V and of the relative numbers (a and b) of the two strains used in the test. Observing the numbers in the two offspring classes enables the parameter combination $F \cdot M \cdot V$ to be estimated, after allowance is made for the numbers used as parents.

			♂ strain	♂ tester
	Frequency		a_m	b_m
	Fertility		M	1
♀ strain	a_f	F	$a_f \cdot a_m \cdot F \cdot M \cdot V$	
♀ tester	b_f	1		$b_f \cdot b_m$

Figure 14.1. Progeny expected in the Competitive Index test of a tested *strain* relative to that of a *tester* stock. V is relative viability.

The hybrids which appear in the offspring are ignored in the test. Although in principle they might provide information about the fertility of the parents, such information is confounded with a viability parameter which is not of interest in the calculation. The tester strain used in both the Knight and Robertson and most of the Latter and Robertson studies involves the second chromosome markers Cy and Pm. However the test gives information about the overall strain fitness rather than any fitness component specifically associated with the second chromosome.

An important feature of the test concerns the raising of parents. Since the two strains are phenotypically distinguishable, they can be bred in the same bottles to eliminate potentially important differences in rearing conditions. Fertilized females of each strain are allowed to lay, and virgin flies of the two strains are collected and separated at eclosion. The design of the test also allows for arbitrarily setting the proportions of the two strains used. From a statistical point of view the best combination is one which gives approximately equal numbers of the two strains in the test progeny. The test also allows any frequency-dependent selection effects to be detected.

One final advantage of the test was not evident until recent years when the manifestations of hybrid dysgenesis became known (see e.g. Engels, 1983). Many tests involve the production of interstrain hybrids, e.g. the viability and balancer equilibration tests (4,5), in which transposons such as the P factor may be activated. If considerable care is not taken to minimize dysgenesis, short-term effects due to sterility and long-term effects due to mutation can lead to misleading conclusions. Although hybrids appear in the competition index test, any dysgenic effects in these genotypes cannot influence the result of the test.

What is being measured?

While the parameter combination $F \cdot M \cdot V$ is a measure which includes contributions from the major fitness components, it is not clear that this measure weights them in the proportions of most interest. This may seem surprising, given that $F \cdot M \cdot V$ is exactly the parameter combination which is responsible for the change of ratio between parents and offspring. However the point at issue is whether the results from the test can be extrapolated to other types of fitness experiments, specifically those at the locus or chromosome level. In these experiments, one is usually interested in the change in gene or chromosome frequency brought about by selection. Then a reduction in either of the female or male fertility components is expected to have only half the weighting of a reduction in the viability component, simply because it affects only one of the two sexes. As indicated by Knight and Robertson, overall fitness as it applies to the change of gene frequency should have a contribution from viability, V, multiplied by a fertility contribution which is expected to be approximately $(M + F)/2$.

Which of the two parameter combinations, $F \cdot M \cdot V$ or $V \cdot (M + F)/2$, provides the more meaningful weighting? The $F \cdot M \cdot V$ combination is only applicable to a rather special case, one in which there are two genotypes, but in which only matings within genotypes contribute to the next generation. This is achieved artificially in the CI by ignoring hybrid offspring. Some interspecific tests discussed below provide a case where the $F \cdot M \cdot V$ combination may be the natural measure to use. In general, however, the $V \cdot (M + F)/2$ combination would be preferable. At the same time it should be noted that if the interest is in ranking a series of strains for fitness, then for realistically achievable levels of accuracy it is unlikely that there will be much difference between the two measures.

Estimating components

The $V \cdot (M + F)/2$ parameter combination cannot be estimated without identifying the frequencies of the different mating types. In their studies on fitness components for single fourth chromosome marker systems of *D.melanogaster*, Prout (1971) and Bundgaard and Christiansen (1972) used a procedure which enabled these frequencies to be estimated. Progeny were raised from individual females from the mating pool, thereby allowing the genotype of the male parent to be inferred. These authors were thus able to estimate all individual components and the overall fitness. In fact Latter and Robertson (1962) carried out a similar procedure on some of their lines to demonstrate that different components could be multiplied together to give the same ranking of strains as measured by the CI. However, they were not concerned, as were Prout, Bundgaard and Christiansen, in combining components to give an accurate overall measure of the fitness. The disadvantage of estimating components is that it involves a great deal of extra labour, and has to date only been applied to artificial marker systems. By comparison, the CI has been used by Latter and Robertson (1962) to give fitness rankings of a wide variety of strains.

Interspecific (IS) competition tests: (2)

The principle of such tests is to compete different strains of one species against a second species. The within-species ranking comes from comparing the results of the between-species competitions. The second species occupies an analogous position to the tester strain in the CI test. There is obviously a close analogy between the IS and CI tests. The IS test is, in general, a multiple generation test, and numerous such tests have been carried out (e.g. Barker, 1963; Mourao and Ayala, 1971).

However the IS test can be restricted to a single generation, and in this case is exactly analogous to the CI test.

While there are advantages in being able to extend the test to multiple generations, the IS test has disadvantages compared to the CI test. First, if the two competing species avoid intermating, then it is not clear that the male mating component is contributing significantly to fitness. All that is required is for each female to be fertilized by a con-specific. Given the apparent importance of the male mating component in intraspecific tests (e.g. Bundgaard and Christiansen, 1972), omission of this component must lead to underestimation of the selection intensity. Secondly, if species are used which differ grossly in resource utilization, it is not clear that the test is very different from a test which just measures productivity within a single species. It can be argued, however, that the fact that one species is usually eliminated in IS tests is a sign that there must be competition for some resource.

The compound-autosome test

A test which was introduced by Hartl and Jungen (1979) serves as a bridge between the CI and IS tests. These authors, working with *D.melanogaster*, used a compound-autosome strain as the tester strain. Strains containing compound autosomes can produce progeny when mated amongst themselves, but lead to inviable progeny when mated to chromosomally normal strains. The one-generation test involving these strains is identical to the CI test as used by Knight and Robertson, except that hybrids do not appear amongst the offspring. The test possesses the advantage that the same combination of competing strains is automatically set up amongst the offspring, so that the test can be extended to multiple generations.

In theory, the multiple-generation compound-autosome test should possess the advantages of taking into account differences in longevity and of producing a more realistic age structure. It is advantageous compared to most examples of the IS test in properly including a male mating component. In practice, Hartl and Jungen found that although the second and third generations of the multiple generation test appeared to give increased sensitivity compared to the first generation, there were problems of establishing a stable age distribution. In later generations there were problems with changes in fitness and with the chromosome configurations breaking down. For overall simplicity and repeatability they recommend the single generation test.

Biomass and productivity: (3)

Little will be said about such noncompetitive tests here. Although they give valid information about adaptation of different strains to particular culture conditions, they do not lead to an estimate of fitness in the same way that competititve tests do. Furthermore the ranking of strains according to biomass and productivity correlates poorly with ranking from the competitive tests (see final section).

The balancer equilibration test (BE): (4)

Sved and Ayala (1970) and Sperlich and Karlik (1970) introduced a method for measuring the fitness of chromosome homozygotes. The method involves setting up population cages containing copies of two chromosomes, the chromosome whose homozygous fitness is to be measured, denoted $\not{+}$, and a balancer chromosome which is homozygous lethal, denoted Cy. The homozygous lethality of the Cy chromosome is expected to lead to its elimination from the population cage in the absence of some counter-selection against the wild type homozygote. If the wild type homozygote has a selective disadvantage s compared to the $Cy/\not{+}$ heterozygote, then an equilibrium is expected at a frequency $1/(1+s)$ for the $\not{+}$ chromosome. Application of the test almost always leads to an equilibrium for the balancer chromosome (even in student experiments). In most individual cages where the balancer chromosome has been eliminated, the behaviour has been so different from other populations as to suggest contamination of the population by one or more extra wild type chromosomes.

The specific measurement being made in the population cage is of the fitness of chromosome homozygotes relative to the fitness of balancer heterozygotes. Balancer heterozygotes are not genotypes whose fitness is of intrinsic interest, but there are appropriate controls for the fitness of these genotypes. If a cage is set up with a mixture of wild-type chromosomes rather than with a single wild-type chromosome, an estimate is obtained of the fitness of balancer heterozygotes relative to wild type heterozygotes, genotypes whose fitness *is* of interest.

In such mixed chromosome cages, the population behaviour of the balancer chromosome is very different from that in the single chromosome cages. The balancer chromosome is usually speedily eliminated. In fact if cages are set up with just two wild-type chromosomes and a balancer chromosome such elimination will occur (Sperlich and Karlik, 1970). This great difference in behaviour between cages initiated with one and two wild-type chromosomes is the reason that the method for measuring homozygous fitness is very sensitive to any contamination by extraneous chromosomes.

The speedy elimination of balancer chromosomes from mixed chromosome cages indicates that balancer heterozygotes are themselves genotypes with lowered fitness. Estimating the value of this fitness with any precision is difficult for what may seem a trivial reason – it is not known how many generations the flies in the population cage have undergone. The best that can be done is to make a 'guesstimate' of generation length as 2.5 to 3 weeks, and to calculate the selective value consistent with the observed rate of loss over the estimated number of generations. In this way, it has been estimated that balancer chromosomes, such as $(SM1)Cy$, have a fitness of only around 50% compared to wild type heterozygotes.

In calculating the fitness of chromosome homozygotes compared to chromosome heterozygotes, what is usually done is to multiply the fitness estimates from the two types of cages. Thus the mean fitness of chromosome homozygotes relative to the $(SM1)Cy$ balancer chromosomes has been estimated to be around 40%. An estimate of the fitness of wild-type homozygotes relative to wild-type heterozygotes is therefore $0.4 \times 0.5 = 0.2$. While it seems unlikely that this procedure would be seriously in error, it is somewhat unsatisfactory that no way has been found to test its validity.

The low fitness of balancer genotypes estimated from the population cage has come as something of a surprise, since these genotypes perform very well in viability tests (Simmons and Crow, 1977). Although it is possible that some reduction in female fertility is caused by crossing over within the inversions, it is known that mechanisms exist in drosophila to minimize such fertility effects. Mandryk (1978) attempted to determine which component of fitness was responsible for the low fitness of $SM1(Cy)$ heterozygotes by measuring viability, female fecundity and male competitive ability. No component was sufficiently reduced to suggest that it might be responsible for the large fitness reduction in the population cage.

Sved and Ayala (1970 and subsequently) have used the BE test specifically to measure the fitness of chromosome homozygotes, but the test can also be applied to measuring the fitness of other genotypes. For example balancer chromosomes can be introduced into an inbred laboratory strain, and their rate of elimination (or possibly equilibrium) compared against the rate of elimination of the chromosome from outbred wild strains to yield an estimate of the fitness of the inbred strain chromosomes. Note, however that this estimate refers specifically to the chromosome in question, e.g. to chromosome II if the Cy balancer is used, rather than to the whole strain as measured by the CI test.

Other approaches: (5)–(8)

The viability of chromosome homozygotes (5) can easily be estimated, and as a result there is a large literature on this approach. While such studies have been of value in defining the effects of mutation on fitness (e.g. Mukai *et al.*, 1972), they do not, in principle, have advantages in the measurement of fitness compared to the more comprehensive BE method.

Naturally occurring chromosome inversions (6) have contributed greatly to the study of selection in natural populations, e.g. in demonstrating temporal changes in frequencies (e.g. Dobzhansky, 1948) and in showing the likely environmental dependence of frequency clines (Oakeshott *et al.*, 1982). However, the special nature of these inversions due to their long evolutionary history means that they cannot be used to measure fitness in the same way as the other measures discussed so far.

Fitness components for single loci (7), as estimated by Prout (1971) and Bundgaard and Christiansen (1972), have already been mentioned. Observations of the fate of mutant genes in laboratory populations (8), without directly estimating fitness components, can also contribute a considerable amount of information on overall fitness. This method bears a close analogy to the BE method at the chromosome level, especially when lethal genes are used (Anderson, 1969).

Comparison between tests

Two principle questions may be asked when comparing different methods for estimating fitness:
(i) When a range of strains is compared using different methods, do the fitness values, or rankings, correlate well between the different methods?
(ii) What absolute estimates of fitness are given by the different methods?

Haymer and Hartl (1983) have given the most comprehensive evaluation of the different methods, principally the CI, IS (single generation) and BE tests. They showed also that the CI and compound-autosome single generation tests correlated very well, although given the similarity of the two tests this is not surprising. Fitness values from these tests correlated, although less well, with those from the single-generation IS test. The lower correlation could be due to a reduction in the male mating component as mentioned above. In agreement with this possibility is the fact that in the IS tests the inbred strains used in the test had a lower absolute reduction of fitness compared to the outbred strains than was the case in the intraspecific tests.

The results from the BE test did not correlate with those of any of the single generation tests. Furthermore, tests on productivity produced results which were uncorrelated with the other tests. Mourao and Ayala (1971) found, however, that multiple generation IS comparisons between *D.pseudoobscura* and strains homozygous for different chromosomes in *D.willistoni* did correlate with fitness as estimated using the BE procedure. Again, neither test correlated with productivity.

In the comparison of absolute fitnesses, the BE test gives considerably lower fitness estimates that the CI test (Haymer and Hartl, 1983). Haymer and Hartl tested a series of strains by both methods, and found average fitnesses of approximately 0.2 for the BE method and 0.6 for the CI method. This difference occurs despite the fact that the CI combines fertility components multiplicatively (see above), which should lower the estimate. There are many differences between the multiple generation BE test and the single generation CI test, including the form of the age structure and possible contributions from longevity and increased crowding levels, all of which appear to combine to increase the sensitivity of the BE test. Which test more accurately reflects the situation in natural populations has yet to be determined.

Chapter 15
Mutation and Maintenance of Quantitative Genetic Variation

William G. HILL

Department of Genetics, University of Edinburgh, West Mains Road, Edinburgh EH9 3JN, Scotland

A fundamental problem in the genetic analysis of quantitative traits is to establish what are the main forces that maintain genetic variability. Mechanisms for the maintenance of variation at the single locus or DNA site have received much more attention and, of course, many of the results are relevant to the analysis of continuous characters. But as there is not sufficient information on the action, number and effects of genes determining quantitative traits to use these results straightforwardly, analysis at the level of the trait is necessary.

Alan Robertson had a continuing interest in these problems and made important contributions. In particular he showed that stabilizing selection led to homozygosity and that its analysis for traits one at a time was a dubious practice because of correlations among traits; with Clayton he obtained among the first estimates of new variation arising in quantitative traits by mutation; he drew attention to the interrelations of natural selection, artificial selection and population size; and collected data on the magnitude of effects and number of genes affecting quantitative traits. Robertson's work pervades an area in which there has been a recent upsurge of interest as the attention of natural population biologists has broadened to quantitative traits and through the stimulus of conflicting theoretical analyses, particularly by Lande and Turelli.

Neutral models

In the absence of selection a balance is expected between the new variation produced by mutation and its loss by genetic drift. Assuming an additive model, the *de novo* variance is $V_M = nuE(a^2)/2$, where n is the number of loci affecting the trait, u is the mutation rate per locus and $E(a^2)$ the mean square difference between homozygotes in effect on the trait. If V_M and effective population size, N_e, are constant the steady-state variance is $V_G = 2N_eV_M$ (Clayton and Robertson, 1955). This value may be approached very slowly, however, if N_e is large: for an initial variance of V_{G0} (zero for an isogenic population, approximately $5V_M$ for a long-standing full-sib line),

$$V_{Gt} = 2N_eV_M + (V_{G0} - 2N_eV_M)\exp(-t/2N_e).$$

Similarly the variance amongst sublines derived without selection from a population is

$$V_{Bt} = 2tV_M + 2(V_{G0} - 2N_eV_M)[1 - \exp(-t/2N_e)]$$

(Chakraborty and Nei, 1982). This shows that the initial rate of divergence of large sublines drawn from an inbred line is very small, or equivalently that mutation does not contribute much to early population divergence, but that the rate of divergence eventually reaches a rate of $2V_M$ per generation.

Alternative methods of estimating V_M, whether from variation among inbred lines, extracted chromosomes, standing variation or selection response have the same expectation if genes are additive. With dominance this is no longer the case, and it is notable that most theoretical and experimental analyses of mutation and quantitative variation have assumed an additive model. It can be shown that V_G is not greatly affected by the degree of dominance (Lynch and Hill, 1986), but if mutants are recessive it takes very much longer for the steady state variance to be reached than with additive or dominant genes. The asymptotic rate of divergence among lines is $2V_M$ as for additive genes, because it depends only on the effects of the homozygotes.

These models formally assume an infinite number of alleles at each locus and an infinite range of effects, with the stepwise difference a having a constant distribution. Cockerham and Tachida (1987) use the 'House of Cards' model (see later) with a finite number of alleles per locus and in which the probability that any allele (effect) sampled is independent of the value of the parental allele. This leads to similar results for variation maintained in the short term, but qualitatively different predictions in the very long term when variation among lines derived from a single founder reaches an asymptote. Their conclusion that mutations are unlikely to yield substantial variation either within or between populations is, however, based on estimates of mutation rate per locus and levels of standing quantitative genetic variation, and does not take into account information on values of the new variation from mutation, V_M.

Estimates of V_M are most extensive for bristle number in drosophila, and these and other estimates have been summarized (Lande, 1975; Hill, 1982b; Lynch, 1988; Mackay, this volume). The typical figure, expressed in terms if the environmental variance, V_E, is $V_M \simeq 10^{-3}V_E$ (a value initially put forward by Clayton and Robertson (1955), although their own estimate was almost an order of magnitude lower!). The heritability, h^2, expected for the neutral additive model is

$$h^2 = 2N_c V_M/(2N_c V_M + V_E)$$

which for $V_M \simeq 10^{-3}V_E$ requires N_e to be in the range 25–500 for h^2 in the range 5–50%. This is another expression of the problem of the neutral model, much discussed at the single locus and molecular level (Lewontin, 1974; Kimura, 1983), that it predicts too close a relation between heterozygosity and population size; and further, $N_e = 500$ is far below population sizes found in nature so there must be selective forces acting to reduce variation. This simple analysis does indicate, however, that in the absence of selection, mutation is sufficient to maintain observed amounts of variation in most populations.

Stabilizing selection

The model of stabilizing selection on quantitative traits in which extremes are assumed to be less fit, has both intuitive appeal and some supporting evidence (e.g. Linney et al., 1971; Kaufman et al., 1977); for further review and discussion see Turelli (1984). It was described by Wright (1935) in terms of a decline in fitness in proportion to squared deviation from the optimum and by Haldane (1954) in the more tractable nor-optimal form, in which the fitness of individuals of phenotypic value X units from the optimum phenotype is proportional to $\exp(-X^2/2w^2)$, where w is an inverse measure of the strength of the stabilizing selection. It follows that the fitness of genotypes of value G units from the optimum is proportional to $\exp(-G^2/V_s)$, where $V_s = w^2 + V_E$ (Lande, 1975). At the individual gene level, V_s also includes variation due to segregation at all other loci.

It was shown by Robertson (1956b) that stabilizing selection leads to an unstable equilibrium at a gene frequency of one-half for populations at the optimum, so that selection leads to homozygosity with an appropiate proportion of loci fixed for the increasing allele. (This *unstable* equilibrium is a property of multilocus models. If only a single locus determined a trait and there was an optimum at the value of the heterozygote, it would lead to *stable* intermediate equilibria gene frequency.) Therefore it is necessary to invoke mutation or other forms of selection to explain variation, and the interactions of mutation and stabilizing selection have received particular attention. Robertson and most others have used an additive model, but it seems unlikely that the results are substantially affected by dominance at individual loci (Slatkin, 1987a).

Infinite populations

Assuming a two allele model, Latter (1960) showed that the equilibrium variance in an infinite population is given by $V_G = 4nuV_s$. An important feature of this equation is that the variance maintained depends only on the number of mutations (nu per haploid genome per generation) and not on their effect (a) on the trait. This is because the mutant has a selective disadvantage proportional to a^2, so the reduction in variance due to selection and increase in variation due to mutation are both proportional to a^2. An alternative viewpoint is to note that the expected total heterozygosity of a mutant in its lifetime is inversely proportional to a^2, so the total variance it maintains is independent of a (Hill and Keightley, 1988). Note that the weaker the natural selection, the larger V_s and the higher the variation maintained. Values of three parameters are required, n, u and V_s, although the total number of mutations nu may be more easy to estimate than n or u separately.

A quite different model in which there is assumed to be potentially infinitely many alleles at each locus with a continual distribution of effects was proposed by Crow and Kimura (1964) and analyzed by Kimura (1965). He obtained the steady state distribution of effects at each locus and showed that this was a normal distribution. The variance contributed by the locus is $[ua^2V_s]^{1/2}$, or summing over loci assuming independence, $V_G = n[uE(a^2)V_s]^{1/2}$. This formulation of Kimura was developed by Lande (1975) and parameter values discussed. Since V_G depends on $n(u)^{1/2}$ rather than nu in Latter's two allele formulation, much larger values of V_G are generally predicted. For example, with $n = 100$, $u = 10^{-6}$, $V_s = 25V_E$ and $[E(a^2)]^{1/2} = 0.2\sigma_E$, the predicted value of V_G/V_E is 0.01 for the two-allele model and 0.1 for the continuum of alleles model.

Turelli (1984) showed, however, that for the normal distribution obtained by Kimura (1965) to hold the variation of mutational effects must be small relative to the standing variation at each locus, which would require unreasonably high mutation rates, in excess of $u = 10^{-4}$. Turelli utilized Kingman's (1978) 'House-of-Cards' (H-of-C) multi-allelic model, in which the effects of the mutant are independent of the effect of the parental allele, and obtained a value for the equilibrium variance equal to that for the two-allele model of Latter. He further illustrated this using a symmetric three-allele model. Subsequently Slatkin (1987a) used a five-allele stepwise mutation model to illustrate the range of applicability of the normal and H-of-C analyses, and further showed the limitations of the normal approximation to very weak selective effects (i.e. values of a^2/V_s) relative to u. More generally Bürger (1986) has undertaken a mathematical analysis of the continuum of alleles model and obtained an upper bound for the variance, which turns out to be very close to the H-of-C approximation. Bulmer (1989) reviews alternative methods of analysis.

Although formally an infinite alleles model, it is not surprising that the H-of-C result is very similar to that for two alleles: essentially mutation rates are sufficiently low that only rarely are more than two alleles contributing substantially to the variance, so a good fit of the H-of-C approximation to a three allele model and to a five allele model, each symmetric about the optimum, have been demonstrated (Turelli, 1984; Slatkin, 1987a).

Theoretical analyses by Lande (1975) for the continuum of alleles normal model and by Bürger (1989) for two alleles, and simulation by Bulmer (1976) for two alleles and by Keightley and Hill (1988) and Bürger (1989) for more general models show that linkage is unlikely to have any significant effect on the conclusions obtained for independent loci.

Finite populations

Although the analysis of Latter (1970) of stabilizing selection in finite populations relied on the Kimura continuum of alleles normal approximation, subsequent analyses have not. Bulmer (1972) used a two-allele model with reverse mutation and Keightley and Hill (1988) an infinite allele model with no reverse mutation. If mutation rates are sufficiently low, these are equivalent. When the population size (N_e) is small and selection is weak, drift predominates so the total variance contributed by a mutant is expected to be $2N_eE(a^2/2)$, i.e. $2N_e$ times its initial variance and so, summing over loci, the expected variance is $2N_eV_M = N_enuE(a^2)$ as shown previously. If selection predominates, such that $N_ea^2/V_s > 10$, approximately, the variance it contributes is close to $4V_s$, and the total is $4nuV_s$ in all (the H-of-C model). However, because of the unstable equilibrium at intermediate gene frequency, the variance exceeds $4V_s$ by about 25% for N_ea^2/V_s around 20. If the latter complication is ignored, and the mutants are regarded as simple deleterious genes, an equation containing the effects of drift and stabilizing selection can be obtained by equating the gain and loss in variance. This is

$$V_G = 2N_enuE[(a^2/2)/(1 + N_ea^2/4V_s)],$$

which reduces to $2N_eV_M$ and $4nuV_s$ at its bounds as $N_e \to 0$ and $N_e \to \infty$ respectively (Keightley and Hill, 1988; Bürger et al., 1989). Simulation has shown this to be an appropriate formula, even where many alleles may be present per locus.

Values of parameters

Latter (1960) was the first to draw attention to the problem of reconciling observed levels of variance and the upper bound for infinite populations of $V_G = 4nuV_s$, and it has received extensive discussion elsewhere, notably by Turelli (1984, 1985, 1988). Whilst observed levels can be accommodated by the Kimura–Lande model (Lande, 1975), these have to be discounted because of the deficiencies of their analysis.

Taking $V_M/V_E = nuE(a^2/V_E)/2 = 10^{-3}$ as discussed previously and $u = 10^{-5}$ (surely a high figure), then $nE(a^2/V_E) = 200$, which for genes of equal effect implies, say, 200 genes of effect σ_E or 3200 of effect $\sigma_E/4$, neither seeming too plausible. A more appealing hypothesis, and one held firmly by Robertson (e.g. 1967a), is that genes and mutants have a distribution of effects, a few having large effect, many small. In this context we might then assume, say, 800 loci influencing the trait, with mutants having $E(a^2/V_E) = 0.25$, but which comprised mutants at, say, one locus with effects as large as $a = 2\sigma_E$, itself contributing 2% of the mutational variance, with 50 or so loci contributing more than half the total. More formal models of distributions of effects are discussed by Keightley and Hill (1988).

Values of V_s have been reviewed by Turelli (1984) who concludes that typical values are approximately $V_s = 20V_E$. These should, however, be regarded as tentative at best, as estimates are invariably noisy and it is difficult, if not impossible, to determine the selection acting directly on any individual trait rather than secondarily through correlated traits. With $n = 800$, $u = 10^{-5}$ and $V_s = 20V_E$, the two-allele and H-of-C models predict $V_A = 0.64V_E$, or a heritability of almost 40%. If these values of n or u are regarded as too high, then lower values of h^2 are predicted.

Values of another quantity are available which are relevant to the calculations but do not enter them directly, namely the number of mutational events detected per generation. Estimates of these in maize are about 0.06 (Sprague et al., 1960; Russell et al., 1963), whereas in the above example $nu = 0.008$ for all mutations, let alone detectable ones. It is difficult to reconcile this value of 0.06 with any model of the genome, for if say an effect of $0.5\sigma_E$ or more can be detected, the detectable mutants alone would contribute to V_M in excess of $0.01V_E$.

Multiple traits

The most important weaknesses of the stabilizing selection model are that selection is assumed to act solely on the phenotype of the trait being observed, and that genes contributing to its variance have no pleiotropic effects on other traits under selection. Robertson (e.g. 1965c, 1974a) frequently pointed out these deficiencies and the problems of interpreting selection at the level of any single trait. Although multitrait analyses (Lande and Arnold, 1983) can take account of selection on all recorded traits, they cannot on any which are unrecorded.

In an analysis of a multivariate model in which effects of mutants are assumed to be sampled from a distribution in which there are no correlations between traits, Lande (1980a) concluded that pleiotropy did not influence the variance maintained. This result is, however, a consequence of the deficiencies of assuming a multivariate normal distribution of allelic effects. Using the H-of-C approximation, Turelli (1985) arrived at a different conclusion, even for mutants with uncorrelated effects. For the case of stabilizing selection acting independently on c characters, each with strength V_s, it turns out that the variance maintained at each trait is $4nuV_s/c$. The simple explanation put forward by Turelli is that all mutants will be deleterious for at least one trait. The result can also be obtained easily for the two-allele model by noting that the expected selective disadvantage of a pleiotropic mutant is proportional to the sum over traits of a^2/V_s, whereas the variance it contributes to each trait is proportional to a^2 for that trait (Keightley and Hill, 1988). (The H-of-C analysis gives somewhat different results in a more general case, however.) If we hypothesize that, in view of the metabolic and developmental interactions in any organism, all genes affect all traits to some degree, it follows that the single trait analysis greatly overestimates the variation likely to be maintained by a stabilizing selection-mutation balance.

Furthermore, it seems unreasonable to assume that traits such as viability and fertility, which are closely associated with fitness have a nor-optimal form of fitness distribution. Therefore most mutants for these traits are likely to be deleterious. See Robertson (1955b) for a classic discussion of the relations between characters and fitness, and Crow (this volume) for a discussion of mutations for viability.

Directional (artificial) selection

With stabilizing selection all mutants are at a selective disadvantage, albeit only in proportion to the square of their effect. In contrast, if there is directional selection as in artificial selection, some mutants are likely to be at an advantage, the proportion depending on the distribution of mutant effects on that trait and on overall fitness. The selective value of a gene affecting only the trait is ia/σ_P for mass selection with intensity i and phenotypic standard deviation σ_P.

Whilst it is clear that if selection is very weak the variance maintained will approach the neutral value of $2N_eV_M$ for additive genes, rather surprisingly this result also holds for additive genes of arbitrarily large effect providing the distribution of mutant effects is symmetric about zero, i.e. increasing and decreasing mutants are equally likely (Hill, 1982a,b). More generally,

$$V_G = 4N_eV_ME^+(a^2)/E(a^2),$$

where $E^+(a^2)$ is the mean square effects of mutants of positive value. There are two ways of obtaining this result, both informative (Hill and Keightley, 1988). The fixation probability of favourable mutants is ia/σ_P, and, if fixed, these give a response of a units; and deleterious mutants are eliminated. With $2N_enu$ mutants expected per generation, $V_M = nuE(a^2)/2$, and the response is therefore $2N_einuE^+(a^2)/\sigma_P$ which, with continued selection can be equated to the response computed in the standard way, iV_G/σ_P. Alternatively, using results of Kimura (1969), the expected total heterozygosity maintained by a favourable mutant gene is $H = 4$; the variance it produces is $Ha^2/2$ and summing over loci gives the final result. Although these arguments strictly apply to two alleles, they seem relevant providing population sizes are not so large that there are often more than two alleles which are not neutral with respect to each other segregating at any time. The prediction of a linear increase in variance with increase in N_e will break down as more mutants segregate and if the range of gene effects are any locus is restricted.

If values of $V_M = 10^{-3}V_E$ are realistic it is clear that mutation can provide substantial useful variation in long term directionally selected populations, providing a proportion of mutants increase the trait. Selection limits are therefore more likely to be the consequence of the absence of favourable mutants or of opposing natural selection than lack of variation (Hill, 1982b; Hill and Keightley, 1988).

With stabilizing selection, genes are typically at extreme frequencies, particularly if their effects are large. Barton (1986) has pointed out that a change from stabilizing to directional selection is expected to lead to an increase in variance as some of the previously deleterious and rare alleles become advantages. In contrast, in the Gaussian model of Kimura and Lande no change in variance is expected if there is a change in mean (Barton and Turelli, 1987). Although a substantial increase in variance is not observed in most selection experiments, the explanation concordant with the stabilizing selection model is that most variation is maintained at loci at which allelic differences have small effect on the trait, because mutants of small effect are most frequent.

Artificial and natural selection

It is a common observation of artificial selection experiments that a plateau is reached at which there remains genetic variability, as can be seen by reductions in response on relaxed or reversed selection (e.g. F.W. Robertson, 1955; Clayton and Robertson, 1957; Yoo, 1980b). One explanation of such plateaux is a conflict between directional and stabilizing selection, in which case the limit in an infinite population depends only on the relative strengths of the two selection pressures (James, 1962; Zeng and Hill, 1986).

In some selection experiments, however, the plateau is largely due to segregation of a gene with a large effect on the trait being selected but which is lethal or nearly so in homozygous form. One such was found by Clayton and Robertson (1957), subsequently suggested to be the *bobbed* locus (Frankham et al., 1978; Frankham, this volume). Theoretical analyses of single locus models have been undertaken. Nicholas and Robertson (1980) showed how the selection limit in a finite population in the absence of mutation would be reduced if genes increasing the trait also had heterozygote superiority for fitness, and Hill and Keightley (1988) described the reduction in rate of response in mutants increasing a trait were deleterious for fitness. In both these models, less response is obtained than would be indicated by the variability in the trait. It would seem worthwhile to further investigate models of both directional and stabilizing selections when mutants have a pleiotropic, invariably deleterious, effect on fitness.

Overdominance for fitness and pleiotropy

The mechanisms for maintaining quantitative variation have been and remain controversial. It has been shown that neutral mutations are sufficient, but introduce a strong dependency between population size and variation. Stabilizing selection with mutation may be an explanation, but the choice of parameters is critical. Further:

We may reach some kind of a solution by pretending to ourselves that the organism consists of independent compartments with no effect on one another. This is essentially what we are doing when we argue from an optimum model. We are saying that genes affecting one particular measurement have their effect on fitness solely through the measurement of that character alone or through some other character which bears a one-to-one relationship to it. The basic assumption spelled out in words like this seems to me to be so unrealistic as to be completely untenable. Can we get around this problem by saying that part of the observed relationship between a character and fitness is due to a direct effect and partly due to a pleiotropic effect? This again is I think merely a hopeful procedure which would break down on detailed logical analysis (Robertson, 1965c).

Bulmer (1973) and Gillespie (1984c) have undertaken analysis of models in which genes show overdominance for fitness and additive effects for metric traits, essentially the converse of the nor-optimal model which shows underdominance for fitness. It is clear that this model can lead to substantial levels of genetic variation in quantitative traits and indeed does not require substantial inputs of variation from mutation, unless population sizes are small (Robertson, 1962a). Despite Robertson's implied criticism of the pleiotropic overdominance model quoted above, it is clear that he took it seriously (e.g. Nicholas and Robertson, 1980) and although consistently critical of the optimum model never put his weight behind alternatives.

The relations of metric traits to fitness and the mechanisms for maintaining variation remained a central interest of Alan Robertson, and indeed his latest research grant was designed to address the problem.

Acknowledgements

This work was supported in part by the Agricultural and Food Research Council and by a NATO collaborative research grant. I am grateful to Peter Keightley and Jim Crow for valuable comments.

Chapter 16
Mutation and the Origin of Quantitative Variation

Trudy F.C. MACKAY

Department of Genetics, North Carolina State University, Raleigh, North Carolina 27695-7614, USA

Mutation creates genetic variation. The evolutionary forces of genetic drift and stabilizing or directional selection both require genetic variation to effect evolutionary change, and deplete variation within the population undergoing that change. We need to understand the way in which mutation affects variation for quantitative traits in order to predict the amount of standing variation for metric traits which can be maintained within natural populations by the balance of mutation and selection and/or drift, to predict the divergence in mean phenotype among subpopulations, and to predict the results of long-term directional selection on a quantitative trait. Alan Robertson has had a longstanding interest in the origin and nature of quantitative genetic variation, and has been responsible for classic experimental and theoretical work on this subject (see also reviews by Frankham and Hill, this volume).

Spontaneous mutation and quantitative variation

Methods for detecting mutations affecting quantitative traits

Although it is difficult to study the effects of mutation on quantitative traits at individual loci, it is possible to assess the overall magnitude of the contribution of mutation to quantitative variation by allowing originally isogenic lines to accumulate spontaneous mutations. The effects of mutation on quantitative variation can be assessed using one of two experimental designs. If artificial selection on the trait of interest is practised on the originally inbred strain, response to selection must be due to the accumulation of spontaneous mutations affecting the selected trait. Alternatively, the inbred strain can be split into a number of sublines which are then maintained separately. Genetic divergence within and among the sublines can occur only if spontaneous mutations accumulate. The advantage of the second approach is that the effects of mutations on several quantitative characters can be assessed simultaneously. A modification of the second approach which has been used in drosophila is to put an originally isogenic, single stem chromosome against a dominantly marked balancer chromosome which suppresses recombination and protects the stem chromosome from recessive deleterious mutations. The stem chromosome is then replicated to found many sublines, and variation among sublines for the quantitative trait of interest is evaluated periodically by making the target chromosome homozygous.

Spontaneous mutation and drosophila bristles

The first experimental demonstration of the accumulation of mutations affecting quantitative traits in drosophila was reported by Mather and Wigan (1942). Divergent selection from a long-inbred strain was practised for sternopleural and for abdominal bristle score, for 21 and 53 generations, respectively. The lines showed clear responses of 0.4-1.0 phenotypic standard deviation. This pattern of very small, but consistent responses to selection from new spontaneous mutations affecting bristle score in drosophila was repeated in subsequent experiments using different inbred strains (Clayton and Robertson, 1955, 1964; Kitagawa, 1967; Hollingdale and Barker, 1971).

Clayton and Robertson (1955) introduced the concepts of mutational variance (V_M, the rate of input of new mutational variance per generation) and mutational heritability (V_M/V_E), and interpreted the response of inbred strains to selection in these terms. Using an analysis later elaborated by Hill (1982a) and Lynch and Hill (1986), Clayton and Robertson (1955) predicted an 'upper limit' to mutational heritability for drosophila bristles to be $10^{-3}V_E$; the actual point estimate from their experiment was a full order of magnitude lower. This prediction was clearly insightful, for when subsequent such experiments are also interpreted in terms of mutational variance and averaged together with the Clayton and Robertson (1955) and Mather and Wigan (1942) results, the overall mean V_M for drosophila bristle traits is $10^{-3}V_E$ (Hill, 1982b; Lynch, 1988). Estimates of V_M from analysis of bristle variation among second chromosomes extracted from a highly inbred strain (Durrant and Mather, 1954) and from accumulation of bristle mutations in replicate lines of a single highly inbred second chromosome (Paxman, 1957) are also consistent with V_M/V_E for drosophila bristles of 10^{-3} (Hill, 1982b; Lynch, 1988).

Spontaneous mutation affecting viability in drosophila

The technique of mutation accumulation among replicates of an originally isogenic stem chromosome in drosophila was applied to the estimation of the spontaneous rate of generation of genetic variation for the viability component of fitness by Mukai and his co-workers (Mukai, 1964; Mukai et al., 1972). Mutations were accumulated for up to 40 generations, and genetic variance of viability was computed periodically from variance among quasinormal (i.e. viability >60%) chromosome lines. The rate of increase of genetic variance of viability was estimated as the regression of genetic variance on time. When these estimates are scaled for the whole genome, V_M for spontaneous viability mutations averages 1.2×10^{-4}. Values of V_M for viability would be much larger if chromosomes bearing more drastic mutations were included in the analysis.

Spontaneous mutation affecting quantitative variation in other species

Lynch (1988) has interpreted the results of the above and other mutation accumulation experiments for quantitative traits of a number of species in terms of mutational heritability. A variety of vegetative and reproductive characters in plant species, pupal weight in tribolium, and life history traits of daphnia have mutational heritabilities that are of the same order as for drosophila bristle traits. Skeletal traits of mice, on the other hand, appear to have rather high mutation rates, with V_M averaging $2 \times 10^{-2}V_E$. However, it is not possible to determine from the composite parameter, V_M, whether the increment in variation was caused by many mutations, each with small effect, or a few mutations with large effects. It should also be noted that the above estimates of mutational variation may be underestimates if new mutations commonly have pleiotropic deleterious effects on fitness.

Spontaneous mutation and long-term response to artificial selection

Indirect evidence about the distribution of effects of spontaneous mutations affecting quantitative traits comes from studying long-term responses to directional selection. Selection theory predicts that continued directional artificial selection on a quantitative trait will eventually lead to a selection limit when all genetic variance initially segregating in the base population has been exhausted by selection. The input of mutations affecting the selected trait will determine the dynamics of long-term response. If there are many potentially mutable loci affecting the trait, then continued response to selection should achieve a steady asymptotic rate (e.g. Dudley, 1977; Yoo, 1980a). If mutational effects are occasionally large, a pattern will emerge of periods of stasis interrupted by sudden jumps in response as the new mutant proceeds to fixation, and there will be variation between replicates in long-term response (Hill, 1982b; Hill and Rasbash, 1986b). Repeated patterns of plateaux followed by sudden, often dramatic, jumps in response, as well as great variation between replicates in response, are typical of the majority of documented long-term responses (e.g. Clayton and Robertson, 1957; Thoday and Boam, 1961; Roberts, 1966; Falconer, 1971; Sheldon and Milton, 1972; Enfield, 1977, 1980; Yoo, 1980a; Sheldon and Evans, 1981).

Additional evidence that mutations affecting quantitative variation may sometimes be of large effect comes from the consistent, albeit infrequent, appearance in selection lines of major morphological mutations affecting the selected trait in the desired direction. In drosophila, the mutation *scabrous* (*sca*), which has a large effect on abdominal bristle score in both homozygotes and heterozygotes, has appeared several times in independent experiments selecting for increased numbers of abdominal chaetae (Hollingdale, 1971; Frankham, 1980a; Yoo, 1980b). In all cases the *scabrous* alleles were not present in the base population, and their appearance in the selection lines can be attributed therefore to mutation. The mutation observed by Yoo (1980b) arose in generation 70, much too late for it to have been segregating in the base population (Robertson, 1978). The *bobbed* (*bb*) mutation has also arisen several times in drosophila lines selected for low abdominal bristle score (Clayton and Robertson, 1957 (presumably, see Frankham, this volume); Frankham *et al.*, 1978; Frankham, 1980a), and Yoo (1980b) also detected a new allele at the *scute* bristle locus in one of his lines. The phenomenon is by no means confined to drosophila bristles: an allele at the *pygmy* locus in mice arose by mutation in a line selected for small body size (Falconer, 1981); other obese and dwarf mutants have appeared in lines of mice selected for large and small body size, respectively (reviewed by Roberts and Smith, 1982); and Bradford and Famula (1984) documented a previously unobserved gene (*hg*) causing rapid postweaning weight gain in a line of mice selected for weight gain. Genes with large effects have also appeared in strains of livestock species selected for meat production (the double muscling gene in cattle (Hanset, 1982) and the halothane gene in pigs (Webb *et al.*, 1982) are well-known examples).

Inferences about the distribution of mutant effects and correlations with fitness

The distribution of effects of mutant genes on quantitative traits may be highly leptokurtic, so that most new mutants affecting quantitative phenotypes are of small effect, but a small proportion have large effects. Although genes with large effects may only constitute a small fraction of the total mutants affecting a quantitative trait, they may nevertheless have a dramatic effect on its variance. Thoday *et al.* (1964) mapped the major determinants contributing to the response of the selection lines described by Thoday and Boam (1961), and concluded that two genes affecting sternopleural bristle number on Chromosome III could account for the response of four bristles. Shrimpton and Robertson (1988a,b) found a minimum of 17 'factors' to be responsible for an increase of 34 sternopleural bristles on a single third chromosome from a line of drosophila selected for high sternopleural bristle score, of which three or four factors contributed one-third of the difference in score of the selected and tester chromosomes.

Although mutant alleles of large effect at quantitative trait loci occur in selection lines, they may not contribute to quantitative variation in natural populations because of deleterious effects on reproductive fitness. Evidence of selection acting to remove mutations of large effect on metric traits from natural populations comes from genetic analysis of long-term selection lines. This approach was pioneered by Robertson (Clayton and Robertson, 1957), who extracted and made homozygous individual chromosomes from drosophila lines selected for high and low numbers of abdominal bristles. Most of their selection lines contained second and/or third chromosomes which were homozygous lethal and had heterozygous effects on the selected trait in the direction of selection. In one line a lethal with a heterozygous effect of 22 abdominal bristles caused a limit to selection for increased score at which genetic variance was maintained. Subsequently lethal chromosomes with heterozygous effects on the selected trait were often found in long-term selected lines of drosophila (Latter and Robertson, 1962; Frankham et al., 1968; Hollingdale, 1971; Madalena and Robertson, 1975; Frankham, 1980a; Yoo, 1980b).

One interpretation of these findings is that mutations affecting bristle score but which were homozygous lethal occurred in the selection lines. Certainly in some cases the lethals in question were not identified in the base population (Frankham, 1980a; Yoo, 1980b), but in other cases evidence concerning their origin is poor. If they were present in the foundation stock, the associated bristle effect must have been reduced, for segregating effects of the magnitude documented by Clayton and Robertson (1957) and Yoo (1980b) would have had perceptible effects on the phenotypic variance of the trait in the early generations of response. Because entire chromosomes were assessed jointly for homozygous effects on viability and heterozygous effects on bristles, it is not possible to say whether a single locus with pleiotropic effects on bristles and fitness or two closely linked loci separately affecting bristle score and fitness were involved. Indirect evidence favours the former interpretation, for there are no reported instances of recombination between the bristle effect and the lethal. Pleiotropic effects on the selected trait and fitness are clearer for the major morphological mutations which have occurred in selection lines. bobbed drosophila females are less viable and have longer development times than wild-type (Ritossa, 1976); the dwarf, obese, pygmy and high growth genes in mice are either sterile or have reduced fertility when homozygous (Falconer, 1981; Bradford and Famula, 1984); and the halothane gene in pigs and double muscling gene in cattle reduce viability and fertility (Hanset, 1982; Webb et al., 1982).

Induced mutation and quantitative variation

X-ray-induced polygenic variation

Genetic variation is necessary for genetic gain under artificial selection and is at the same time exhausted by artificial selection. Given the (what was perceived to be) sluggish rate of accumulation of spontaneous polygenic mutations, it was natural to query whether accelerated mutation rates, such as those caused by X-irradiation and chemicals, could produce genetic variation for metric traits utilizable by selection. If so, this would be of obvious practical importance to animal breeders. The results of selection for bristle traits from X-irradiated, initially isogenic populations of drosophila are mixed. Scossiroli and Scossiroli (1959) achieved remarkable responses in sternopleural bristle sore when they irradiated both sexes of a nearly isogenic ($h^2 = 0.06$) Oregon population of drosophila every generation or every alternate generation. Heritabilities calculated from response of the irradiated populations to selection were 0.35 (every generation) and 0.11 (alternate generations). These large responses have not been repeated by other workers (Clayton and Robertson, 1955, 1964; Kitagawa, 1967; Hollingdale and Barker, 1971). The effect of X-irradiation was in all cases judged by response to selection of abdominal bristle score. When interpreted by the theory of Hill (1982a, 1982b) and scaled to 1000r/generation, these responses give an average $V_M = 2.9 \times 10^{-3} V_E$, or three times higher than the spontaneous rate (Mackay, 1987). A small but demonstrable increase in response was reported by Roberts (1967) after irradiating a line of mice at a selection limit for body weight, but irradiating chicken sperm did not improve the response to selection for increased egg number (Abplanalp et al., 1964). Therefore X-irradiation does generate genetic variation for quantitative traits above the spontaneous mutation rate, but the magnitude of the increase is so small that it is not a practical way to further selection response in animals.

Induced mutations affecting viability in drosophila

One reason irradiation has not been successful in creating significant genetic variation for quantitative traits is the associated wide-scale genetic damage and induction of recessive lethals it causes. Clayton and Robertson (1964) extracted third chromosomes from drosophila populations that had received a cumulative dose of 270 kr and found that 82% were homozygous lethal. From the rate of production of X-ray-induced variation for bristle traits they calculated a cumulative dose of 500 kr would be necessary to reconstitute from an initially isogenic population the level of genetic variation for bristle traits typical of natural populations. Clearly the population would expire from the deleterious effects of irradiation on fitness long before variation for traits like bristle score reached appreciable levels. The detrimental effects of X-irradiation and chemical mutagenesis on homozygous and heterozygous viability in drosophila have been well documented (see reviews by Mukai, 1979 and Simmons and Crow, 1977). The effect of induced mutagenesis on total fitness is even more pronounced; chromosomes harbouring EMS-induced mutations have heterozygous effects on fitness as great as their homozygous effects on viability (Mitchell, 1977; Mitchell and Simmons, 1977; Simmons et al., 1978).

The nature of genetic variation for quantitative traits

Quantitative variation and variation at 'regulatory' loci

There are currently only speculative answers to questions about the nature of the genes causing quantitative variation. Since any aspect of the phenotype that is measurable can be defined as a quantitative trait, the total number of traits expressed by an organism is very large. It would be most unusual to define a trait for which there is no genetic variation, as judged by resemblance among relatives, response to selection and differentiation of inbred lines. Clearly most loci must simultaneously affect more than one trait, and their effects on quantitative variation are likely to be secondary to their major function.

Consider the *hairy* locus in drosophila. This locus was originally identified by mutant alleles with major morphological effect on numbers of microchaetae. Some mutant alleles of *hairy* disrupt embryonic segmentation, some are lethal, and the *hairy* locus has been implicated in the regulation of the unlinked *achaete–scute* complex (Botas et al., 1982). The *hairy* locus is a complex gene consisting of two regulatory regions and one structural region (Ingham et al., 1985). Mutations in the different regions produce the variety of phenotypes. If this pattern of gene organization is common (and it appears the *cut* locus has similar properties (Johnson and Judd, 1979)) then quantitative variation may result from mutations in the regulatory regions, yielding a spectrum of direct effects and pleiotropic effects on other traits and on fitness.

There are several other lines of evidence implicating regulatory mutations as a source of quantitative genetic variation. Mukai and Cockerham (1977) have suggested that regulatory mutations are responsible for variation in drosophila viability, based on a comparison of the spontaneous mutation rates of viability polygenes and isoenzyme structural loci on the second chromosome of *Drosophila melanogaster*. Variation in the level of activity at structural enzyme loci has been shown to be partly determined by unlinked modifier loci (McDonald and Ayala, 1978; Laurie-Ahlberg et al., 1982). Finally, Wilson (1976) has noted that the rate of morphological divergence among mammals is much too great to be accounted for by the rates of divergence of structural loci, and has argued therefore that regulatory mutations which alter the expression of genes are instrumental in morphological evolution.

Highly repeated DNA and quantitative variation

Studies of DNA renaturation kinetics have revealed that eukaryotic DNA can be separated into sequences that are unique, moderately repeated and highly repeated. There is evidence that repeated DNA can contribute to variation for quantitative characters. Where DNA is organized into tandemly repeated multiple copies, or tandons, variation in copy number can be generated by the process of unequal crossing over. Many 'loci' are organized in this fashion, including rRNA, 5S RNA, wool proteins, actin, keratins, collagens, chorionic proteins, histones, antibody genes, haemagglutinins and serum esterases (Frankham, 1980a). Selection for decreased abdominal bristle number in drosophila has on several occasions resulted in the occurrence of *bobbed* mutations at the rRNA tandon (Clayton and Robertson, 1957 (presumably); Frankham *et al.*, 1978; Frankham, 1980a, this volume). Similarly, resistance to organophosphate pesticides by *Culex* mosquitoes has been attributed to the increase in copy number by at least 250-fold of an esterase locus (Mouches *et al.*, 1986), a mechanism that might prove to be general in the evolution of resistance to organophosphate pesticides in a number of insect species.

Transposable elements and quantitative variation

Much of the moderately repeated component of the eukaryotic genome is composed of transposable element (TE) sequences. Such sequences have been found at all levels of biological complexity, from yeast to mammals, and they can be grouped into broad categories which share common structural features across their different taxonomic hosts (reviewed by Shapiro and Cordell, 1982; Finnegan, 1985). TE sequences can comprise a significant fraction of the genome; for example, in *D.melanogaster* the *copia*-like sequences account for roughly 10% of the total DNA (Rubin, 1983). Transposable element sequences are mutagenic – in fact, most TEs have been identified from the mutations, often unstable, that they have caused. It is thought that most 'spontaneous' mutations in drosophila have been caused by TE insertions (Rubin, 1983), and many insertional mutations have been described in maize and yeast. The *dilute* coat colour mutation in the mouse has also been shown to be caused by an insertion of an ecotropic retrovirus (Jenkins *et al.*, 1981).

Insertions of TEs at or near a functioning gene may have manifold effects on that gene's expression, which have been documented by detailed genetic and molecular analysis of TE-induced mutations and their reversions in maize, drosophila and yeast. TE insertions may increase or decrease the amount of transcript produced (Tsubota and Schedl, 1986), or alter the developmental timing or tissue specificity of expression (Chia *et al.*, 1986). The severity of the mutational effect depends both on the position of the inserted TE within the gene and on the structure of the inserted sequence. Imprecise excisions of elements which leave pieces of the TE behind often yield wild-type pseudorevertant phenotypes (Rubin, 1983; Tsubota and Schedl, 1986); and insertion of a second TE at or beside the first may ameliorate the severity of the mutant phenotype (e.g. Rubin, 1983; Engels, 1988; Mount *et al.*, 1988). Insertional mutations can be modified by unlinked genes which suppress (restore to wild type) or enhance (create a more extreme phenotype) them (Modolell *et al.*, 1983; Rubin, 1983; Fassler and Winston, 1988; Mount *et al.*, 1988). Other phenotypic effects of TEs are drastic phenotypes resulting from imprecise exisions extending into a coding region, and position effect variation arising from TE-induced inversions and translocations.

Clearly the mutational repertoire of TEs is so vast, and they are so common, that it would be surprising if they did not cause mutations affecting quantitative characters. Not all TEs transpose sufficiently frequently to make experimental investigation of their effects on quantitative variation feasible, although transposition is frequent enough to cause differentiation between strains and between individuals within strains in the sites they occupy (Montgomery and Langley, 1983; Leigh Brown and Moss, 1987; Montgomery *et al.*, 1987). However, some families of TEs do transpose at high rates when individuals of strains containing them are crossed to strains without them, e.g. the *P* (Kidwell *et al.*, 1977; Bingham *et al.*, 1982), *I* (Bucheton and Picard, 1978; Bucheton *et al.*, 1984), and *hobo* (Blackman *et al.*, 1987; Yannopoulos *et al.*, 1987) elements of drosophila, and some murine ecotropic retrovirusus (Jenkins and Copeland, 1985). Using crosses of drosophila strains containing

P elements to those without *P* elements, Mackay (1984, 1985b, 1987, 1988) showed that there was a correlation between increased variation for bristle traits and *P* element activity, and estimated the mutational heritabilities resulting from the crosses to be of the same order as heritabilities of these traits from natural populations – two orders of magnitude greater than previous estimates of mutational heritability for drosophila bristles. *P* elements even transpose relatively frequently within strains containing them (Preston and Engels, 1984). Mackay (1988) estimated spontaneous mutational heritabilities for two bristle traits from response to selection from an inbred *P* strain, and found mutational heritabilities about twice as great as those estimated from X-ray-induced variation. TE insertions also cause deleterious effects on viability (Yukuhiro *et al.*, 1985; Mackay, 1986; Cooley *et al.*, 1988) and total fitness (Fitzpatrick and Sved, 1986; Mackay, 1986).

P element mutagenesis in drosophila, when refined by controlling the effects of segregation and instability introduced by crossing different strains, should prove valuable in generating mutations affecting quantitative variation which can then be studied with the full range of genetic and molecular techniques that have been applied to insertional mutations at other drosophila loci. The answers to questions about the distribution of mutant effects, dominance, epistatic and pleiotropic properties of polygenic loci may well be accessible using these techniques. TEs may prove at least partially responsible for the quantitative variation we observe in natural populations, and also provide the means by which such variation can finally be studied at the *genetic* level.

Acknowledgements

This work was supported in part by NIH Grant GM11546 and by a NATO collaborative research grant. Paper number 11822 of the Journal Series of the North Carolina Agricultural Research Service, Raleigh, NC 27695-7601.

Chapter 17
Selection Experiments and the Nature of Quantitative Variation

D.S. FALCONER

Department of Genetics, University of Edinburgh, West Mains Road, Edinburgh EH9 3JN, Scotland

What can one learn about the genetics of quantitative characters by doing selection experiments? Many experiments, particularly the earlier ones, were done with the purpose of finding answers to purely empirical questions such as: Can the population mean of a particular character be changed by selection? The answer is, almost always it can. Can continued selection take the mean beyond the original range? The answer is that it can, often in only a few generations. Are there limits to the change that can be made by long-continued selection? The answer here is less clear; changes amounting to many times the standard deviation can be made within twenty or so generations, but mice cannot be made the size of rats nor drosophila the size of house flies within the time that experiments have been carried on. Is selection equally effective in increasing the character as in reducing it? The answer is, sometimes but often not. Some insight into the genetics of the character selected can be obtained from the answers to questions such as these.

Another purpose of some experiments has been to test current theory, particularly whether the response obtained agrees with the prediction from the heritability in the base population estimated from correlations between relatives. The observed responses have on the whole agreed with the predictions well enough to uphold the theory. We cannot, however, infer much about the nature of the genetic variation from the agreement between observed and predicted responses. The reason is that the prediction requires only that the genetic variation is caused by more than a very few genes; the properties of the genes, their dominance or epistasis or their mode of action, play no part in the prediction. Most experiments, however, have disagreed with the prediction in one way or another, particularly in the later generations, and some inferences about the genetics of the character can usually be made from these discrepancies. The discrepancy most often encountered is an asymmetry of response, the realized heritability being different for selection in opposite directions. Unfortunately, asymmetry has so many possible causes that it cannot be diagnostic of any particular genetical feature and the best the experimenter can usually do is to point to what seems the most probable cause.

The following necessarily brief summary of the conclusions that have been drawn from selection experiments is far from complete and omits much of interest.

Number of 'genes'

The first selection experiments established the fact that the mean of the character selected can be changed to far beyond the limits of the original variation. This fact, incidentally, still surprises many undergraduate students. Two examples from the earlier experiments, which are summarized by Wright (1977, Chaps 7 and 8), will suffice. In the Illinois selection for oil content of maize seeds, started in 1896, the means of the high and low lines were both outside the original range after 5 generations, and after 61 generations they differed by 34 times the original phenotypic standard deviation (see Wright, 1977, p.186). In the experiment of MacArthur (1949), selection of mice for body weight took the means beyond the original limits in, again, 5 generations, and after 21 generations the high and low lines differed by 11 phenotypic standard deviations.

These early experiments confirmed the 'multifactorial' theory of the inheritance of continuously varying characters. They showed that the characters were controlled by many genes, or at least by more than a few. But how many? By comparing the total response with the original genetic variation it is possible to estimate the number of 'genes', more properly called effective factors, or the segregation index. The selection for oil content of maize seeds for example gave a segregation index of about 20 (Wright, 1977, p.190; Lande, 1981). Gene numbers so obtained are notoriously unrealistic. The main reason is that this is the number that there would be if all the genes had the same magnitude of effect on the character and were unlinked, a situation that it is impossible to believe could be true. There must in reality be some genes with large effects and increasingly more genes with ever smaller effects. What we need to know is the form of the distribution of gene effects. Hill and Rasbash (1986a) showed that the shape of the distribution does not have a great effect on the response to selection, so it cannot be determined directly by selection experiments. Instead, it is necessary to identify genes, or chromosome segments, by their linkage with markers, a method that is practicable only with drosophila. The genes identified are those by which a selected line differs from a tester stock containing the markers. There are several difficulties inherent in the method (Robertson, 1967a; McMillan and Robertson, 1974), the chief of which is that a segment of chromosome identified as a 'gene' of large effect may in reality be several genes each with smaller effects which could be separated by crossing-over if the experiment were continued for longer. Leaving the difficulties aside, however, the identification of genes in this way has shown conclusively that for bristle characters in drosophila a large part of the variation is due to only a few genes. For example, Spickett and Thoday (1966) found that 5 effective factors accounted for 87.5% of the difference between a selected line and the tester. Robertson (1967a), reporting preliminary results, concluded that three-quarters of the difference between a selected line and the tester could be accounted for by 10 effective factors. Shrimpton and Robertson (1988b) identified 18 effective factors on the third chromosome affecting sternopleural bristles. The factor with the largest effect had a homozygote difference of 1.7 phenotypic standard deviations, and the distribution of effects was clearly skewed, with the smaller effects being the more numerous.

Mutation

It used to be thought that mutation was too slow a process to have appreciable effects on selection responses. The expectation, ignoring mutation, was that the rate of response would gradually diminish as gene frequencies moved toward fixation, and when all the genes affecting the character selected were fixed the response would cease and the population would reach a selection limit. Robertson (1960c) developed a theory of limits from which the maximum limit attainable could be predicted under certain conditions. Observations of two sorts, however, showed that selection responses did not conform to the simple expectation of limits. First, lines that appeared to have reached limits were not in fact fixed because they responded rapidly to reversed selection. And, second, some selection lines did not reach limits but went on responding for as long as selection was continued; the Illinois selection of maize mentioned earlier was one of these (Dudley, 1977).

Then Hill (1982b) argued that when selection is being applied, mutation may not be negligible. Responses must be expected to continue indefinitely as a result of new variation produced by mutation. To decide whether the response attributable to mutation would be fast enough to be detectable we need to know the rate at which mutation produces new variation. This can be estimated by applying selection to a line previously made isogenic by inbreeding. Clayton and Robertson (1955) found by this means that the genetic variance of abdominal bristle number in drosophila produced by mutation in one generation was about one thousandth part of the genetic variance in a random breeding population. This rate of origin of new variation is enough to give detectable responses to selection when none of the original variation remains. Hill (1982b) calculated the response expected from mutation for abdominal bristle number and compared it with the rate observed by Yoo (1980a). The predicted rate was 0.4 bristles per generation and the observed rate over generations 50 to 80 was 0.3 bristles per generation, thus showing very good agreement.

Knowledge of the rate of origin of new variation is of great importance for understanding responses to selection and also for models of the maintenance of variation in natural populations subject to stabilizing selection (see for example Lande, 1988; Turelli, 1988).

The mutational variance can be increased by mutagenic agents such as X-rays (Clayton and Robertson, 1964), and is enormously increased by transposable elements in drosophila (Mackay, 1987).

When variance generated by mutation is brought into consideration responses to selection are expected to continue indefinitely without limit. Yet selection responses often do seem to reach limits at which no further response is obtained. For example, five of Yoo's (1980a) six lines appeared to have reached limits. Eisen (1980), reviewing long-term selection in mice, lists 16 experiments in all of which there was an apparent limit when response had ceased. The reasons for these unexpected limits are not always known but there are several possible reasons of which the following seem the most likely (Hill, 1982b). The limit is not real and continued selection would have shown the response to be continuing; the mutational variance is less than it is thought to be; all the useful mutations that are possible have already occurred and been fixed; natural selection opposes the artificial selection.

Gene frequencies

Selection experiments cannot tell us much about the gene frequencies in the initial population. Hill and Rasbash (1986a) examined theoretically the consequences for the selection limit, in the absence of mutation, of three different distributions of gene frequencies – all genes at frequencies of 0.5, a uniform distribution with all frequencies equally probable, and a U-shaped distribution with most genes at frequencies near 0 or 1 which would be expected from neutral mutation. They found that the limit expected was not much affected by the gene frequency distribution. It seems unlikely therefore that studies of selection limits would be able to distinguish between different hypothetical distributions of gene frequencies.

The distributions considered by Hill and Rasbash (1986a) were all symmetrical, with no correlation between gene frequency and gene effect. If there were such a correlation it would lead to an asymmetrical response to selection in opposite directions (Falconer, 1954). If alleles that reduce the character were at low frequencies and alleles that increase it at high frequencies, then upward and downward selection would produce markedly different responses. The upward response would gradually diminish, but the downward response would increase to a maximum rate when the gene frequencies reached intermediate values and would subsequently diminish. Then when (ignoring mutation) limits were reached the total response downwards would be much greater than the total response upwards. A very clear example of this pattern of asymmetrical response is the selection for wing-length in drosophila by F.W. Robertson and Reeve (1952, Figure 3). The upward and downward selected lines both reached limits showing no response from about generation 40 to 50. The total advance upwards was about 15 units and downwards about 48 units. If all the genes were additive and at the same initial frequency this would imply a frequency of 0.76 for the genes that increased wing-length. Unfortunately, however, the same pattern of asymmetrical response would result from directional dominance, and the asymmetrical limits could also result from natural selection opposing the upward selection but not hindering the downward selection. There was some evidence of natural selection acting in this way, because when the selection was relaxed the high line went back half-way to the control level but the low line did not change. So the asymmetrical response is not diagnostic of asymmetrical gene frequencies in the base population.

Dominance

Dominance, like the gene frequencies, does not have characteristic effects on responses to selection. The short-term responses are predicted by the heritability in the base population whether there is dominance or not. Directional dominance, i.e. the dominant alleles acting in the same direction, however, could produce asymmetrical responses in long-term selection. But the asymmetry is indistinguishable from that resulting from asymmetrical gene frequencies and is therefore not diagnostic. Directional dominance is unambiguously identified by inbreeding depression.

The properties of lines at selection limits may be affected by directional dominance. Recessive alleles are more easily fixed by selection than dominant alleles because selection becomes less effective as the frequency of the recessive allele becomes rarer. If all, or most, of the recessive alleles reduce the character then downward selection will lead to fixation of the recessives at the limit; but the responses to upward selection may become so slow as to be undetectable while many recessive alleles are still present at low frequencies. A selection line at the lower limit will therefore not respond to reversed selection or to inbreeding, but a line at the upper limit will respond to reversed selection and will be depressed by inbreeding. The contrast between high and low lines when subjected to reversed selection was very clear in drosophila selected for thorax-length (F.W. Robertson, 1955). Two-way selection was applied to three different populations derived from wild flies. After all the lines had reached limits, reversed selection was applied and it was effective in all three high lines but in none of the low lines. Furthermore, relaxation of selection in the high lines had little effect, proving that natural selection was not the main cause of the limits. Two examples of lines, apparently at limits, being depressed by inbreeding are the mice selected for high litter size by Falconer (1971) and Eklund and Bradford (1977). In both cases the genetic variation at the limits that was revealed by the inbreeding depression was attributed to recessive alleles at low frequencies.

Selection experiments can give some evidence about overdominance. If some genes were overdominant with respect to the character selected, selection in one direction would favour heterozygotes and lead to equilibrium gene frequencies. The response would then cease, but the line would respond to reversed selection. This possibility was not excluded in the drosophila experiment of F.W. Robertson (1955). But in the two mouse experiments it was excluded on the grounds that inbreeding the selected populations produced one or more highly inbred lines with no decline of litter size.

Another form of overdominance, however, is frequently found, particularly in drosophila experiments. This involves fitness. The effect on the character may be additive, with the heterozygote intermediate, but one homozygote has an adverse effect on fitness and is often lethal. The combination of artificial and natural selection therefore favours the heterozygote. The effect on the character is usually large. A lethal found by Clayton and Robertson (1957) had an effect of 5.8 phenotypic standard deviations of the base population; fifteen lethals tested by Yoo (1980b) ranged in effect from 0.3 to 3.6 standard deviations of the nonlethal homozygotes at the time they were tested. Consequently the variance is much increased so that the effectiveness of selection on the other genes is diminished and the response is much reduced. Examples are the experiments, all with drosophila, of F.W. Robertson and Reeve (1952), Clayton and Robertson (1957), Frankham et al. (1968) and Yoo (1980b). These lethal genes that are found so often in drosophila experiments may have been present in the base populations, but if they were they must have been at very low frequencies. They are more likely to have arisen by mutation during the course of the selection, as was proved in the experiment of Yoo (1980a,b).

Epistasis

Analyses of selected lines have shown epistatic interactions to be widespread. This was highlighted by the assays of chromosomes from drosophila lines selected for body size, summarized by F.W. Robertson (1955). The prevalence of epistasis in selected lines does not, however, necessarily mean that much of the genetic variance in an unselected population is due to epistasis because the variance generated by the interactions found has not been estimated. Other evidence of interaction in drosophila comes from the effect on the character selected of a lethal gene being greater in the selected line than in the base population (F.W. Robertson, 1955; Clayton and Robertson, 1957). On the other hand, Rathie and Nicholas (1980) found no evidence of nonadditive effects on abdominal bristles in an experiment which compared responses in subdivided and undivided populations.

A clear case of epistasis in mice is the recessive dwarfing gene studied by Fowler and Edwards (1961). This gene, called midget, was found in a line selected for small size. Its effect on body size was large enough to make it easily classified and it gave good Mendelian ratios when in the background of the line in which it was found. But when crossed to two different unrelated small lines, its effect was much reduced and about 60% were misclassified as normal. There must have been a gene present in the original line, but not in the other two lines, which enhanced the effect of the midget gene.

Scale

It is common practice to transform the scale of measurement so as to make the distribution of phenotypic values normal. But what happens if the genetic and environmental variances require different transformations? This is an aspect of quantitative variation that has received little attention. In terms of gene action, a difference in the distributions of genetic and environmental variation would represent genotype×environment interaction. If the distributions of breeding values and environmental deviations were skewed in opposite directions, or if one were normal and the other skewed, this would result in a nonlinear offspring–parent regression and an asymmetrical response. The asymmetry of response would be seen immediately in the first generation, whereas with asymmetrical gene frequencies or directional dominance the asymmetry does not appear till later when the gene frequencies have been changed.

The results obtained by Meyer and Enfield (1975) selecting for pupa weight in tribolium are consistent with the interpretation of the distributions being skewed in opposite directions. Single-generation selection was applied upwards and downwards with three intensities of selection in nineteen replicates. The base population was a cross of two highly inbred lines, so no asymmetry could be attributed to asymmetrical gene frequencies. Selection downwards gave a higher realized heritability than selection upwards at all intensities. The ratio of downward to upward realized heritability was 2.6 at the highest intensity (10% selected) and 1.5 at the lowest intensity (50% selected). Offspring–parent regressions in the selected individuals and their progeny were consistent with the realized heritabilities. Thus the asymmetry of response was associated with a nonlinear offspring–parent regression in the base population. These results point to the distribution of breeding values being negatively skewed (the tail toward low values) and the environmental variance being positively skewed. This difference in the distributions would result in the additive genetic variance being greater when pupa weights were low and the environmental variance being greater when pupa weights were high. Figure 12.2 in Falconer (1981) shows distributions which would result in heritabilities roughly the same as those found in the tribolium experiment.

Fitness

It is an almost universal experience that fitness is in some way impaired in lines that have been subjected to artificial selection, whatever the character selected. This is what would be expected if the character selected had been previously subject to stabilizing selection, natural selection favouring individuals with intermediate values. For characters subject to inbreeding depression it would also be expected from random fixation unless the selected population was very large; but the loss of fitness from this cause can be determined by other means (Latter and Robertson, 1962).

The question of interest is whether the change of mean in the selected line itself causes the loss of fitness. Do the genes causing the variation of the character selected also affect fitness? In other words, is the variation caused mainly by neutral genes? The evidence comes mainly from the effects of relaxing the artificial selection, letting natural selection operate alone. If stabilizing selection is operating in the laboratory population then a selected line will regress to the unselected level when the artificial selection is relaxed, provided there is still additive variance in it, which can be shown by a response to reversed selection. On the whole, relaxed selection has been found to change the mean only a small way toward the control level. For example, the three drosophila populations selected upwards for thorax-length (F.W. Robertson, 1955) returned at most half-way to the control level, though reversed selection brought the means all the way in a few generations. Drosophila lines selected for high abdominal bristle number returned less than one quarter of the way under relaxed selection (Robertson, 1967a). Experiments such as these lead to the conclusion that most of the genes responsible for the variation of the character in the base population must be nearly neutral with respect to fitness (Robertson, 1967a). This conclusion is supported by chromosome assays by Mackay (1985a). The effects of different third chromosomes on bristle numbers (abdominal and sternopleural) and on fitness were measured. Bristle numbers and fitness were not associated.

A greater effect of natural selection was found by Enfield (1980) in a comprehensive study of tribolium selected for high pupa weight. Selection was relaxed after 51 generations and the mean declined steadily for 68 generations with no sign of reaching a stable level. In the selected lines there was a high incidence of sterility (about 50%), and the sterility decreased in the relaxed lines. This suggested that sterility was caused by high pupa weight, but another population with a high pupa weight did not decline in mean when maintained with no artificial selection. The conclusion reached was that the sterility associated with high pupa weight was due to the pleiotropic effects of some genes and not to the pupa weight itself.

Recessive lethals with pleiotropic effects on drosophila bristles have already been mentioned. Genes with adverse pleiotropic effects on fitness must, of course, be at low frequencies in unselected populations and, unless there are a large number of them, cannot contribute much to the variance of a character that is not an important component of fitness.

Conclusion

The theory by which short-term responses can be predicted is very robust because it does not depend on the properties of the genes concerned, nor on their numbers if more than a very few. The strength of the theory, however, means that comparisons between observed and predicted short-term responses do not allow us to make deductions about the nature of the genetic variation. It is only when other features, mainly connected with long-term responses, are considered that deductions can be made.

The character most studied by selection is the number of abdominal bristles in *Drosophila melanogaster*. Most of the genetic variation is caused by not very many genes, perhaps 5 to 20. The genes are mainly additive, with little dominance or epistasis, and are nearly neutral with respect to fitness. There are, however, some genes which affect bristle number and are severely deleterious, often lethal, in homozygotes. Their effect on bristles can be large – up to 6 phenotypic standard deviations. Some of these genes may be present at low frequencies in unselected populations, but most probably arise by mutation during the course of selection. The features of any particular selection line depend to a large extent on the particular genes of large effect that it contains.

Few generalizations can be made about other characters and other organisms. Many experiments have features peculiar to themselves, but all of these can be reasonably well accounted for by dominance, epistasis, and pleiotropic effects on fitness. It seems likely that, as with drosophila bristles, the peculiar features are often due to particular genes of large effect which have arisen by mutation during the selection.

Chapter 18
Tests of Theory by Selection Experiments

Carlos LÓPEZ-FANJUL

Departamento de Genética, Facultad de Ciencias Biológicas, Universidad Complutense, 28040 Madrid, Spain

The classical theory of quantitative genetics generally assumes that variation in traits is due to independent segregation at many neutral genes, all with small additive effects between loci (the infinitesimal model), and that populations are of infinite size. Predictions can then be made of the rates of response to selection pressures, but they are strictly only valid for a single generation when applied to finite populations. Alan Robertson's main contribution to this field has been the extension of the theory to populations of finite size, under the assumption of additive gene action but including the effects of linkage.

This chapter analyses selection experiments specifically carried out to examine the validity of the current theory on the effect of population size on short-, medium- and long-term selection response. It concentrates on predictions with potential practical application, i.e. extrapolation at the statistical level from laboratory work to breeding schemes.

Short-term: the effect of selection on effective population size

With artificial selection, genetically superior families will contribute more offspring to the selected group in the next generation, even though all parents contribute equally to the scored group, thereby reducing the effective size of a selected line. Robertson (1961c) found that the ratio of the variance effective size (N_e^V) to the actual size (N) progressively declines with increasing intensity of selection (i) and heritability of the selected trait (h^2). The ratio is also dependent on the shape of the distribution of the contribution of offspring to the scored group, whether multinomial or constant, the reduction being greater for the former. For inbreeding effective size (N_e^F) an exact expression for N_e^F/N has been given by Burrows (1984), which, however, is only valid for the first generation of selection and does not differ appreciably from that obtained by Robertson, unless selection is very intense. These predictions refer to neutral loci only, i.e. those not affecting either fitness or the selected trait.

The simplest experimental test of this theory should aim at comparing the expected and observed values of N_e^V/N_R in a set of lines started independently from the same base population, all with the same number of individuals scored and selected per generation, where N_R is the size expected with random sampling. This can only strictly be done for the first generation of selection because there is no available theory for the effect on N_e^V of the increasing complexity of the genealogy of selected individuals after several generations. Alternatively, N_e^V has been estimated by N_e^F, which may however differ in populations undergoing selection, but which takes into account the increase in the average relationship among individuals as selection progresses.

Several experimental checks of the theory have been carried out in lines of *Drosophila melanogaster* selected for high abdominal or sternopleural bristle number (i values ranging from 0.8 to 1.7, h^2 values ranging from 0.05 to 0.59) (Jones, 1969; Yoo, 1980c; Gallego and García-Dorado,

1986). Good agreement was generally found between observed and expected values of N_e^V/N_R after one generation of selection, and the fit between the expected N_e^V/N_R and the observed N_e^F/N_R was still acceptable after 7–9 generations. In one case (Gallego and García-Dorado, 1986), a high proportion of the genetic variation of the selected trait appears to be due to the segregation of a small number of loci of large effect and at appreciable frequencies in the base population, and the lines essentially reached fixation at those loci after two generations of selection. The genetic situation in this case is clearly different from that assumed in the infinitesimal model so a large discrepancy was noted between expected and observed effective sizes.

Medium-term: optimizing selection response

Immediate response to individual selection is maximized when the intensity of selection is highest, but only at the expense of reducing the effective size of the selected line. Consequently, the available genetic variance will decline, jeopardizing future gains. The maximum advance at the limit is, however, expected from selecting the best half of the population every generation (Robertson, 1960c), so short-term response is sacrificed to ultimate gain. There will therefore be an optimum proportion selected (p_o) that maximizes the genetic advance achieved in a fixed number of generations (t). When selection is carried out from a constant number (T) of individuals scored per generation and the sex-ratio among scored and selected individuals is equal to one, Robertson (1970c) showed that p_o is a function of t/T only: p_o is low when t/T is small and approaches 0.5 as t/T increases. For unequal numbers of males and females in the scored and selected groups, maximum response after t generations is obtained when the number of males and females scored and selected is the same; but when the sex ratio among scored and/or selected individuals is not one there are different optimum selection intensities in the two sexes (Jódar and López-Fanjul, 1977). These results apply to any unimodal and symmetric phenotypic distribution (Cockerham and Burrows, 1980).

Two experimental checks of the theory for a sex-ratio of one have been reported. Ruano et al. (1975) carried out individual selection to increase egg laying of virgin females of Tribolium castaneum for 32 generations and Frankham (1977) applied mass selection for increased abdominal bristle number in D.melanogaster for 50 generations. Good agreement with the predicted ranking was obtained in both experiments for values of t/T up to 0.18, corresponding to predicted optimum proportions up to 20%, but none of the comparisons made above that value of t/T showed the expected ranking of treatments.

Theoretical predictions for unequal numbers of each sex were tested by García-Dorado and López-Fanjul (1985) by carrying out mass selection for high sternopleural bristle number in D.melanogaster for 20 generations. As predicted by theory, higher responses were obtained for a sex ratio of one. Good qualitative agreement was also found between observed and expected rankings of the different types of selected lines over time. Lines selected with a sex ratio of 10 for maximum response after 10 generations proved the only exception, and only one replicate survived to generation 20 (probably because these lines had the lowest effective size, and suffered most from fitness deterioration). This replicate also showed a large accelerated response to selection from generation 8, related to the presence of a recessive lethal associated with the terminal inversion $3R^c$.

There are several factors not considered in the model which may affect the results:
(i)   linkage (Robertson, 1970c), likely in drosophila,
(ii) alleles of large **effect on the** selected trait, found in both drosophila experiments (Frankham et al. 1968; García-Dorado and López-Fanjul, 1987); and
(iii) effective sizes lower than assumed, by accident in mass cultures and/or from the practice of artificial selection (Robertson, 1961c).

All of these will increase the predicted values of both the optimum proportions to be selected and the between-replicate variance, introducing more noise in the comparisons as selection continues. Progressive fitness deterioration will also result in increasing the sampling variance about the treatment means. All this concords with results of Frankham (1977) where discrepancies are concentrated in later generations ($t > 25$).

When the intensity of selection and the number of individuals scored per generation have already been decided, further optimization can still be obtained by choosing the maximum possible effective size compatible with them. This can be achieved by intentionally making unequal the contributions of selected individuals to the next generation (Toro and Nieto, 1984). The effectiveness of this proposal has been confirmed by computer simulation (Nieto et al., 1986). Likewise, an experimental check was carried out by selecting *D.melanogaster* for 13 generations for low sternopleural bristle number (Nieto et al., 1986). Similar responses were obtained in two sets of lines which had equal ($N_e = 7.8$) or unequal ($N_e = 9$) parental contribution of progeny. It was hypothesized that the second set was not superior due to segregation of genes of large effect, although fitness deterioration was considerably smaller in these lines.

Long-term: selection limits

The theory of limits to artificial selection (Robertson, 1960c, 1970e) provides the basic conceptual framework for the interpretation of long-term selection results (see review by Robertson and Hill, 1983b). The theory has been developed in terms of the chance of fixation of alleles initially present in the population, the limit being eventually reached when the original variation has been exhausted by incessant selection. It has been extended to include the influence of nonobservable variables, such as the number of loci affecting the trait, as well as the distribution of allelic effects and frequencies (Hill and Rasbash, 1986a). In practice, a population is said to be at a selection limit when continued selection pressure no longer results in further gain. The genetic nature of lines at the limit has been found to differ even among selected replicates, and only occasionally has fixation been reported. Nevertheless methods for overcoming the limit are heavily dependent on the knowledge of its nature. These aspects have been comprehensively reviewed by Al-Murrani (1974) and Eisen (1980).

Here experimental tests of predictions from limits theory are discussed. This analysis has been deliberately restricted to those experiments lasting no more than 20–30 generations, so that the effect of new mutations on selection response will not seriously affect the conclusions (Hill, 1982b, 1986). As Robertson (1960c) pointed out 'this theoretical investigation [on selection limits] has probably its real value not in predicting exactly what is going to happen in reality but in enabling one to design experiments on selection limits and to interpret them when we have done it'.

The effect of linkage

Robertson (1970e) considered k biallelic loci on a chromosome, all with the same effects and initial frequencies, contributing a heritability h^{*2} to a trait. The final advances from selection are: L_f (for independent segregation), L_0 (with no crossing-over), and L_1 (when loci are equally spaced along a chromosome of length l). Starting from linkage equilibrium, it was concluded that:
(i) L_f/L_0 tends to $2N_e ih^*/3$ as k increases (for $N_e ih^* > 5$);
(ii) L_1/L_0 approaches L_f/L_0 asymptotically as $N_e l$ increases, and
(iii) the effect of suppressing crossing-over on cumulative selection response will be unlikely to appear before $2/ih^*$ generations.
The theory has been extended to an infinite number of loci (Robertson, 1977a).

Two experiments with *D.melanogaster* ($l = 0.5$) (McPhee and Robertson, 1970; Thompson, 1977) bear specifically on these predictions. They both compared the limit achieved with the usual level of crossing-over with that attained where crossing-over on chromosomes II and III was severely reduced by balancer chromosomes. McPhee and Robertson carried out divergent selection for sternopleural bristle number ($N_e \simeq 10$, $i \simeq 1$, $h^* \simeq 0.4$ for the two autosomes) for 17 generations. Crossing-over suppression reduced the limit by about 25%, and the observed and expected half-lives with no recombination coincided ($2/ih^* = 5$ generations). Thompson selected for positive phototaxis ($N_e \simeq 50$, $i \simeq 1$, $h^* \simeq 0.1$ for the two autosomes) for 21 generations. Similar advances were made by both sets of lines, which were still responding when the experiment was concluded (as would be expected from the predicted half-life of $2/ih^* = 20$ generations). The values of $N_e ih^*$ in the two experiments were nearly equal; but that of $N_e l$ was much larger in Thompson's so the absence of recombination could scarcely be a limiting factor to the selection response obtained in this study. Some reduction would be expected in that of McPhee and Robertson whose computer simulation of their experiment was compatible with a small value of k, which implies that the loci involved were in approximate linkage equilibrium in the base population.

The effect of population structure

For additive loci, Robertson (1960c) showed that selecting to the limit m replicate lines with proportion N/M followed by crossing them and reselecting the cross with proportion mN/mM (single-cycle structure), was expected to lead to the same final gain as a single line L selected from the start with proportion mN/mM. Maruyama (1970) generalized the result by showing that crossing can be done at any time. The best of the m lines is, however, likely to exceed L for a short time (Madalena and Hill, 1972). Simulation shows that between-line selection will always result in a lower limit, as will a repeated-cycle structure in which selection is carried out in m replicates and, every t generations, the best n are crossed and a new set of m lines started from the cross and reselected (Madalena and Hill, 1972). These authors concluded that subdivision and crossing schemes (single-or repeated-cycle) are unlikely to be useful except for the elimination of deleterious recessives. Wright (1939) suggested that a repeated-cycle structure may be more efficient than selecting a single large population if epistatic gene action is important. This has been confirmed by simulation studies covering additive×additive epistasis (Enfield and Anklesaria, 1986).

Four experiments have been carried out to test these predictions. Two of them used the same design, started from the same base population of *T.castaneum*, and differed only in the number of generations of selection for increased 21-day pupal weight, either 24 (Goodwill, 1974) or 41 (Katz and Enfield, 1977). The responses obtained from a repeated-cycle structure ($m = 6$, $n = 2$) were compared to that from an undivided population, the proportion selected being the same in all lines. After 6–10 cycles of 3 generations or 3–5 cycles of 7 generations either no significant differences were detected between the responses (Goodwill, 1974), or the response of the large line exceeded the others (Katz and Enfield, 1977).

Rathie and Nicholas (1980) selected for increased abdominal bristle number in *D.melanogaster* for 17 generations and compared repeated-cycle structures of 6 generations with ($m = 10$, $n = 5$) or without ($m = n = 10$) between-line selection to a large line. A linear response to selection was observed in each case, but the large line exceeded the other two from the start.

Finally, Madalena and Robertson (1975) selected *D.melanogaster* for low sternopleural bristle number for 28 generations. They studied:

(i) single-cycle structures with ($m = 8$, $n = 2$; $m = 8$, $n = 5$) or without ($m = n = 8$) between-line selection, and two periods of 6 or 12 generations;

(ii) a three-cycle structure ($m = 8$, $n = 4$–5, 6–9 generations);

(iii) a large line.

The interpretation of the results is complicated by the presence of a recessive lethal affecting bristle score which, at the end of the experiment, was at a frequency of one-third in six out of eight types of structure considered. Nevertheless, there was general agreement with the predictions as single-cycle structures with no between-line selection did not surpass the limit reached by L, and single-cycle structures with between-line selection generally attained a limit close to that of L. The repeated-cycle structure clearly surpassed L, and this was interpreted in terms of favourable recessives present at low frequencies in the base population.

The experimental results agree on the whole with theoretical expectations. The traits considered have genetic variation mainly determined by additive gene action (Carbonell *et al.*, 1985; Robertson, 1955b) and are peripheral with respect to fitness (Kaufman *et al.*, 1977; Robertson, 1955b, 1967a). Obviously, it is desirable that experiments of this kind be carried out for epistatic traits in order to test Wright's hypothesis.

The limit attained by within-family selection

When the same number of individuals are scored and selected each generation, Robertson (1960c) concluded that equal limits will be attained by individual and within full-sib family selection, because in the latter the effective size is doubled but only half the additive variance is used. Dempfle (1974) has shown, however, that within full-sib family selection implies a relatively lower rate of decay of the genetic variance than individual selection and may therefore result in a higher limit, particularly when selection is intense, families are large and the heritability is high. Nevertheless, this prediction may not hold when the genetic variation of the selected trait is controlled by a small number of loci of large effect, and could be reversed if initial gene frequencies are low and the effective size is small (Young and Skavaril, 1976).

Gallego and López-Fanjul (1983) carried out individual and within full-sib family selection for low sternopleural bristle number in *D.melanogaster*, with the same number of individuals scored and selected per line and generation in the two treatments, and similar limits were achieved. The response to selection and the additive variance were exhausted very quickly, although the heritability of the trait in the base population was high, and genetic differences among replicates at the limit were small. These observations are consistent with the presence of a small number of genes of large effects and at appreciable frequencies in the base population which determine a high proportion of the genetic variation of the selected trait. Thus the factors which would theoretically determine the long-term superiority of within-family selection could only be shown to be operating in the short term.

Butler *et al.* (1984) carried out 18 generations of individual and within full-sib family divergent selection for 8-week body weight in two mouse populations. A plateau was reached by individual selection, while the lines selected within families continued to respond, but only 60–80% of the response achieved by individual selection was obtained by within-family selection. Based on the genetic similarities of selected replicates at the limit, Falconer (1973) suggested that the genetic variation of body size in mice is determined by a small number of loci. If so, the conditions required for the completion of Dempflé's effect will only be met in early generations. It is therefore possible that the experimental results can be explained by a violation of the assumption of a large number of loci.

Conclusions

The study of the interrelation between effective size and selection response is generally based on the assumption of a large number of additive neutral loci. The relevance of this theory has been assessed experimentally and on the whole good qualitative agreement between expectation and results has been found. Although different organisms and characters were used, all traits were largely additive and not strongly connected with fitness. Traits showing substantial nonadditive variance and closely related to fitness have not been studied, even though the theory (e.g. sublining and selection) makes specific references to them.

Discrepancies between observations and predictions arose from opposing natural selection in the long-term and/or genes of large effect on the selected trait. Fitness always deteriorated in the selected lines but this only decreased the power of resolution of the experiments without seriously affecting the expectations. However, the presence of major genes and lethal genes with considerable effect on the selected trait were found in most drosophila selection experiments, and caused deviation from prediction. These genes were not detected in tribolium and mouse experiments, perhaps because they were not sought.

Chapter 19
The Genetic Basis of Selection for Growth

J.C. McCARTHY

Faculty of Agriculture, University College, Dublin 4, Ireland

and

R.C. ROBERTS

Chief Scientists' Group, Ministry of Agriculture, Fisheries and Food, Horseferry Road, London SW1P 2AE, England

Introduction: The Robertson model of the effects of selection for growth

Although Alan Robertson made a deep and lasting impact on our understanding of the genetics of growth in animals, his written contribution to the subject is limited to a few reviews (Robertson, 1973d, 1980a and 1982a). The first and most important of these was read to the British Poultry Breeders' Roundtable meeting at Birmingham in 1973. There, in a little-known paper, he reviewed two related topics, the observed effects of selection for growth rate (or body size) and the predicted effects of selection for efficiency. In conclusion he suggested 'that selection for growth rate or the various related characters ... can be mostly considered in terms of genetic variation in intake and in partition (the proportion of energy used for growth which is used for fat deposition)'. Here we shall first recollect Robertson's basic model of the genetics of growth. Next, we shall describe how the model was elaborated in the context of selection for growth rate in mice. Finally we shall review how the model stands in the light of more recent studies of the effects of selection for different aspects of growth, including efficiency.

It was clear to Robertson in 1973, from the results of several experiments with mice and poultry, that selection for increased body weight or growth rate had two predictable side-effects: genetically large animals consumed more food and also became fatter. But crucially, he concluded from the results of Falconer and Latyszewski (1952) and Clarke (1969) that selection could affect the level of fatness independently of the rate of food intake. In the first of these studies, a line of mice selected for increased 6-week body size on a fixed amount of food, restricted daily to about 75% of the normal intake of unselected mice, was found to be leaner when fed *ad lib.* than a line of comparable size selected on *ad lib.* feeding. In Clarke's study of replicated lines of mice selected for high and low 6-week weight on *ad lib.* feeding, mice in the large lines showed no increase in fatness up to the age of selection but after that they became fatter. The large mice did not grow fatter at young ages; rather, according to Clarke, they were 'relatively leaner', i.e. leaner at a given body weight.

Robertson postulated the action of two different sets of genes from these results, one affecting food intake and the other the partitioning of energy between lean and fat. He went on to suggest that the same kinds of genes might determine responses to selection for body size in pigs and poultry and would have predictable effects if selection were carried out for increased feed efficiency in those species. We shall return to those aspects of his paper later but first, we shall consider his model in more detail in the light of later studies of selection for body size in mice.

Elaboration of the model

The patterns of fat deposition in mice selected for high and low weight at 5 and 10 weeks of age were described by Hayes and McCarthy (1976). These results were remarkably similar to those of Clarke's. Selection had little effect on fatness up to the age of selection but the large mice became much fatter at later ages. It appeared from Allen and McCarthy's (1980) further study of the development of individual fat depots in these mice that the pattern of fat deposition was altered rather similarly at both ages, but to a greater extent in those selected at 10 weeks. Large mice were leaner relative to body weight prior to the age of selection. The small mice showed the symmetrical change and were relatively fatter prior to the age at selection. Prompted by the Robertson model, Hayes and McCarthy interpreted these results in terms of changes in energy utilization and conversion, and accepted the two causes of genetic variation in growth rate. One was associated with the supply of energy for growth and the other with the efficiency with which that supply is used for weight gain. The source of variation in energy supply was assumed to reflect differences between animals in their rate of food intake. The variation in the efficiency of energy conversion was assumed to reflect differences between animals in the composition of their weight gain, i.e. in the relative proportions of lean and fat in the body. They proposed that differences in the tissue composition of weight gain led to variation in the efficiency of energy conversion, because weight gain in fat is more costly than weight gain in lean. Hayes and McCarthy suggested that fat was about seven times as costly as lean – a figure estimated from perusal of basic nutrition texts at the time. Later authoritative reviews of the energetic costs of tissue deposition, for example by Webster (1977), indicate that the relative costs for fat and lean may in fact be nearer 5 to 1, but this would not greatly alter the interpretation.

The model, so elaborated, provided a more general explanation of the effects of selection for body weight in mice. Selection for large size favours mice which are capable of ingesting and metabolizing greater amounts of energy. But these animals must also utilize that energy sparingly for weight gain, by increasing their relative leaness during early postweaning growth. The effects of selection for low weight are explained simply as the converse of this. Small mice combine a reduced intake with increased relative fatness in early growth. The elaborated model also clarified the puzzling results of Hull (1960) who had found that selection for high weight at 3 weeks of age had resulted in higher levels of fatness at later ages than had selection at 4.5 or 6 weeks. Because the proportion of fat increases with age, there is more scope for selection for efficient partitioning at later ages and, therefore, lines selected at older ages are relatively leaner.

Soon afterwards Roberts (1979) reviewed the overall effects of selection for body weight in mice and concurred broadly with this general interpretation of the effects of selection on body composition. Several experiments with mice showed that selection for increased size led to increased adult fatness including the then recent work of McPhee and Neill (1976). However, Roberts expressed some reservations about the completeness of the model, in two respects, and took it upon himself to suggest amendments which he referred to as 'speculative'.

The first amendment was in regard to the genetic relationship between growth rate and appetite. One of the two major elements in the model postulated by Robertson (1973d) and by Hayes and McCarthy was a set of genes affecting intake. Robertson raised the question whether genetic differences in intake are generated by differences in growth rate or *vice versa*, but dismissed it as irrelevant to the model. Hayes and McCarthy's paper was ambiguous on this point – their diagrammatic representation of the model clearly indicates that selection was seen to affect intake as a secondary consequence of screening for growth rate. But their textual statement was that differences in the rate of growth are 'caused by genetic differences in the rate of food consumption'. Prompted by the work of Webster and his colleagues (e.g. Radcliffe and Webster, 1976) with the

Zucker rat, Roberts adopted their suggestion that the increased impetus for protein deposition in large mice was the principal determinant of their increased of food intake. The increased appetite, if it remained high after the age of selection, would, he claimed, inevitably lead to the high levels of fat observed in adult animals. The implication was that genetic variation in the rate of food intake observed after selection was the effect rather than the cause of genetic difference in the rate of growth. This was an extension of the Hayes and McCarthy version of the model, since it supposed that selection acted on genes whose primary action was on a determinant of intake and not on intake itself.

A second amendment to the model suggested by Roberts was more speculative still. He reviewed the overwhelming evidence that mice selected for increased size or growth rate convert their food to weight gain more efficiently. But does this imply that there are genes which affect efficiency other than those which affect body composition? Roberts suggested that there were, on the grounds that a large proportion of total energetic intake is expended as heat – heat generated either in the processes of metabolism or heat otherwise generated to maintain body temperature. He suggested that fat deposition and heat expenditure might be alternative forms of energy utilization. But this does offer an explanation why direct selection for gross efficiency often leads to an increase in fatness, in direct contrast to what might be expected if the energetic contents of fat and lean were the only factors at work. If all this is true, it is a fairly radical departure from the simpler considerations of earlier models.

The model now

The broad basis of the model described above, particularly the version of Hayes and McCarthy, was readily incorporated into the conventional wisdom of animal breeding in the context of selection for body size (e.g. Barlow, 1978; McCarthy, 1983). However, those of us actively involved in mouse work found it perhaps less adequate to explain the effects of selection in mice. This was particularly so in the case of selection on restricted feeding. Under restricted feeding, any response would demonstrate the effect of 'partitioning' genes in the absence of variation in growth rate associated with appetite. However, the outcome was not always so clear cut.

There have been many selection experiments for growth rate or body size on *ad lib*. feeding in mice (about 20 were reviewed by McCarthy, 1982) but far fewer in which selection was carried out on restricted feeding. Even these few are not consistent among themselves. Estimates of the realized genetic correlation between growth rates on different feeding levels have varied widely. Pooled estimates from individual studies range from about 0.3 (Falconer and Latyszewski, 1952; Hetzel and Nicholas, 1986) to about 0.6 (Falconer, 1960b; McPhee and Trappett, 1987). These intermediate values for the correlations indicate that different combinations of genes are involved in the responses on *ad lib*. and on restricted feeding, as would be predicted from the exclusion of appetite genes. In each of those studies, the mice selected for increased growth rate on restricted feeding were leaner than those selected on *ad lib*.

However, these outcomes are not obtained consistently. In two studies (McPhee *et al*., 1980; Yuksel *et al*., 1981) the genetic correlation between growth rates on different levels of feeding was close to unity, and the mice selected under restriction were fatter than those selected on *ad lib*. The possible reasons why the results of selection in these two studies differed from the previous four are discussed below.

As stated earlier selection for growth on restricted feeding would be expected to produce leaner animals because the differences between animals in food intake would be mostly removed by the restriction. Thus, differences in the efficiency of conversion to weight gain would be the major source of variation in growth rate. Accordingly, animals with a high proportion of lean relative to fat should be the ones selected because, other things being equal, they should be more efficient. However, other things may not be equal, and it is not totally clear how restriction affected variation in food intake in the experiments quoted. In only one case (Falconer and Latyszewski, 1952) were mice continuously restricted in food supply by the daily feeding of a limited amount of food. In his second experiment, Falconer (1960b) restricted intake by diluting the food with indigestible fibre, thus allowing mice to vary their intake of a less nutritious diet. In the other four studies quoted the mice were restricted by feeding a limited amount of food every two days, allowing mice to eat freely on the first day but restricting them to what may be left on the next. This method of restriction may have special consequences for the outcome of selection. One of us (JMcC) has just completed an experiment which shows this.

Briefly, this experiment selected for increased 3–6 week gain on 3 feeding regimes:
A-line: *ad lib.* feeding;
I-line: intermittent feeding, on alternate days;
L-line: limited feeding on a daily basis.

Both the I and L lines were restricted to about 75% of the intake of the A line. The I-line allowed full expression of appetite every second day, with possible days of hunger in between. The L-line accommodated no difference in appetite at all, and all the food was eaten daily. The two methods of restriction led to different genetic correlations between growth under restriction and on *ad lib.* The correlation was large and positive between A and I but small and negative between A and L. The reasons why these correlations are so different can be deduced from Table 19.1 and Figure 19.1, but the nature of the interactions will not be explored further here. It is sufficient for now to note that in the I-line, the role of appetite was not excluded, so the mice became fatter when fed on *ad lib.* In the L-line, selection had all been for efficiency, at the expense of fat (as predicted by the Hayes and McCarthy model) and even at the expense of appetite when fed *ad lib.*

Table 19.1 The growth and food intake from 3–6 weeks on *ad lib.* feeding of mice selected on *ad lib.* (A) and restricted feeding (I and L) relative to their unselected control (C). (Wade, K. and McCarthy, J.C., unpublished data)

Line	C	A	I	L
Gain (g)	17.0	+1.8[1]	+1.0[1]	–0.9[1]
Intake (g)	106.7	+6.2[1]	+5.0[1]	–6.9[1]

[1]Different from C-line, $P < 0.05$.

What implications for a genetic model of growth can be drawn from selection under restricted feeding in mice? It is difficult to be dogmatic because of the different methods and the different degrees of restriction used. Where restriction has been applied in a rigorous fashion on a daily basis, there is clear evidence that partitioning genes come into play. When the mode of restriction was less rigorous and allowed some effect of intake genes, the results are less easy to relate to the model. The most difficult results in this regard are those of McPhee *et al.* (1980) and of Yuksel *et al.* (1981), which are reminiscent of the somewhat ambiguous effects of selection in mice for efficiency on *ad lib.* feeding. It would appear from these results and from those of a recent but similar study by McCarthy (unpublished data) that mice selected for feed efficiency on *ad lib.*:
(i)   do not exhibit an increase in intake;
(ii) tend to be smaller but fatter at the beginning of the test period; and
(iii) still remain relatively fat throughout their growth period.
The model outlined earlier is not fully adequate to describe the effect of selection for efficiency under *ad lib.* feeding.

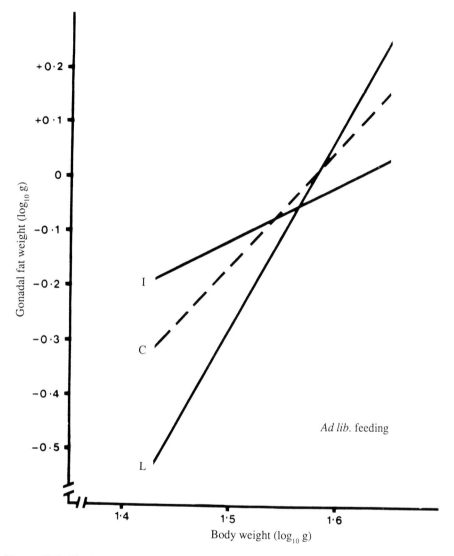

Figure 19.1. The logarithmic relationship between fat weight and body weight on *ad lib.* feeding of mice selected on restricted feeding (I and L) and of unselected controls (C). (Somers, C. and McCarthy, J.C., unpublished data.)

The evidence reviewed briefly here suggests that different strategies of energy utilization come into play, depending on which trait is at a premium. When selection is simply for weight gain, mice typically increase both their intake and their efficiency (Roberts, 1981). However, the increase in efficiency is at least partly a direct function of the increase in intake, because of the proportionate reduction in the fraction of that intake needed for maintenance. Maintenance requirements as such may not alter much, and the main response comes from the increase in intake. The situation is quite different when selection puts a direct premium on efficiency. Mice then seem to respond by reducing maintenance, and as shown above, they may do this in a variety of ways not readily expected of the Robertson model. The few results on selection for efficiency show the need for more studies on genotype×feeding regime interactions. We have some glimmerings of how the system may work, but at the moment it seems rather complex. Apparent complexity usually means a lack of understanding.

It was noted earlier that unelaborated versions of the Robertson model seemed to work well when applied to selection for body size, when the criteria of selection were growth over a given period or weight for age. Our later discussion of selection for efficiency seems to call for modifications of the model, at least as suggested by studies on the laboratory mouse. The mouse appears capable of adopting a range of strategies to balance the energetic demands of the selection criterion with the supply of energy in the feeding regime. We did not explore these modifications in larger animals, and it is too large a topic to incorporate within the compass of this review. It could be that the mouse is not a good model for the efficiency of energy utilization in larger animals, for the following reason. The food intake of the growing mouse is only about 1.2 times what it requires for maintenance, leaving only one-sixth of its total intake for growth. The pig, in contrast, requires only about one-third of its intake for maintenance. Differences of this magnitude in relative food intake may cast doubt on the relevance of studies on efficiency in the mouse to larger animals.

Alan Robertson's lasting impact on the genetics of growth was to provide a satisfying intellectual framework to address the issues in a way that reduces the apparent complexity to a few simple, internally coherent, concepts. We may not yet have all the answers, but he showed the way to ask sensible questions on the subject.

Chapter 20
Identification of Genes which Influence Quantitative Traits

O. MAYO

Biometry Section, Waite Agricultural Research Institute, The University of Adelaide, Australia

Individual genes which influence quantitative traits were probably recognized before recorded history began. Major genes producing gross deficiencies in human height and intelligence, for example, have been recognized for many centuries, well before the particulate nature of inheritance was elucidated. The problem of interest here, however, is the identification of genes underlying quantitative traits, since quantitative traits are normally analyzed by statistical methods which depend upon the effects of individual genes being largely unrecognizable (Fisher, 1918). The overall problem has at least three separate components: first, the demonstration that a small number of genes has a major effect on a trait; second the mapping of such genes; and third the identification of gene products, i.e. the identification of gene function, given the existence of a gene.

Determination of gene number

There are separate methods for species which may be treated solely observationally or experimentally, i.e. in general, humans and other species.

The oldest method of determining the minimum number of genes influencing a trait is the Castle–Wright index (Castle, 1922; Wright, 1952). For two inbred lines, P_1 and P_2, this segregation index (i.e. estimate of the apparent minimum number of genes) is $(\bar{P}_1 - \bar{P}_2)^2/[8(V_{F2} - V_{F1})]$, where $\bar{P}_i$ denotes the line mean and V_{Fi} is the variance of the ith filial generation. This method assumes equal gene effects, independent gene action, directional distribution of alleles, and other inherently unlikely but useful properties.

Wright (e.g. 1959, 1968) suggested other indices for estimation of apparent gene number and for the detection of a single factor if backcross generations have also been obtained. This latter index is $(V_B - V_{P1})/(\bar{B} - \bar{P}_1)^2$. It takes the value 1 if the gene is segregating and is otherwise less. Such methods are discussed further in the next section. The Castle–Wright index is still being investigated and, in principle, enhanced. For example, Lande (1981) and Cockerham (1986) suggested a number of improvements. However, these improvements require the estimation of many more parameters and therefore larger samples, and possibly the estimation of parameters for different generations (as for the backcross index), so that they may yield no improvement in practice unless these other generations are usually available.

Furthermore, when the Castle–Wright index was applied, suitably modified for autotetraploids, to tetraploids made from certain diploids (in grasses), the diploid and tetraploid estimates differed widely (Mayo and Hopkins, 1985). This emphasizes that mode of gene action is ignored in making such estimates and allowance for dominance and other complicating factors can be made only in very arbitrary ways.

Newer methods, such as genotype assay (Jinks and Towey, 1976), have similar defects. This method depends upon the detection of the minimum number of factors still segregating in F_t. Each individual of F_t is selfed, then two random progeny are taken from these offspring and selfed. These progeny means and variances can differ only if their parent was heterozygous at one or more loci. Then bounds can be obtained for the relationship between the number of genes involved and the proportion of progeny still segregating, and the gene number estimated therefrom. The method lacks power and can be biassed (Hill and Avery, 1978).

Detection of major genes underlying a continuous distribution

If a quantitative trait has a skewed distribution, tests exist to determine whether it is merely a skewed distribution which may be normalized by transformation or if it is made up of a mixture of distributions. In the latter case, a population sample may yield evidence of several genes. Distributions may be split by very archaic methods, such as Pearson's method of moments, or by more modern methods (e.g. Maclean et al., 1976). The methods developed by Morton, Lalouel, Elston and others (e.g. Morton and MacLean 1974; MacLean et al., 1976; Lalouel et al., 1983; Demenais et al., 1986) are very adaptable and have been very widely used. They now allow the detection of segregation of a major gene underlying a quantitative trait with a distribution exhibiting any degree of skewness, and the partition of the determinants of a trait among major and minor genes. The statistical properties such as robustness and power have been thoroughly investigated (e.g. Demenais et al., 1986) and if these methods are used carefully they will continue to be most successful for detecting major genes which influence quantitative traits.

All of these methods, however, assume normality of the underlying distributions about the means of single genes, the effect given by one major gene and very many minor genes. Less distribution-dependent methods have therefore been developed, under the heading 'Structured Exploratory Data Analysis', by Karlin et al. (1981). These methods tend to be rather unsatisfactory in practice, however. They rely on ratios of powers of departures of offspring measurements from that of the midparent to differences between parents or separate departures of offspring from the two parents. These quantities are very sensitive to measures of skewness and kurtosis, whereas in other methods these measures are themselves used as major gene detectors (Mérat, 1968; O'Donald, 1971), or skewness is removed by a power transformation, either ad hoc (Morton and MacLean, 1974) or systematic (Box and Cox, 1964). Furthermore, the different statistics suggested by Karlin et al. can give inconsistent or misleading results when gene effects are very unequal but gene numbers are not very large (Mayo et al., 1983, 1985).

Related in spirit are the intrafamily correlations of Matthysse et al. (1979), which should detect genes whose effects decline approximately as a geometric progression, a model originally suggested by Fisher (McMullen, 1962). Unfortunately they do not (Mayo et al., 1985).

Famula (1986) does not discuss the problems of Karlin's major gene index from the same point of view, but has indicated that it can be combined with Best Linear Unbiased Prediction (BLUP) to remove maternal, seasonal and other effects when the method is applied to domestic animals. That is, sire, dam and offspring phenotypic values are replaced by their predicted values, and Karlin's statistics are calculated from these predictions. The method can then be applied to many different kinds of data. Famula (1986) has shown that the major gene index then detects a major gene for rapid postweaning weight gain in mice in families in which it is segregating (Bradford and Famula, 1984; Famula et al., 1986). However, in experimental populations, multiple generation data and prior estimates of genetical parameters are usually available, and without these the problems mentioned above for human populations will persist. It would be of great interest to apply Famula's methods to the Booroola ovulation rate gene in sheep (Piper and Bindon, 1982).

The above-mentioned methods of Mérat (1968) and O'Donald (1971) and other related methods, have been tested in practice and have been shown to be rather ineffective (Hammond and James, 1970, 1972), as might be expected for the reasons discussed earlier.

Mapping of genes

There are two major sets of methods, those based on populations, and those based on test crosses. The former essentially apply to human populations.

In human populations, association of trait differences with marker genes should mean, in the absence of linkage disequilibrium, direct effects on alleles of the marker gene on the trait. This should be testable with distinct separate marker genes in linkage disequilibrium with the original markers (Mayo et al., 1971). If, however, there is linkage disequilibrium, there may be a quantitative trait locus segregating in the population or a direct effect of the marker locus. The basic model assumes a marker gene, M, with two alleles, M_1 and M_2, which has no effect on the trait of interest, and a quantitative trait locus, A, with two alleles, A_1 and A_2. The three genotypes have means and variances: A_1A_1, μ_1, σ_1^2; A_1A_2, μ_2, σ_2^2; A_2A_2, μ_3, σ_3^2; and there is recombination fraction θ between A and M. If either $\theta = 0$, or if M actually has an effect, then there will appear to be a direct effect of M, as opposed to a linked effect from A. Family studies should distinguish between these possibilities.

However, in some cases there are a priori reasons to assume that M has a direct effect, and in such cases the six μ_i, σ_i^2 parameters could be estimated directly. Alternatively, one could assume a mixed model:

$$\hat{Phenotype} = \mu_{ij} + G_j + E_j, \text{ where } G_j \sim N(0,\sigma_G^2), \text{ and } E_j \sim N(0,\sigma_E^2), \text{ and } \sigma_1^2 = \sigma_2^2 = \sigma_3^2,$$

and then partition the variance directly into that attributable to M, that attributable to other genetic influences (σ_G^2), and a residual (σ_E^2). This approach has been used by Boerwinkle and Sing (1987) to assess the influence of four alleles of the apolipoprotein E gene on plasma lipids. They found that σ_G^2 absorbed 56% of the total variance in total cholesterol, apo E 8%. Further, specific apo E alleles raised or lowered cholesterol level by as much as 4%. (Four alleles would require the estimation of ten σ_i^2 parameters if they were not all assumed equal.)

Several family study methods have been suggested for distinguishing linkage from direct effects, or for estimating θ. A.P. Hill (1975) has given a method based on the hierarchical partitioning of total variance in a trait in a set of sibships into that between sibships, between marker genotypes within sibships, and a residual within marker genotypes. Like many other human methods, this depends on constancy of trait variance over sibships (Mayo et al., 1980). Hill suggested full maximum likelihood for estimation. Because this method is so demanding in its assumptions, Smith (1975) suggested a nonparametric variation on this theme, making fewer assumptions about the distribution of the trait, but requiring three-generation information and making no use of families smaller than three in size. Part of the method derives from Galton's 'Law of Ancestral Heredity', making its pedigree even longer than most of the other methods. This method lacks power, and its one- and two-generation derivatives are even weaker. Despite this lack of power, however, these methods have been used successfully, for example to demonstrate linkage between the Kell blood group system and dermatoglyphic patterns on the sole of the foot (e.g. Bener, 1982).

Complex segregation analysis (Lalouel and Morton, 1981; MacLean et al., 1984) is a general approach to the analysis of traits which are not inherited in a simple Mendelian fashion, i.e. modifiers, pleiotropy and epistasis may mask the effects of major genes which determine or influence discrete traits. They are particularly applicable to those diseases which have high heritability but are not determined by single genes, such as heart disease and diabetes. These methods currently allow examination in one analysis of the influence on a trait of a major disease susceptibility gene possibly linked to a known marker (in linkage disequilibrium), pleiotropic effects, an unlinked modifier and environmental effects (Rich *et al.*, 1987). Applied to insulin-dependent *diabetes mellitus*, these methods detected a major gene in significant gametic association with the HLA region. The gene detected had a frequency of about 0.14 and penetrance in homozygotes of 0.12. This was a large study of a system showing very tight linkage ($\theta = 0$); in smaller studies, the large number of parameters to be estimated might mean that looser linkages would not be detected.

Human data have many problems from an analytical point of view; they are observational, family sizes are small, generation intervals are long, genotypic effects may be age-dependent, and so on. At the opposite end of the scale of tractability lie results from self-fertilizing species of plant. It is then possible to analyze large numbers of offspring of known homozygosity with genotypes well-characterized for many marker loci, to repeat crosses, and to set up complexes of crosses yielding interconnected results.

These general observations may be illustrated with perhaps the simplest case, two genes with alleles homozygous in different parental lines. Suppose that there is the same situation of a marker gene, M, and a quantitative trait locus, A, as set out above. There will be again three genotypes A_1A_1, A_1A_2 and A_2A_2, and one may assume as before that the trait is normally distributed with a different mean and variance for each genotype, and recombination between A and M of θ. Then the distribution of the trait for each marker genotype in the F_2-generation is simply given by the weighted sum of the separate Gaussian distributions for the three trait genotypes with means and variances as before. Thus, seven parameters have to be estimated and maximum likelihood may be used. Because of the complexity of the problem, however, Weller (1986) recommends using as a starting point moment estimates (Pearson, 1902; Zhuchenko *et al.*, 1979) of some of the parameters, approximate maximum likelihood for others. By this method, Weller was able to show that if $\theta = 0.2$ and the difference between μ_1 and μ_3 is more than 1.63 phenotypic standard deviations, a sample of size 2,000 will yield accurate estimates of the seven parameters.

Weller (1987) has described the application of these methods to a cross between two tomato species. This wide cross was likely to increase the chance of detecting genes because one species (*Lycopersicon esculentum*) had been strongly selected as a crop plant, the other (*L. pimpinellifolium*) not. Weller used 10 markers and measured 18 quantitative traits, and identified 48 significant associations (putative 'quantitative trait loci'), compared with the 9 expected. In one case, 8 traits were associated with a particular marker. These results illustrate the promise of the method and some of the problems. It will be very difficult to separate statistical artifacts, real but irrelevant results arising from gene interaction and diffuse regional effects from the desired major gene effects.

Somewhere between plant genetics and human genetics lie the methods of drosophila genetics (Breese and Mather, 1957; Thoday, 1961; Thoday *et al.*, 1964; Davies, 1971). These are similar to those of plant genetics, with the exceptions that lines must be maintained of known genotype (i.e. clones and seeds cannot be obtained), and that normal recombination occurs only in females, so that informative backcrosses cannot be made with heterozygous F_1 males. (Conversely, this property can also be used to transfer blocks of unrecombined genetical material between strains.) Hence, multiple generations must be raised to assess any chromosome segment.

The basic principle requires that the genotype to be tested is crossed to a stock homozygous for at least two recessive markers on a chromosome segment of interest. Heterozygous females are then backcrossed into the marker strain, and their heterozygous male offspring are scored for the quantitative trait of interest. Then the number of different modes observed on the trait within the recombinant marker phenotypes give a minimum estimate of the number of quantitative trait loci between the marker genes concerned, and the relative frequencies of the different classes depend on the map distances between loci. These methods have been quite widely used, despite the elegant demonstration by McMillan and Robertson (1974) that they can both detect genes which do not exist and overestimate the effects of real genes by confounding their effects with those of adjacent genes. The former effect arises when a gene which increases the trait of interest is close to one which decreases it, and both these genes lie between the two markers. Thoday (1961) originally developed the method for detecting and mapping genes in lines strongly selected for increased and decreased levels of certain traits; under these circumstances, the method is less unreliable. Perhaps the most comprehensive demonstration of the utility of this approach is the investigation of the influence of the third chromosome on sternopleural bristle score (Shrimpton and Robertson, 1988a, b). This supports the hypothesis that both major genes and more numerous minor genes influence the trait.

Identification of gene products

As noted initially, many major genes have been identified by their gross effects on traits of interest. Similarly, their gene products have later been identified because the differences were so substantial. However, when major genes have underlain a continuous trait segregation but have not been individually detected, their primary effects have been established with much less ease. For example, several genes were shown quite some time ago to have significant effects on the blood levels of cholesterol and other lipids (e.g. Mayo *et al.*, 1969 and many later papers by many authors; see e.g. Fröhlander, 1987 for references). However, while major genes for lipid levels have been identified and while this work is very important (Goldstein and Brown, 1973; Hegele and Breslow, 1987), they have not been associated with the marker genes of original interest, nor with a substantial proportion of the genetic variance in the trait (Sing *et al.*, 1975). The genes which make such a substantial contribution, referred to in the previous section, form another group (Boerwinkle and Sing, 1987). There is much work yet to be done in this area (George and Elston, 1987).

A related problem is the partitioning of the variance in a trait primarily determined by a single gene. Various methods may be used, from the very simple, like the use of multivariate analysis of variance to partition out the influence of the structural gene on red cell acid phosphatase activity (Bishop *et al.*, 1987), to the complex, like the similar use of maximum likelihood to partition variation in α-1-antitrypsin level between the PI polymorphism and various physiological and environmental factors (Martin *et al.*, 1987). Maximum likelihood is the method of choice when the relevant exogenous factors can be specified.

Several hundred genes affecting intelligence of man have been detected. This is because intelligence, however defined, is a quintessentially human trait, and it may be damaged in so many different ways in the complex process of human development. Thus, while an understanding of the lesion which produces the retardation called Tay–Sachs disease is important for genetical counselling and possibly, in the long term, for treatment, it does not yield a lesson of general importance. Major genes which influence normal intelligence (e.g. Ashton and Borecki, 1987) remain controversial and difficult to identify unequivocally. In contrast, in other areas of applied genetics, identification of a gene with particular effects may be useful in species other than the original one and in different circumstances. For example, several dwarfing genes are known in cereals which take effect by producing gibberellic acid insensitivity. In addition to dwarfing, these genes allow increases in yield. It may be that DNA probes for these genes will allow the identification and manipulation of genes influencing yield in species where dwarfism or semi-dwarfism is irrelevant, unlike the cereals.

Discussion

Demonstration that one or a few genes at most influence a trait is an important problem in at least three different applications of genetics.

In human genetics, identification of a single gene influencing a trait may aid in disease prediction in a given individual (e.g. in the case of the association between serum cholesterol and cardiovascular disease), in genetic counselling (e.g. some of the diseases which may be regarded as threshold manifestations of underlying continuous traits; Curnow and Smith, 1975), and in understanding other continuously varying traits with thresholds (e.g. Kalmus and Seedburgh, 1975).

In plant and animal breeding, the methods used for single gene incorporation are much more direct than those for truncation selection, and will become even more direct as genetic engineering becomes more practicable and routine. Thus the single genes which are the raw material for such engineering will become more important. In ordinary directional selection, the limit to which selection is directed depends on how many genes affect the trait (Robertson, 1970e) and therefore knowledge of this number will allow more efficient selection.

In evolution, contending theories emphasize selection on many genes rather than few, so that resolution of such controversies may depend on knowledge of the numbers of genes involved (Lande, 1981). While it is not universally agreed that this dichotomy is necessary or real (e.g. Mayo and Hancock, 1985), a knowledge of gene number is still important in all approaches to the problem.

Given the importance of all of these problems, it is unfortunate that all the methods discussed in this review have rather low power and efficiency, and therefore to be effective, require very large, well defined samples, usually involving two or more generations. This results from their necessary dependence on variances and higher moments of distributions, so it is not likely that the better statistical methods which will no doubt be developed will be much more powerful or efficient. However, recombinant DNA technology and the maximum likelihood methods will certainly result in the mapping of many genes of the types discussed in man and in economically important species of plants and animals in the near future.

Chapter 21
The Quantitative Effects of Genes Which Influence Metric Traits

L.R. PIPER

CSIRO, Division of Animal Production, Armidale, NSW 2350, Australia

and

A.E. SHRIMPTON

Department of Genetics, University of Edinburgh, West Mains Road, Edinburgh EH9 3JN, Scotland

In classical genetic models for continuously distributed or quantitative traits, the phenotypic values are the sum of independently and normally distributed environmental and genetic components, the latter assumed to arise from segregation at a large number of independent loci each of small effect. Such models are the basis of the statistical theory of quantitative genetics initiated by Weinberg (1909, 1910), Fisher (1918) and Wright (1921) and greatly extended by Lush, Alan Robertson and others into a predictive theory of animal breeding. This theory has provided animal breeders with reasonably reliable short-term predictions of direct response to selection but with somewhat less reliable predictions of correlated response (Bohren *et al.*, 1966). However, the theory has several major deficiencies (Robertson, 1966a): it cannot predict the likely limits to selection except under very simplifying conditions, and it is unable to predict the changes in reproductive fitness that might be expected to accompany selection for any metric trait.

It has been recognized for some time (Robertson, 1967a) that to overcome the deficiencies of the classical variance components approach, it would eventually be necessary to provide a description of the genetic variation in any trait in terms of the contributions of the individual loci affecting that trait. This description would not necessarily be complete but would attempt to specify the numbers of loci, the effects, type of gene action and frequencies of alleles for those loci controlling the major part of the observed genetic variation.

It will be some time before such descriptions are available for most economically important traits of domestic animals. However some evidence is available from laboratory species and the purpose of this chapter is to briefly review the evidence in relation to one aspect of this description, namely the distribution of the effects of genes influencing metric traits in animals.

Possible distribution of gene effects

At the level of the biological activity of individual enzymes or hormones it seems reasonable to suppose that there will be a distribution of allele effects. For any individual locus with alternative alleles, there will be a limited range of activities, but over all the loci contributing to the variation in any metric trait there will be some distribution of allelic effects. What will be the effect of segregation at these loci on the expression of traditional metric traits such as body and tissue growth, milk production, wool production, etc?

The work of Kacser and Burns (1973, 1979, 1981) shows that in linked biochemical pathways there is considerable buffering of the flux or flow through that pathway against genetic change in the activity of component enzymes. A decrease in the activity of a particular enzyme may be compensated by an increase in pool size of its substrate with little or no observable change in the flow rate or flux. Such effects were clearly demonstrated in the studies of Flint *et al.* (1981) with *Neurospora crassa* where very substantial reductions in enzyme activity (to less than 10% of normal) were required in order to significantly influence the rate of synthesis of arginine. Under the Kacser and Burns theory, any distribution of individual enzyme activities might then be transformed into something approaching a negative exponential distribution of absolute (+ve and -ve) gene effects at the production trait level. That is to say, the vast majority of gene substitutions would be expected to have small effects on the production trait but a few genes may have substantial effects. Because currently available procedures for detecting segregation and estimating the numbers and effects of quantitative trait loci have low precision (O. Mayo, this volume), only those genes with large effects are likely to be detected. The biological expectations and the limitations of genetic analysis combine to create a situation where a few major genes may be detected but where in most cases the genetic variation will be ascribed to the effects of segregation at a large but indeterminate number of loci each of small effect.

Evidence from drosophila

Drosophila is the most studied organism with regard to locating the genes controlling metric traits. This is probably due to the formidable background of work, largely using *Drosophila melanogaster*, which provided a vast array of marker genes and balancer chromosomes.

Morphological traits, such as bristle number (sternopleural, abdominal and scutellar), body size, wing length and wing vein patterns have been most frequently investigated, but behavioural, reproductive and stress characters have also been studied.

The first gene location study of a metric trait in drosophila was reported by Warren (1924), who showed that all four chromosomes carried 'egg size' genes. Karp (1936) used a multiply recessive marked third chromosome marker stock to analyze the third chromosome for abdominal bristle genes. He postulated at least six genes and suggested that increasing effects alternated with decreasing effects along the chromosome in a balanced distribution. Breese and Mather (1957) analyzed a selected chromosome using a multiple marker stock and also showed that the third chromosome carried at least six factors controlling abdominal bristles.

Thoday (1961) developed this method to map more precisely the location and effects of quantitative bristle genes. A chromosome with a minimum of two recessive markers is used to analyze a homologous chromosome with a different average score for the metric trait. Recombinant males are used to establish lines for progeny testing. The number of recombinant lines analyzed and the number of offspring recorded per line determine the sensitivity of the assay. Subsequently Thoday and co-workers made a number of investigations of chromosomes from a line selected for sternopleural bristle number (Thoday and Boam, 1961; Wolstenholme and Thoday, 1963; Thoday et al., 1964; Spickett and Thoday, 1966). These studies led them to conclude that relatively few genes were required to explain the responses to selection, with a total of nine major sternopleural bristle effects being located over all three major chromosomes. Schnee and Thompson (1984) found that gene effects determining the response to selection for sternopleural bristle number were not the same at all temperatures. Davies (1971) used a multiple recessive marker stock to find at least eight sternopleural bristle effects on a high third chromosome.

This subject has been a long term interest of Alan Robertson (1966a, 1973c, 1977a) and, with a series of co-workers, he developed the stocks and techniques for genetic analysis of drosophila chromosomes (Da Silva, 1961; Mostafa, 1963; Louw, 1966; Piper, 1972; Shrimpton, 1982). The final investigation in this series contrasted a third chromosomes from each of two lines selected for high and low sternopleural bristle count. Using the Studentized Range Test at least 17 factors were required to account for the difference in score of 24 sternopleural bristles (Shrimpton and Robertson, 1988a,b). The largest sternopleural effect detected was nearly $2\sigma_p$ and was at or close to *hairy* (*h* III:26.5). Other effects ranged down to around $0.5\sigma_p$ but there was an ascertainment problem for effects of less than $0.6\sigma_p$. Presumably the scoring of more flies would have reduced this limit. Only a conservative estimate of the minimum number of effects and their approximate locations could be given. The results were compatible with the hypothesis that quantitative characters are under the control of a few major genes supported by numerous genes of smaller effect. Epistasis between factors was detected.

McMillan and Robertson (1974) investigated the sensitivity and precision of this form of genetic analysis. Their concern was to devise a strategy which would maximize the joint probability of determining the number and location of the quantitative trait loci and of accurately estimating gene effects. They showed that, for a given total number of flies measured, the optimum balance between the number of lines created and the number of flies measured per line, depends on the *a priori* distribution of gene effects which is unfortunately unknown. A further limitation to the precision of the method was the possibility that genes of opposite effect to that expected from their average scores were fixed in the tester and tested chromosomes.

An alternative approach to the problem of detecting genes controlling metric traits was proposed by Wright (1952). His method requires the existence of lines differing in some metric trait and involves repeated backcrossing to one of the genotypes, accompanied by appropriate selection to isolate major genes. He suggested selecting backcross individuals from the middle of the backcross distribution, where all large alleles have probabilities of one-half of being present, and selecting the progenies according to their coefficients of variability. Mostafa (1963) studied the sternopleural bristles in drosophila using this procedure but found that large blocks of chromosome, rather than single major effects, were isolated.

A third approach used on drosophila is to re-define the metric character into discontinuous components. Spickett (1963) was able to separate sternopleural bristle number differences into more specifically defined discontinuous characters. He divided the sternopleurite into four regions and counted the bristles in each region. This enabled him to ascribe a different morphological effect to each of three previously located loci and to discover a fourth. This approach combined with that of Thoday (1961), forms a powerful technique in drosophila.

More recently, advances in molecular genetics have been turned to this problem. Mackay (1985b) used P-element hybrid dysgenesis to provide additional mutational variation which could be exploited to enable a greater response to selection for abdominal bristle number. These lines have been analyzed (A.E. Shrimpton, T.F.C. Mackay and A.J. Leigh Brown, unpublished) and shown to contain a large number of new P-element insertions. At least one of these insertions has been shown to be at a known bristle gene, *smooth* (*sm*II:91.5), enabling its subsequent cloning through P-element homology. The implication of these experiments is that there are many genes which can influence bristles and presumably any other metric trait. The drawback with this type of experiment is the lack of control over the amount and timing of the P-element transposition. By the time flies can be examined there are numerous new insertions obscuring the bristle mutant. Other experiments involved P-element contamination of chromosomes for a single generation (Mackay, 1986) then screening for changes in the metric trait. This reduces the number of candidate insertions but requires the examination of many more flies.

Evidence from mice

Several loci (*dwarf*, *pygmy*, *adipose*, *obese*, *diabetes*) with alleles causing major changes in mature size and/or body composition have been described in mice (Green, 1981). However, variation in growth traits has generally been found to be quantitative in nature with responses to selection consistent with the classical genetic model of many loci each of small effect.

An exception to the established pattern has recently been reported by Bradford and Famula (1984). Mice with weight gains of three or more standard deviations above the mean were noted from around generation 25 in a line selected for high 21–42 day weight gain. The very high growth rate of these exceptional mice has been postulated to be due to homozygosity for an autosomal recessive gene, *hg*, with *hg hg* mice having weight gains 3.5–$5.0\sigma_p$ greater than their *Hg* contemporaries.

No other reports of genes with major effects on metric traits in mice have been reported, though a phase of rapid response in a long-term 6 week weight selection line observed by Roberts (1966) could have been due to a major gene affecting 6 week weight.

Evidence from domestic animals

Genes causing major discontinuities in traits such as coat colour, hornedness and fleece quality have been known for some time. In recent years, however, there have been a number of reports of major genes at loci influencing classical production traits in domestic animals. Examples of such genes include the dwarf gene in poultry (Mérat and Ricard, 1974), the halothane sensitivity gene in pigs (Smith and Bampton, 1977), the double muscling gene in cattle (Rollins *et al.*, 1972) and the *F* gene which greatly enhances prolificacy in Booroola Merino sheep (Piper and Bindon, 1982). Genes with effects on prolificacy similar to that of the *F* gene have also been postulated to be segregating in Javanese thin and fat tail sheep (Bradford *et al.*, 1986), Icelandic sheep (Jonmundsson and Adalsteinsson, 1985) and in the Cambridge breed which was developed as a synthetic based on highly prolific founder stock from a variety of breeds (Hanrahan and Owen, 1985). There is some indication that the postulated gene in the Icelandic sheep may cause a smaller increase in litter size than that caused by the Booroola *F* gene (0.6 lambs compared with 1.0 lambs; Piper *et al.*, 1988) but more data are required for Icelandic sheep before this indication could be confirmed.

At least two genes with major effects on resistance of sheep to helminth parasites have been postulated (Albers *et al.*, 1984; Outterridge *et al.*, 1985) but the existence of these genes has not formally been confirmed. In general, most production traits of domestic animals appear to behave under selection in a manner consistent with classical quantitative genetic models. It is probably fair to speculate, however, that until recently there has been little or no interest in the systematic examination of data sets for evidence of the segregation of major genes affecting production traits.

Discussion

In classical models of quantitative genetics, the genetic variation in any metric trait results from segregation at a large number of independent loci each of small and essentially equal effect. The biochemical literature certainly lends support to the notion of genetic variation at a large number of loci, in this case enzyme loci (e.g. Harris, 1975; Robertson, 1970a), but it is clear that not all mutants have equal effects (e.g. Yuhas *et al.*, 1967). Across all loci in any biochemical pathway contributing to the variation in a metric trait, it is therefore reasonable to expect that there will be a range of gene effects. However, the buffering capacity of linked biochemical pathways is very strong (Kacser and Burns, 1981), and very large changes are required in the activity of individual enzymes in the pathway if the flux through the system is to be significantly altered (Flint *et al.* 1981). The consequence of this buffering for expression at the level of the metric trait is that most gene substitutions will have small effects, but a few may have substantial effects.

The evidence reviewed in this chapter from drosophila, mice, and domestic animal species supports the above expectation. However, the power of currently available statistical procedures is such that the only genes likely to be detected in domestic animal species are those with effects greater than one to two standard deviation units. There may be genes with intermediate effects on quantitative traits in domestic species but they will remain undetected until the genetic marker maps become much more detailed and the quantitative methods developed in human and plant genetics become available (e.g. Hill and Knott, 1988; Elston, 1988).

C QUANTITATIVE APPROACHES TO ANIMAL BREEDING

Chapter 22
Alan Robertson's Contributions to Theory and Application of Animal Improvement

J.W.B. KING

CAB International, Bureau of Animal Breeding and Genetics, West Mains Road, Edinburgh EH9 3JX, Scotland

Although the two topics of theory and application would appear to be quite distinct, in terms of development of ideas they have frequently been interlinked in animal breeding as in many other applications of scientific method. Alan Robertson came to the field of animal breeding with a background of wartime operational research, so that it should be even less surprising that the two branches of enquiry overlap and intermingle in a complex fashion. Inevitably, early investigations into practical animal breeding problems revealed deficiencies in the methodology of the time and led to new theoretical developments. Likewise new ideas, coming from a theoretical base, about how genetics might be employed to create changes in populations led to quite radical innovations in practical breeding programmes. Despite this tangled history, it does seem worthwhile attempting a review of the two lines of enquiry under separate headings since they both represent fields of endeavour which Alan Robertson pursued with great success.

Theory of Animal Improvement

Early collaboration with J.M. Rendel into the problems of dairy cattle improvement were particularly rewarding. Two early papers (Rendel and Robertson, 1950b; Robertson and Rendel, 1950) set the scene for much that was to follow. The first paper demonstrated the limited progress that might be expected within a closed herd even if selection were devoted entirely to the improvement of milk yield. The second, on 'The use of progeny testing with artificial insemination in dairy cattle', showed the way ahead in a clear and concise manner and must be one of the most quoted papers in the animal breeding literature. Building on earlier formulations of Dickerson and Hazel (1944), it was shown how the pathways for improvement and generation intervals could be broken down into four categories; those for bulls used to breed bulls, bulls to breed cows, cows to breed bulls and cows to breed cows. This formulation led to estimates of expected progress dependent on the selection differentials obtained and the heritability of the character under selection. The innovation of artificial insemination (AI) raised new possibilities which were demonstrated to depend upon the size of the breeding unit. Dickerson and Hazel (1944) had shown that progeny testing within a closed herd was unlikely to be useful but with artificial insemination progeny testing could be carried out in many herds, thus radically changing the nature of the breeding operation. Some of the complications which arise are dealt with in the applications section in this review and are therefore postponed for later discussion.

Parameter estimation

One of the assumptions in predicting progress through selection by alternative means was that the heritability of the character under selection was known. The estimates available at that time were so variable as to call into question their reliability. Recognizing the deficiencies of many sets of records, a number of papers were produced showing how better estimates might be obtained where it was possible to pre-arrange the disposition of records. The results of these investigations were made public at the Symposium of Biometrical Genetics held in Ottawa in 1958 (Robertson, 1960d) and published in more detail later (Robertson, 1959b,c). As a young research worker attending one of my first international meetings I well remember the general buzz of approval with which these revelations were received by the Ottawa audience. For many animal breeders who were working with existing sets of records the message was not encouraging in that the results showed clearly that family sizes of the order of two or three were very inefficient and that for half-sib analyses 20 or 30 would be more appropriate. A later publication on the estimation of variance components in an unbalanced single classificiation (Robertson, 1962b) showed how basically unsatisfactory data sets might be used were it necessary to do so.

Optimizing progress by selection

Family structure was important not only for the estimation of parameters but also in determining the amount of progress to be expected from any form of family selection. This problem had been encountered at an early stage in designing AI progeny-testing schemes given that there was a choice between testing a limited number of bulls on a large number of daughters or a greater number of bulls on a lesser number of daughters. The results were different from those for parameter estimation in that the optimum was mostly dependent on the total size of the population and only to a small extent on heritability. It was shown always to be desirable to progeny test more than four times the number of bulls than would ultimately be required for general use.

When attention was switched from short-term gains to long-term progress different results were obtained again (Robertson, 1960c). These important conclusions are described elsewhere but are no less relevant in the present context for the improvement of farm livestock where there is a long-term commitment to the use of particular breeds.

Concern with inbreeding

The formulation of improvement patterns based on mating a large number of cows to a comparatively small number of sires through artificial insemination raised almost immediately the problem of expected rates of inbreeding. Fears were expressed that a rapid decline in genetic variation might result (e.g. Rabesa, 1950) but Rendel and Robertson (1951) were able to show that the expected decline would be slow, although still important in the long term. A few breeders professed to favour inbreeding as a means of creating uniformity but calculations showed that inbreeding did not inevitably lead to uniformity as, for example, when variation was due to rare recessive genes (Robertson, 1952). With selection the problem of inbreeding might be rather more serious than originally thought, in that the most outstanding sires in the parental generation were likely to produce the most outstanding sons in the next which would in turn be selected, so increasing the expected rate of inbreeding (Robertson, 1961c). Other papers reflect the continuing interest in the effects of inbreeding but are dealt with in other sections of the present volume. A related problem was tackled in calculations of the appropriate sizes for control populations where drift was to be minimized (Gowe et al., 1959).

The variability of animal populations

Early work by animal blood groupers showed that there was a large amount of polymorphism present in most populations of farm animals. The extent of this variation was further demonstrated by the discovery of protein polymorphisms, mostly through the application of electrophoretic techniques. Those working with such polymorphisms were anxious to explain the basis for so much variation and many effects and hypotheses were reported. These findings were put into perspective in the 10th European Conference on Animal Blood Groups in a wide-ranging review of the relevance of biochemical polymorphisms to genetic improvement (Robertson, 1966c). The big advantage of protein pclymorphisms was that many more genotypes were ascertainable and these seemed to show an excess of heterozygosity. The extent of this excess was, however, shown to be no more than might be expected from the limited sample of parents in most domestic animal populations (Robertson, 1965b).

Interest in the variability of populations continued to be fuelled by new discoveries in molecular biology and revelations about the architecture of the genome. Alan Robertson maintained a detailed interest in these developments and their evolutionary significance. He also intepreted the new findings and their relevance to animal breeding for many less up to date in their assimilation of the new knowledge (e.g. Robertson, 1982b, 1984, 1986a). Here we find perceptive questions relating not to how genetic transformation might be effected but how we might choose genes for transfer, comments about the uniqueness of each transgenic animal and the problems of analysis of all the effects of such an unique event.

Application of genetic ideas to animal improvement

For the applied research worker concentration on one species of farm animal has advantages in giving a depth of understanding and knowledge of feasibility which is not possible if the effort is more widely spread. Dairy cattle breeders should rate themselves fortunate that Alan Robertson concentrated on their animal for so much of his work. Early reviews of previous experiments on inbreeding and crossbreeding (Robertson, 1949a,b) revealed many shortcomings in experimental design and set standards for future work. General methodologies for the analysis of field data were worked out at an early stage, and some special techniques elaborated to deal with some special problems such as all-or-none traits (Robertson and Lerner, 1949). Statistical techniques gave rapid dividends and some notable landmarks in the accumulation of new knowledge on dairy cattle improvement will be noted as headings to this section of the review.

Progeny testing

Although progeny testing had generally been 'seen to be a good thing', it had not previously come under detailed scrutiny except by Dickerson and Hazel (1944). An important contribution of Alan Robertson in this field was to show how direct estimation of breeding values might be obtained by working from first principles (Robertson, 1955a). Up to this time most of the formulae in use had been obtained by the use of Sewall Wright's path coefficients. These led to the same answers but the direct methods were often much simpler and less prone to error in their derivations. With the use of this method calculations of the accuracy of progeny tests with varying numbers of daughters became very straightforward. The practical complication was that with artificial insemination daughters of individual bulls were scattered over many herds, often with not very many individuals in each. Some other countries, such as Denmark, sought to solve the problem by gathering together groups of daughters from particular bulls into special stations in order to make direct comparisons. Although initially such methods seemed attractive in appearing to give higher values of heritability (Robertson and Mason, 1956) it was shown that field records could be used as satisfactorily and at lesser cost. The key to the use of field records was the innovative introduction of the 'contemporary comparison' in which the heifers from a particular bull were compared with animals of the same age lactating in the same herd. By appropriate weighting an average difference could be obtained and at the same time an effective number of daughters calculated for substitution in the progeny-testing formulae. The contemporary comparison method was shown to be superior to other methods of evaluation that had been used up to that time. The idea was a new one and had to be explained in a series of meetings (Johansson and Robertson, 1952; Robertson, 1953b; Robertson, Stewart and Ashton, 1956). The contemporary comparison method was adopted in 1954 by the Milk Marketing Board of England and Wales and for a long period of time remained the standard method of evaluation for dairy bulls. Comparisons of the progeny tests of bulls and those of their sires showed the expected degree of concordance (Robertson, 1960a; Ødegård and Robertson, 1967).

The use of contemporary comparisons ran into difficulties when it could no longer be assumed that the contemporaries were a random sample of the breed. As had been foreseen by Alan Robertson, the extensive use of progeny-tested bulls which increased rapidly, especially when the freezing of semen made it possible to nominate particular bulls, necessitated the introduction of more sophisticated methods.

Artificial insemination (AI) and changing breed structure

Analysis of pedigrees had shown that a breed was not an amorphous whole but showed a definite structure, being dominated by a small number of fashionable herds which supplied the majority of breeding stock to other pedigree breeders. These breeders in their turn sold stock to other breeders and to commercial producers, thus establishing a definite hierarchy. This type of structure was well recognized and a useful innovation was the introduction of a method of describing the degree of concentration in numerical terms (Robertson, 1953a).

The early bulls used in AI came for the most part from fashionable herds in the anticipation that these bulls, coming from herds with high levels of milk production, would create improvements in the herds in which they were used. In the event, analysis of records showed that AI bred heifers were not different in milk yield from their naturally bred contemporaries and only slightly superior in fat percentage of the milk (Robertson and Rendel, 1954). The absence of large genetic differences between bull-breeding herds was confirmed by more direct analysis of progeny records (Robertson and MacArthur, 1955). These initial results might have been considered the end of AI as an improvement method had it not been equally clear that there were large differences between individual bulls coming from bull breeding herds. As the theoretical analysis (Robertson and Rendel, 1950) had shown, the possibilities were there provided the correct testing structures were implemented and discriminating selection of parent stock was made. The testing of many young bulls for progeny-testing purposes was an innovation and Alan Robertson obtained the support of the Agricultural Research Council to set up a pilot scheme in East Anglia and run from Cambridge, which would progeny test four young bulls each year. This prototype ran about for about fifteen years and, while no formal analysis of results has been reported, one estimate of the response obtained was 6% (Robertson, 1969b). The exact response obtained from such a small scheme was, however, irrelevant because it served its purpose in demonstrating the system and was rapidly overtaken by the introduction of large scale progeny testing of young bulls by the Milk Marketing Board in 1961. Although not going to the length of some similar breeding schemes in Europe in that the herd of origin of young bulls still counted for something, a very radical change in breeding structure was introduced. The basic cycle of selecting bull dams and mating them to the very best progeny-tested sires to breed a new crop of young bulls for subsequent testing has formed the basis of many similar schemes around the world. The principles laid down by Alan Robertson in the UK and by Henderson in the USA formed a platform for quite revolutionary changes in dairy-cattle breeding practices. Furthermore, subsequent statistical analyses have shown that, although the annual rate of genetic change with such schemes may be small, it is cumulative and of long term significance to the dairy industry. Only very recently with the introduction of embryo transfer has there been any suggestion that the basic progeny testing structure might once again be changed with advantage.

Some of the complexities of dairy cattle breeding

Theoretical calculation of the improvements to be made by selective breeding have tended to be couched in terms only of milk yield. This simplification is almost guaranteed to raise the immediate ire of all practical breeders who recognize many more elements in the complex process of dairy production. Clearly one important aspect is the useful working life of dairy cows and an early attempt to analyze this was made (Rendel and Robertson, 1950a). Dairy herds as kept in practice are subject to a continuing culling process of animals of all ages making analysis difficult. The problem in describing this process satisfactorily and analyzing any genetic element in it represents a considerable intellectual challenge. A mathematical approach to the problem was described (Robertson, 1966b) and in an analysis of records an attempt made to correlate milk production and longevity (Robertson and Barker, 1966). The heritability of survival was found to be low but significantly different from zero and to be genetically correlated with milk yield. As many subsequent analyses have shown, the connection between milk production and longevity is a favourable one. The extent of variation in longevity that is unconnected with production is more problematical and the general feeling would be that the last word has yet to be said on this difficult topic.

Other aspects of milk production have proved more amenable to analysis. For example, the variation of parameter estimates with the level of herd production was investigated (Mason and Robertson, 1956). In addition investigations of milk composition were also made in response to some national concern that there might be an undetected decline in milk quality (Robertson, Waite and White, 1956). Apart from such variations in output, there has also been concern over the inputs required for milk production. In the field situation these are not readily measured and therefore they were the subject of some theoretical and practical analysis (Mason et al., 1957) leading to suggestions that body size should be included in dairy cow selection criteria. The concern with the ratio of outputs to inputs has clearly continued over the years having been expressed in more recent times in a series of investigations in mice reported elsewhere. The efficiency of milk production remains a key area of investigation for the future.

Use of marker genes

The early interest in the variability of populations led to an opportunity of analyzing records for associations between markers and major genes affecting production. The results from two collaborative investigations were clear. With blood groups (Neimann-Sørensen and Robertson, 1961) it was shown that, while individual associations might be significant, the total amount of the genetic variation in milk yield which was associated with such blood group markers was a very small proportion of the total and thus of no practical consequence. An analysis of transferrin polymorphisms (Jamieson and Robertson, 1967) reached a similar conclusion.

The advent of molecular biology and the possibility of detecting many new restriction-fragment-length-polymorphisms has engendered some new optimism in this field.

Advice – formal and informal

Over the years Alan Robertson has given of his knowledge and insight to many committees and commissions, not all of which are well documented. What are also largely unrecorded are the many instances of advice given to many different students, breeders and breed societies. This help was freely given in a public or confidential manner as appropriate.

Special mention must be made of the Cattle Breeders Club in this connection. The Club had been started by George Odlum and John Hammond as a means of exchanging ideas on cattle breeding. Alan Robertson came to the early meetings with some trepidation as he records, but gradually contributed more and more either as a speaker or as someone who could persuade the right people to take the platform. It was therefore natural that in due course he should succeed to the post of President. His talks to breeders were models of their kind, conveying simple or complex messages with studious avoidance of any unnecessary jargon. At the same time he acknowledges having learnt much from breeders.

Concluding remarks

Among the domestic species, the dairy cow must have seemed one of the most unlikely for the early application of ideas of quantitative genetics. The low reproductive rate, long generation interval and sex limitation of milk production makes the task far from easy. The catalyst for change was the introduction of artificial insemination combined with the insight of a research worker able to see how changed breeding structures could be used to good effect. The fact that artificial insemination features in the improvement programmes for dairy cattle in all developed dairying countries is a tribute to that foresight. Although animal breeding may be entering a new era there seems no doubt that there will continue to be a need for investigators with the kind of devotion and inspiration which Alan Robertson has given to animal breeding over the years.

Chapter 23
Estimation of Genetic Parameters

Karin MEYER

Department of Genetics, University of Edinburgh, West Mains Road, Edinburgh EH9 3JN, Scotland

In animal breeding, knowledge of the genetic properties of the traits we are interested in is the first prerequisite in establishing a selection programme. Unless we are concerned with traits controlled by single or few genes, in which case we are generally more interested in gene frequencies, estimation of genetic parameters is synonymous with the estimation of variance components. In this context, variance comprises not only the variance of an observation for a particular trait and individual but also covariances between traits as well as covariances between individuals for the same or different traits.

Estimation of genetic parameters then involves partitioning of observational components, i.e. phenotypic covariances between relatives, into causal components such as variances due to additive genetic effects, dominance, epistasis and permanent and temporary environmental effects (Falconer, 1981). This utilizes the known degree of relationship between animals and the resulting expectations of covariances between them.

Analysis of variance

Traditionally, phenotypic covariances between relatives have been estimated using analysis of variance (ANOVA) or analogous procedures. In general, these require that individuals can be assigned to groups with the same degree of relationship for all members. Family structures considered most often are, for instance, paternal half-sib groups or parents and their offspring.

Using the ANOVA, the covariance among members of a family or group of relatives is usually determined as the variance component between groups. This involves partitioning the sum of squared observations (SS) due to different sources of variation in the model of analysis, groups of relatives being one of them, and equating the corresponding mean squares, derived as the SS divided by the associated degrees of freedom, to their expectations. The same principle applies for multivariate analyses but considering sums of cross-products between traits instead of SS. For balanced data, the partial SSs are orthogonal and their expected values are simple linear combinations of the variance components between groups so that calculations are straightforward, even for multiple cross-classifications, and estimators are unique.

Data arising in animal genetics are usually not balanced but methods analogous to the ANOVA have been developed for unbalanced data. In particular, Henderson's (1953) method 3 of 'fitting constants' has found extensive use. This approach replaces the SS in the balanced ANOVA by quadratic forms involving least-squares solutions of effects for which variances are to be estimated. Its widespread application was greatly aided by the availability of a 'general' least-squares computer program tailored towards applications commonly arising in animal breeding (Harvey, 1960, 1977).

Consider a mixed linear model for one trait, represented by

$$\mathbf{y} = \mathbf{Xb} + \mathbf{Zu} + \mathbf{e} \tag{23.1}$$

with $\mathbf{y}$, $\mathbf{b}$, $\mathbf{u}$ and $\mathbf{e}$ representing the vectors of observations, fixed effects, random effects and residual errors, respectively, and $\mathbf{X}$ and $\mathbf{Z}$ the corresponding design matrices. Assume all levels of $\mathbf{u}$ pertain to the same source of variation, for example sires, and that $V(\mathbf{u}) = \sigma_u^2 \mathbf{I}$, $V(\mathbf{e}) = \sigma_e^2 \mathbf{I}$ and $\mathrm{Cov}(\mathbf{u}, \mathbf{e}') = 0$. The normal equations for (23.1) are

$$\begin{pmatrix} \mathbf{X'X} & \mathbf{X'Z} \\ \mathbf{Z'X} & \mathbf{Z'Z} \end{pmatrix} \begin{pmatrix} \hat{\mathbf{b}} \\ \hat{\mathbf{u}} \end{pmatrix} = \begin{pmatrix} \mathbf{X'y} \\ \mathbf{Z'y} \end{pmatrix}. \tag{23.2}$$

Absorbing fixed effects reduces (23.2) to

$$\mathbf{Z'MZ\hat{u}} = \mathbf{Z'My} \tag{23.3}$$

with $\mathbf{M} = \mathbf{I} - \mathbf{X'}(\mathbf{X'X})^- \mathbf{X'}$. Method 3 estimates of variance components are then (Searle, 1971)

$$\hat{\sigma}_e^2 = [\mathbf{y'y} - \hat{\mathbf{u}}'\mathbf{Z'y} - \hat{\mathbf{b}}'\mathbf{X'y}]/(N - r(\mathbf{X}) - r(\mathbf{Z}) + 1) \tag{23.4}$$

$$\hat{\sigma}_u^2 = [\hat{\mathbf{u}}'\mathbf{Z'My} - (r(\mathbf{Z}) - 1)\hat{\sigma}_e^2]/\mathrm{tr}(\mathbf{Z'MZ}) \tag{23.5}$$

with $r(\mathbf{X})$ and $r(\mathbf{Z})$ denoting the column rank of $\mathbf{X}$ and $\mathbf{Z}$, respectively, N the number of observations, and tr the trace operator. This ignores any covariances between the levels of $\mathbf{u}$. More generally, $V(\mathbf{u}) = \sigma_u^2 \mathbf{A}$, where, if the levels of $\mathbf{u}$ pertain to animals, $\mathbf{A}$ is the so-called numerator relationship matrix. An extension of method 3 to account for $\mathbf{A}$ not equal to the identity matrix has been considered by Sørensen and Kennedy (1986).

Maximum likelihood

Typically, animal breeding data originate from selection experiments or are field records from livestock improvement schemes and thus are seldom a random sample. ANOVA-type methods of analysis, however, are based on the assumption of random sampling and consequently the resulting estimates are frequently biased by selection. In contrast, maximum likelihood (ML) methods of estimation under certain conditions account for selection. In essence, it is necessary that all information which has contributed to selection decisions is included in the analysis, unless it is totally uncorrelated to the traits(s) analysed. Even if these conditions are only partially fulfilled, ML estimators are often considerably less biased by selection than their ANOVA counterparts (Meyer and Thompson, 1984).

ML estimation is conceptually very simple. It requires assumptions about the distribution of the data. For a given model of analysis, parameters to be estimated and data with a specified distribution, we can calculate the likelihood of particular numeric values of the parameters, i.e. how likely it is that the data have been sampled from a population with these parameter values. This is analogous to probability calculations where we determine the probability of observing a specific set of data for given parameter values, but with 'cause' and 'effect' reversed. ML estimates are then, by definition, the parameter values for which the likelihood is maximized.

In estimating variance components by ML, data are generally assumed to have a multivariate normal distribution. Consider two traits with truncation selection on the first, i.e. whether or not an individual has a record for the second trait depends on the magnitude of the first record. As shown, for instance, by Curnow (1961), ML then exploits two properties of the multivariate normal distribution to obtain unbiased estimates of the covariance between traits and the variance of the second trait.

Firstly, the regression of the second, indirectly selected trait, on the first, directly selected trait is unbiased. Secondly, the conditional variance of the second trait given the first, i.e. the variance about regression, is not affected by selection.

While early work on the use of ML in the analysis of animal breeding data has been stimulated by concern about selection bias, more recent interest has focused on the desirable statistical properties of the estimators and the flexibility of the method. As reviewed by Harville (1977), ML estimators are consistent, asymptotically normal and efficient, i.e. all information available is utilized in an optimal way. Moreover, they are well defined for cases which cannot be accommodated by standard ANOVA models. For example, ML allows both the covariance between parents and offspring and the covariance between sibs to be utilized simultaneously to estimate the additive genetic variance or heritability (Hill and Nicholas, 1974).

Restricted maximum likelihood

A major drawback of ML estimation in a mixed model is that fixed effects are treated as if they were known, i.e. the loss in degrees of freedom due to fitting these effects is ignored. For example, the ML estimator of the phenotypic variance would be the total SS divided by the number of observations, while the unbiased estimator would have the denominator reduced by one to account for the fact that the overall mean has been estimated. If the model of analysis comprises many fixed effects, as is almost invariably the case for animal breeding data, this can yield estimates considerably biased. In particular residual variances can be severely underestimated.

Fortunately, a modified ML procedure, the so-called Restricted Maximum Likelihood (REML) as described by Patterson and Thompson (1971), overcomes this problem by maximizing only the part of the likelihood which is independent of the fixed effects. Conceptually, this is achieved by replacing the data by linear functions thereof, 'error contrasts', with an expectation of zero. These can be viewed as the observations adjusted for generalized least-squares estimates of the fixed effects. For balanced data, REML estimates are equal to those from an ANOVA. Considering the estimation of variance components for an unbalanced one-way classification, Robertson (1962b) derived optimal weights for family means which were effectively those used in REML estimation.

Even more than ML, REML estimation of variance components is computationally highly demanding and this has limited practical applications. However, over the last decade considerable research effort has concentrated on the development of specialized and efficient algorithms. This has been closely linked to advances in the genetic evaluation of animals by Best Linear Unbiased Prediction (BLUP). In particular, the so-called Animal Model (AM) has influenced the use of mixed model methodology in the statistical analysis of animal breeding data considerably.

The AM includes a random effect for the additive genetic merit or breeding value of each animal, both for animals with records and animals which are parents only, incorporating all known relationship information in the analysis. This requires the inverse of the numerator relationship matrix $\mathbf{A}$. Hence it gained practical importance with the availability of procedures to obtain $\mathbf{A}^{-1}$ directly from a list of pedigrees (Henderson, 1976; Quaas 1976) which made the AM computationally feasible for larger data sets. Kennedy et al. (1988) discuss genetic properties of AMs, outlining how the AM can account for changes in genetic means and variances. Thus the AM allows an optimal analysis of data involving multiple generations arising, for instance, from selection experiments (Sørensen and Kennedy, 1986; Kennedy, 1988).

In terms of variance component estimation, the AM has changed thinking from the interpretation of covariances between relatives to a linear model framework where we determine variances directly by fitting corresponding random effects in the model of analysis. Covariances between random effects for relatives are now taken into account by specifying the variance matrix of random effects accordingly. With the AM, the additive genetic variance is estimated as the variance of animals' additive genetic merit instead of, for example, four times the variance between sires or twice the covariance between parents and offspring. Analogously, nonadditive genetic components can be estimated by fitting a corresponding random effect, such as a dominance or maternal genetic effect, for each animal.

Advances in theory, in particular the development of specialized and efficient algorithms, together with an increase in the general level of computing power available have led to progressive use of REML. As for the least-squares analysis of unbalanced data, this has been furthered by the availability of appropriate software. Widely distributed statistical packages like SAS and GENSTAT now provide options for REML analyses. In addition, a number of researchers have developed specialized programs, in animal breeding (e.g. Van Raden and Freeman, 1986; Meyer, 1987a, 1988a) as well as in other genetically orientated branches of biology (Shaw, 1987), which have been made available to the scientific community.

REML algorithms

As emphasized by Harville (1977), ML estimation requires the numerical solution of a constrained nonlinear optimization problem. Standard numerical analysis textbooks (e.g. Gill *et al.*, 1981) classify iterative procedures to determine the optimum of a function, i.e. minimum or maximum, according to the amount of information from derivatives of the function which is utilized.

Method of scoring

Methods which use both first and second derivatives, i.e. geometrically speaking information on slope and curvature of the function, have been found to converge quickest. A special case is Fisher's Method of Scoring which requires expected rather than actual values of the second derivatives which are often easier to calculate. This has been used by Patterson and Thompson (1971) and, for animal breeding applications, in REML algorithms described by Thompson (1973), Schaeffer *et al.*, (1978), Meyer (1983, 1985) and Cue (1986).

REML algorithms are often formulated in terms of Henderson's (1973) mixed model equations (MME). For the example (23.1) considered above, the MME differ from (23.2) only by a term $(\lambda \mathbf{A}^{-1})$ added to $\mathbf{Z}'\mathbf{Z}$, where $\lambda = \sigma_e^2/\sigma_u^2$ is the variance ratio for assumed values of σ_u^2 and σ_e^2. Method of Scoring estimates can then be found iteratively as solutions to

$$\begin{pmatrix} q - \lambda \operatorname{tr}(\mathbf{A}^{-1}\mathbf{C}) + \lambda^2 \operatorname{tr}(\mathbf{A}^{-1}\mathbf{C})^2 & \operatorname{tr}(\mathbf{A}^{-1}\mathbf{C}) - \operatorname{tr}(\mathbf{A}^{-1}\mathbf{C})^2 \\ \operatorname{tr}(\mathbf{A}^{-1}\mathbf{C}) - \operatorname{tr}(\mathbf{A}^{-1}\mathbf{C})^2 & \mathrm{NDF}/\lambda^2 + \operatorname{tr}(\mathbf{A}^{-1}\mathbf{C})^2 \end{pmatrix} \begin{pmatrix} \hat{\sigma}_u^2 \\ \hat{\sigma}_e^2 \end{pmatrix} = \begin{pmatrix} \hat{\mathbf{u}}'\mathbf{A}^{-1}\hat{\mathbf{u}} \\ \hat{\mathbf{e}}'\hat{\mathbf{e}}/\lambda^2 \end{pmatrix}, \tag{23.6}$$

where $\mathrm{NDF} = N - \mathrm{r}(\mathbf{X}) - q$ denotes the degrees of freedom for residual, q the number of levels of $\mathbf{u}$, and $\mathbf{C} = (\mathbf{Z}'\mathbf{M}\mathbf{Z} + \mathbf{A}^{-1})^{-1}$ is the inverse of the coefficient matrix for random effects after absorbing $\mathbf{b}$. The efficient matrix in (23.6) is proportional to the information matrix for σ_u^2 and σ_e^2, i.e. its inverse supplies estimates of their lower bound sampling variances.

Expectation-Maximization (EM) algorithm

Even for simple models, calculation of (expected) second derivatives has proven to be computationally highly demanding if not prohibitive. Hence, to date, most REML applications are based on the so-called Expectation-Maximization (EM) algorithm (Dempster *et al.*, 1977) which, implicitly, requires first derivatives of the likelihood to be evaluated. The resulting estimators then have the form of quadratics in the vector of random effects solutions, obtained by BLUP for the assumed values of variances to be estimated, which are equated to their expectations.

For (23.1), REML estimates of variance components using the EM algorithm can be obtained as

$$\hat{\sigma}_u^2 = \hat{\mathbf{u}}' \mathbf{A}^{-1} \hat{\mathbf{u}} / [q - \lambda \ \text{tr}(\mathbf{A}^{-1}\mathbf{C})] \tag{23.7}$$

$$\hat{\sigma}_e^2 = \hat{\mathbf{e}}' \hat{\mathbf{e}} / [\text{NDF} - \lambda \ \text{tr}(\mathbf{A}^{-1}\mathbf{C})] \tag{23.8}$$

with $\hat{\mathbf{e}} = \mathbf{y} - \mathbf{X}'\hat{\mathbf{b}} - \mathbf{Z}'\hat{\mathbf{u}} = \mathbf{M}(\mathbf{y} - \mathbf{Z}'\hat{\mathbf{u}})$. Alternative forms are (Harville, 1977)

$$\hat{\sigma}_u^2 = \hat{\mathbf{u}}' \mathbf{A}^{-1} \hat{\mathbf{u}} + \sigma_e^2 \ \text{tr}(\mathbf{A}^{-1}\mathbf{C})] / q \tag{23.9}$$

$$\hat{\sigma}_e^2 = [\mathbf{y}'\mathbf{y} - \mathbf{y}'\mathbf{X}\hat{\mathbf{b}} - \mathbf{y}'\mathbf{Z}\hat{\mathbf{u}}] / [N - \text{r}(\mathbf{X})]. \tag{23.10}$$

The EM algorithm has been found to be slow to converge, especially for low heritabilities. However, modifications have been suggested which improve its performance without increasing the computational burden considerably. These included a reparameterization, estimating variances of family means rather than variances between families, as described by Thompson and Meyer (1986) and Harville and Callanan (1988) who referred to it as 'linearization'. For a multivariate analysis, Robinson (1988) used a so-called secant algorithm which approximates the matrix of second derivatives iteratively. Meyer (1986) combined an EM step to estimate the residual components with a Method of Scoring step to estimate covariances for random effects. Schaeffer's (1979) 'common intercept approach' attempted to reduce the number of iterations required by predicting changes in estimates.

Though computationally less demanding than the Method of Scoring the EM algorithm in general requires the direct inverse of a matrix of size equal to the number of levels of all random effects, in each round of iteration. This imposes severe restrictions on the kind of analyses feasible, especially for multivariate analyses. Simplifications have been described for special cases.

As noted by Patterson and Thompson (1971), REML estimators can be expressed in terms of latent roots of matrices in the MME. For a model with one random effect, as in (23.1), Dempster *et al.* (1984) described the use of a singular value decomposition of the MME to estimate variances via the EM algorithm. With this strategy, the major computational burden is the calculation of eigenvalues and eigenvectors. However, this is required only once per analysis, and subsequent iterations to solve (23.7) and (23.8) (or (23.9) and (23.10) are fast. Along the same lines, Smith and Graser (1986) advocated the use of a Householder transformation to reduce the coefficient matrix in the MME to tridiagonal rather than diagonal form, which is computationally less demanding. Thompson and Meyer (1988) extended this approach for a Reduced Animal Model.

For models with two random effects, partitioned matrix results have been utilized to reduce the size of the matrix to be inverted in each round of iteration to the number of levels of one of the random effects (Meyer, 1987b). Another approach has been to employ a nested two-step procedure, combining an EM-step to estimate two variance components, as described above, with a direct search for the maximum of the likelihood function for the third component (Smith and Graser, 1986).

Derivative-free approach

Finally, the minimum or maximum of a function can be located without knowing its derivatives. Methods range from a direct search, based on mere comparisons of function values, to procedures which approximate first and even second derivatives using numerical techniques (Gill *et al.*, 1981). The use of a derivative-free approach for REML estimation of variance components has been considered first by Graser *et al.* (1987) for an AM with animals as the only random effect and a univariate analysis. For this case, the REML log likelihood is

$$\log L = -\tfrac{1}{2}[\text{const} + q \log \sigma_u^2 + \text{NDF} \log \sigma_e^2 + \mathbf{y}'\mathbf{P}\mathbf{y} + \log|\mathbf{B}| + \log|\mathbf{A}|] \tag{23.11}$$

with $\mathbf{P} = \mathbf{S} - \mathbf{SZCZ'S}$ and $\mathbf{B} = \mathbf{C}^{-1}$. They showed that the SS of residuals, $\mathbf{y'Py}$, and the log determinant of the coefficient matrix, $\log |\mathbf{B}|$, can be evaluated simultaneously by augmenting $\mathbf{B}$ by the vector of right hand sides and the total SS, $\mathbf{y'y}$, and absorbing all rows and columns into the latter. In addition, the residual variance can be estimated directly as $\mathbf{y'Py}/\text{NDF}$ so that $\log L$ can be maximized with respect to one parameter only, the variance ratio λ, estimating $\hat{\sigma}_u^2$ subsequently as $\hat{\lambda}\hat{\sigma}_e^2$.

This has been extended to models including additional random effects, such as an environmental effect due to litters or a maternal genetic effect, and to multivariate analyses (Meyer, 1988b, c). Im and Gianola (1988) considered the ML analysis of binomial data. The derivative-free approach is highly flexible, accommodating a wide range of models of analysis, and thus provides a powerful tool, in particular facilitating analyses under an animal model.

Canonical transformation

As discussed above, a reparameterization of the model or a transformation applied to the MME can often reduce computational requirements substantially. For multivariate analyses, a transformation of the data with corresponding effects is available for special cases.

When all traits are recorded for all individuals at the same or strictly corresponding time(s), design matrices $\mathbf{X}$ and $\mathbf{Z}$ are equal for all traits. Consider p traits for a model with one random factor, and let $\mathbf{T}$ and $\mathbf{E}$ (of size $p \times p$) denote the covariance matrices of random effects and residuals, respectively. As outlined by Hayes and Hill (1980), a canonical decomposition of $\mathbf{T}$ and $\mathbf{E}$ then yields a transformation to new traits, so-called canonical variables, which are both genetically and phenotypically uncorrelated.

This reduces the multivariate analysis to a series of corresponding univariate analyses. REML analyses exploiting the canonical transformation have been described for the Method of Scoring (Meyer, 1985), the EM-algorithm with tridiagonalization (e.g. Taylor et al., 1985) or diagonalization (Van Raden and Freeman, 1986; Lin, 1987) of the coefficient matrix, and the derivative-free approach (Meyer, 1988c; Thompson and Juga, 1988b). For the latter algorithm, it also proved useful for models involving more than one random effect by reducing the number of nonzero off-diagonal elements in the coefficient matrix and thus the computational effort to evaluate $\log L$.

A special algorithm for diagonal $\mathbf{E}$, i.e. zero residual covariances, has been described by Schaeffer et al. (1978). This is the case when different traits are measured on different sets of animals, for example growth rate of male and female calves in beef cattle. In special cases, this constellation can also be achieved through a transformation of the data. For instance, when traits are recorded sequentially, i.e. there are missing records but each animal with trait t has all records 1 to t, the inverse of the Cholesky decomposition of $\mathbf{E}$ provides a transformation to traits with uncorrelated residuals (Schaeffer, 1986a).

Discussion

This review has concentrated on the analysis of animal breeding data by REML. A variety of other methods of variance component estimation, related to those presented has been omitted in this paper. These include noniterative procedures, like minimum variance (MIVQUE) and norm (MINQUE) quadratic unbiased estimation (Rao, 1972), and Bayesian approaches to the estimation of genetic parameters (Gianola and Fernando, 1986). Various approximations to REML which are computationally simpler have been suggested, for instance 'a simple method' (Henderson, 1980), 'pseudo REML' (Schaeffer, 1986b), 'the tilde-hat' approach (Van Raden and Jung, 1988) and 'REML' approximating the inverse of the coefficient matrix (Wright et al., 1987). However, some of these have been shown to be biased by selection of the data when REML is not (Ouweltjes et al., 1988). Special procedures for noncontinuous data have not been discussed here. Categorical traits are considered by McGuirk in this volume.

Methods of estimation of genetic parameters in animal breeding have changed markedly over the last decade. Increasing use of mixed model methodology in genetic evaluation (BLUP) has been followed by progressive use of procedures to estimate variance components, which fit into the same framework. Models of analysis have become more detailed and appropriate. The animal model especially has helped to change thinking to a linear model concept where we directly estimate variances due to random effects fitted. Because of its statistical properties, REML is now considered the optimal procedure for a wide range of analyses. It is computationally demanding but the development of specialized algorithms, together with a continuing increase in the general level of computing power available, are making applications easier and easier.

Future advances are expected to improve the situation even further, in particular with respect to multivariate analyses, the model which can be fitted and the size of data set which can be analysed, so that, eventually, REML will become the standard method for estimating genetic parameters. When all is said and done, however, we have to bear in mind that – even with the best method of analysis – our estimates of genetic parameters are only as good as the data on which they are based. In particular, we need to ensure that we have sufficient information on all effects we want to fit and parameters we want to estimate in order to obtain meaningful estimates. Even then sampling correlations between parameter estimates may be large. This is becoming all the more important as interest increases in the estimation of nonadditive genetic effects, for example dominance or cytoplasmic effects or epistasis, and as modern algorithms, such as the derivative-free approach, appropriate software and fast computers make it easy to carry out corresponding analyses.

Chapter 24
Design of Experiments to Estimate Genetic Parameters Within Populations

Robin THOMPSON

AFRC Institute of Animal Physiology and Genetics Research, Edinburgh Research Station, West Mains Road, Edinburgh EH9 3JQ, Scotland

The genetic parameters considered will be heritabilities, additive genetic variances, covariances and correlations. These parameters are of interest:
(i)   as a quantitative summary of inheritance in traits,
(ii) in deciding if there is enough genetic variation in a trait to make selection effective,
(iii) in considering alternative selection schemes such as individual or progeny testing,
(iv) in optimization of a selection scheme by choice of population structure and incorporation of information from relatives and other traits using selection indices or best linear unbiased prediction (BLUP).

Poor parameter estimates lead to imprecise prediction of progress. Indices using family information are robust to errors in heritability in that little progress is lost by using incorrect heritabilities. Indices combining information from several traits are less robust (Sales and Hill, 1976a,b).

In some areas of livestock breeding, most parameter estimates come from data not specifically collected for this purpose. In dairy cattle, for example, most estimates are derived from records collected mainly for management reasons. The number of animals involved in field data can be many more than the number of animals in designed experiments, but depending on the structure of the population and nature of measurements taken, field data may not be able to give relevant or precise estimates.

It is important to define the population of interest, the parameters of interest and the use made of the parameters in order to make a rational choice of experimental design. In some cases there has to be a compromise between getting a precise answer to a specific question and getting less precise answers to more general questions. For example if one wishes to know if selection on trait X or on Y makes more change in trait Z, then selection on the difference, X–Y, should give a precise answer. But this design is inappropriate if one wishes to know how quickly Z will be changed by selection on X (or Y).

This chapter reviews some of the designs used to estimate genetic parameters in farm livestock, including:
(A) designs based on collateral relatives (one generation),
(B) designs based on noncollateral relatives (between two generations),
(C) a combination of (A) and (B),
(D) longer experiments for (S) single traits and (M) multiple traits.

Some of the early work, especially on A and B, was carried out by Alan Robertson.

Designs based on collateral relatives for a single trait (AS)

If there are data on a single trait from s half-sib families of size n, then an obvious parameter of interest is the heritability, h^2, the ratio of additive variance to phenotypic variance. An analysis of variance can easily be constructed to give mean squares between sires, B, and within sires, W, the heritability is estimated as $[4(B-W)/n]/[W+4(B-W)/n]$ and its variance is

$$32[1+(n-1)h^2/4]^2(1-h^2/4)^2/[n(n-1)(s-1)].$$

For a given total number of measurements, sn, the variance is minimized if $n = 4/h^2 + 1$ or approximately $4/h^2$ as given by Robertson (1959b).

In a sense there is a compromise between knowing W more precisely, its precision increasing as n increases, and knowing B more precisely, its precision decreasing as n increases. The optimal size depends on h^2, the parameter being estimated. The design is robust in that discrepancies of n from the optimum have little effect on the efficiency, although small families should be avoided especially at low heritabilities, when the optimal size can be large (e.g. $n = 80$ for $h^2 = 0.05$) (Robertson, 1959b). Robertson also considered hierarchical structures with full- and half-sibs. For h^2 estimated from half-sibs, then $4/h^2$ dams per sire and one progeny per dam is optimal. To give equal information on both full- and half-sib correlations he suggested using $2/t$ progeny per dam and 3 or 4 dams per sire, where t is the half-sib correlation.

Designs from collateral relatives for multiple traits (AM)

If several traits are measured there may be interest in the heritabilities of all traits and of genetic correlations. In the former case Taylor (1976a) suggested that a minimax criterion might be appropriate to give a method for combining information on several different parameters. One chooses a design that minimizes the maximum possible variance. This is often thought of as a pessimistic criterion in that the worst case that is minimized might have a small probability of occurring. The family size is chosen to minimize the maximum value of the variance of heritability $V(h^2)$. In the half-sib case the maximum variance is when $h^2 = 2(n-2)/(n-1)$ and $V(h^2)$ is minimized when $n = 4$, which leads to smaller family sizes and less efficient designs than suggested in the univariate designs above, especially if the heritabilities are low. Taylor (1976b) also considered hierarchical designs.

Robertson (1959c) and Tallis (1959) discussed designs for estimating genetic correlations. The formula for the variance of genetic correlations is complicated (Tallis, 1959) and Robertson, by considering two traits with equal heritabilities and the sum and difference of these traits, was able to give a perceptive alternative derivation of the variance for this special case. He suggested that the optimal size is $4(1+r^2)^{1/2}/[h^2(1-r^2)]$, where r is the genetic correlation. More generally Tallis (1959) has tabulated designs for a range of values of heritabilities and genetic and phenotypic correlations.

When estimating genetic correlations it can be useful to use a canonical transformation, essentially working with linear combinations of the original traits that are uncorrelated both genetically and phenotypically (Thompson, 1977; Meyer, 1986). Transforming Tallis' parameter sets to uncorrelated traits suggested that the family sizes are approximately related to the heritabilities of these uncorrelated traits, which with low heritabilities imply large optimal family sizes as in the univariate case.

Minimax designs for genetic correlations can be derived from results given by Taylor and Hnizdo (1987). As in the univariate case the family sizes are small.

Designs based on non-collateral relatives (between two generations) (BS) for single traits

In these designs x will be used as a mid- or single-parent value with variance σ_x^2, and y, a possibly different trait measured on full- or half-sib offspring, with variance σ_y^2. The regression b, of y on x, can be related to heritability, if x and y are the same trait, and to genetic regressions, if not. Let r and t be parent–offspring and sib correlations respectively. If s families of size n are measured then

$$V(b) = [(t-r^2) + (1-t)/n]\sigma_y^2/s\sigma_x^2. \tag{24.1}$$

Increasing n reduces both the variances of the family mean about its predicted value, included in the numerator of (24.1), and the sum of squares of parental values, the denominator of (24.1).

Hence the optimal value of n strikes a balance between these two conflicting terms. Latter and Robertson (1960) showed that if the cost of recording parents is k per family and the rearing cost of a family is c relative to recording an offspring, then the total cost is proportional to $(k + c + n)s$. Hence the optimal family size is $n = [(k + c)(1-t)/(t-r^2)]^{1/2}$. Optimal family sizes are generally larger for the sib methods than the regression method. Sib methods tend to be more efficient than regression methods at low heritabilities and less efficient at high heritabilities. Falconer (1960a) gives an example when sib covariance and regression methods give equal variances for a heritability of 0.25.

If the regression of offspring on parent is linear (as is commonly assumed) then $V(b)$ can be reduced by up to a half by assortative mating of parents (Hill, 1970). $V(b)$ can also be reduced by only raising offspring from the extreme animals, and mating between animals within the high and low groups.

Hill and Thompson (1977) considered the case when the ratio, R, of the number of animals measured in the two generations, was fixed and Hill (1970) considered the case when R could be optimized. Hill and Thompson (1977) gave a graph to find the optimal proportion of animals to use as parents. The optimal proportion is low, usually less than 20%, and the optimal family size larger than when there is no selection of parents. $V(b)$ can be halved compared to when there is no selection of parents. Estimates of heritability from regression on sire with selection are more efficient than half-sib covariance estimates for all values of heritability (Hill, 1988).

Combination of single trait designs (CS)

Two types of relationships have been used above to estimate genetic parameters, sib covariances and parent–offspring covariances. Sometimes designs generate both types of relationship and it is appropriate to combine the estimates. For example, if several offspring are raised from each sire, half-sib covariance and sire–offspring covariances could be combined. These estimates are correlated (Hill and Nicholas, 1974) and weighted estimates can be derived. However sib covariance estimates are biased with selection of parents (for example Robertson, 1977c). Maximum likelihood (ML) techniques can be used to correct for this bias and give efficient estimates. These can be based on a multivariate analysis of variance (Thompson, 1976a).

Optimal designs are given by Hill and Nicholas (1974) for no selection of parents and by Thompson (1977) for assortative mating and selection of parents in a variety of hierarchical mating structures with measurements on parents and offspring. The general conclusions are that when using half-sibs and no selection of parents, ML gives the most gain, about 20%, in combining estimates at intermediate heritabilities ($h^2 = 0.2$). At high values of h^2 most of the information comes from offspring–parent regression and at low values most of the information comes from sib covariances, and in both these cases the gain from using ML is of the order of 5%.

With selection of parents, ML is more useful at low heritabilities, giving gains of 33% at $h^2 = 0.05$ compared to using parent–offspring regression alone. At high values of heritability ($h^2 = 0.4$ to 0.6) there is very little gain from using ML. The optimal family sizes are again of the order of $4/h^2$ for low h^2 and the designs are reasonably robust against not knowing the heritability value.

For practical experimental reasons it is natural to try to use equal numbers of offspring from each animal selected and symmetry arguments suggest that equal sizes are optimal when only within and between family sib information is being used. In more complicated hierarchical cases the balanced designs are not always optimal (Anderson, 1981). However when parents are measured it may be better to relate the numbler of offspring to the phenotypic (or genetic) worth of the parent. In a sense, selection of extreme animals is a crude form of this.

The minimization of the sampling variance of the regression estimate suggests raising n_i offspring from a parent with value x_i about a mean of 0, with n_i proportional to $|x_i|/v_i$ where v_i is the variance of the offspring mean about the regression line. Then if $r = h^2/2$ and $t = h^2/4$,

$$n_i = [k|x_i| - (1 - h^2/4)]/[h^2(1 - h^2)/4]$$

where $k = [nh^2(1 - h^2)/4 + (1 - h^2/4)]\bar{x}$, $\bar{n}$ is the mean number of offspring and $\bar{x}$ is the mean value of the $s/2$ selected high parents. Hence the optimal number is linearly related to the parental value. In practice, the formula for family size will need iterative refinement of s to ensure n_i is positive for all parents included in $\bar{x}$. A numerical example of 30 animals with 120 progeny in all was examined with $h^2 = 0.4$ using expected order statistics to give parental values. The optimal value of s was found to be 12, and 23, 15, 10, 7, 4 and 1 offspring should be taken from the top six and bottom six ranking animals. This gives a reduction of the order of 5% in the sampling variances from the optimal balanced design of keeping 20 progeny from the three best and worst parents. There is little advantage in raising a variable number of offspring per family, given that h^2 is needed to determine n_i. Hill (1988) found similar numbers and came to the same conclusion from a simulation exercise.

Designs from noncollateral relatives for multiple traits (BM)

There has been less interest in developing optimal designs for multivariate parameters from parent and offspring data, probably because of the difficulty of reducing the $p(p + 1)$ genetic and phenotypic parameters in p traits to one relevant design criterion. The relevance of genetic correlation as an important parameter, and hence designing experiments to minimize its variance, is not immediately obvious. In some cases it is reasonable to assume that the phenotypic variances are known, certainly they are of an order of magnitude easier to estimate than genetic variances. It seems reasonable to concentrate on the genetic parameters of the standardized traits (traits with unit phenotypic variance) and a natural criterion is the generalized variance of these parameters (the D-optimal criterion, Silvey, 1980). This criterion is invariant to the transformation of the standardized traits.

A common method of estimating multivariate parameters is to adapt the idea of selection of parents in the univariate case. If two traits are of interest, then two experiments are carried out in parallel. In one, extreme animals for X are selected and offspring measured on X and Y and the differences between the two groups of offspring are compared with their parental difference. The experiment is similar with parents selected on Y. Cameron and Thompson (1986) showed how to use individual parental values rather than mean parental values and to predict offspring values using both traits measured on parents. They investigated the use of linear indices to select animals as parents and found that although the most efficient design depended on the genetic parameters, a simple design was based on deriving phenotypically uncorrelated traits, L_1 and L_2, and selecting parents on the basis of these two indices.

An alternative to carrying out essentially two independent selection experiments is to use one scheme selecting extreme animals on a quadratic index $L_1^2 + L_2^2$, which select parents as far from the mean as possible. This scheme can reduce the variance of genetic parameters by a factor of approximately two, compared to classical designs for uncorrelated traits, and with bigger gains for correlated traits. Again the designs are fairly robust against not knowing the genetic parameters a priori.

Cameron and Thompson (1986) consider the case when parents were selected in only one sex, but there will be further gains from selecting the parents of the other sex and mating the animals in some assortative or 'minimizing distance' manner.

These analyses are based on traits for which there is equal interest, cost of recording and measurement. In some cases one trait is relatively cheap to measure, e.g. milk yield, and other traits are more expensive or difficult to measure e.g. blood metabolites or carcass measurements on live animals, and Hill (1988) has compared alternative designs.

Designs for more than two generations for single traits (DS)

Sometimes a control group is kept and parents are randomly selected to produce animals in the next generation. Little work has been done on the efficiency of parameter estimates from such designs, but Table 24.1 gives results for some simplified designs (K.D. Atkins and R. Thompson, unpublished). In these designs there are s individuals, each individual has n offspring and one of these offspring is chosen at random to have n offspring, the process being repeated for g generations to gives $(gn + 1)$ recorded animals. In practice there are other branches in the pedigree and many links providing information are ignored. Female pedigrees in species such as pigs or sheep are modelled by small values of n, larger n is more appropriate to male pedigrees. Table 24.1 shows asymptotic variances of ML estimates of h^2 and an estimate derived from parent–offspring covariances and phenotypic variances. These results show the gain from increasing generations in reducing the variance of h^2 per observation, especially at small n. In one sense there are s families of size $gn + 1$, but unequal covariances between members of the family. For example there are $gn(n + 1)/2$ half-sibs and gn offspring–parent pairs. Qualitatively these results mimic Robertson's results for one generation. The parent–offspring information extracts a large proportion of the information, especially for heritabilities near 0.5. When $g = 1$ the information on heritability comes from the

Table 24.1
Asymptotic variance per observation for maximum likelihood (ML) and regression (Reg) estimates of heritability, for various heritabilities (h^2), family sizes (n) and generations (g)

h^2		0.05		0.10		0.4		0.6	
g	n	ML	Reg	ML	Reg	ML	Reg	ML	Reg
1	2	5.50	6.07	5.64	6.12	5.89	6.12	5.43	5.82
1	3	4.53	5.46	4.70	5.65	5.53	5.76	5.19	5.49
1	5	3.60	5.02	3.98	5.22	5.49	5.76	5.27	5.52
1	9	2.76	4.86	3.32	5.23	6.01	6.40	5.94	6.18
1	17	2.10	5.04	2.88	5.75	7.48	8.13	7.63	7.92
2	2	4.19	5.11	4.46	5.22	5.26	5.41	4.92	5.13
2	3	3.63	4.83	3.98	4.98	5.21	5.37	4.91	5.10
2	5	3.02	4.64	3.51	4.40	5.44	5.65	5.18	5.40
2	9	2.43	4.68	3.10	5.10	6.20	6.56	5.98	6.32
2	17	1.94	4.97	2.86	5.75	7.88	8.61	7.76	8.38
4	2	3.57	4.63	3.91	4.75	4.95	5.07	4.64	4.83
4	3	3.14	4.52	3.62	4.68	5.06	5.20	4.75	4.97
4	5	2.74	4.48	3.31	4.74	5.44	5.65	5.12	5.43
4	9	2.28	4.59	3.03	5.04	6.32	6.73	5.99	6.51
4	17	1.88	4.93	2.89	5.77	8.09	9.01	7.81	8.79

regression of offspring on parent and the deviation about the regression line which has a variance function of $h^2(1-h^2)$ (Curnow 1961; Robertson, 1977c). As $h^2(1-h^2)$ is relatively flat over the heritability range 0.4 to 0.6, it will be hard to make precise inferences about h^2 from this type of information.

The case where several generations can be measured and parents can be selected has been discussed thoroughly by Hill (1971b). These designs are in a sense a natural extension of one round of selection of parents, and there can be substantial reductions in variance by using several rounds of selection.

Discussion

In this brief review, I have only highlighted the basic points in relatively simple cases with simple models, for example solely additive trait. Overall it has been shown that careful study of experimental design is well repaid.

It is hoped that these points are useful in more complicated cases. Designs for estimating maternal genetic parameters using methods similar to AS and CS are given by Eisen (1967) and Thompson (1976b). These are essentially covariance methods and I wonder if an adaptation of the designs of section BM might be more useful. In this context response is a function of several genetic parameters and of selection differentials for direct and maternal components (Mueller and James, 1985). Paralleling the multivariate designs above one might select parents on the basis of these different components and not just on the 'direct' component as usual.

One important area where design has not often been considered is selection experiments with overlapping generations, but possible analyses are becoming clearer by showing how to construct more appropriate cumulative selection differentials (e.g. James, 1986; Thompson and Juga, 1989a). In principle ML analysis can be carried out to provide estimates, but it can be difficult to know exactly where all the information comes from, for example differences between selection lines, half-sib, full-sib and offspring covariances, how efficient are simple estimators, and hence how a design might be improved.

There is also a need to know if BLUP methods for predicting breeding values can be used with advantage in selection experiments. BLUP should lead to greater response which should help to reduce sampling variances of estimates, but BLUP requires more assumptions about parameters and might lead to an increase of genetic drift.

Acknowledgements

My initial interest in becoming an agricultural statistician came about from reading about Alan Robertson's life in the *Farmer and Stockbreeder* when I was fourteen. For this and for his encouragement throughout my career I am very grateful.

Chapter 25
The Estimation of Genetic Parameters for All-or-none and Categorical Traits

B.J. McGUIRK

Premier Breeders, Vallum Farm, Stamfordham, Northumberland, NE18 0LL, England

While many of the characters of interest to animal breeders are normally distributed, or almost so, there are other economically important traits which show a discrete distribution, sometimes with as few as two classes. Many characters associated with reproduction and survival are of the all-or-none variety. Where more than two discrete classes are used, this may again reflect the true biological nature of the character, for example, litter size in sheep, or it may be for reasons of convenience and simplicity, as for type characteristics, or to indicate degrees of severity to some disease or other condition. For all such characters which are not normally distributed, it is important to have available appropriate statistical procedures for both the estimation of genetic parameters and the prediction of breeding values. This chapter will attempt a brief review of procedures to estimate the heritability of characters where there are only two possible classes, all-or-none traits, and to indicate their extension for characters where more than two classes are possible.

The theory and practice of genetic analysis of all-or-none or binomial traits in an animal breeding context was effectively begun in three related papers which examined disease and mortality data in poultry (Lush et al., 1948; Robertson and Lerner, 1949; and Dempster and Lerner, 1950, in which Alan Robertson also contributed an Appendix). The essential difficulty posed by all-or-none traits which these papers both recognized and tried to accommodate is that means and variances are correlated, thus violating one of the assumptions of the analysis of variance. Subsequent research has been directed to the development of appropriate and efficient forms of analysis, which do not require prior transformation of the data.

Half-sib estimates of heritability

Robertson made two contributions to this history. With Lerner (Robertson and Lerner, 1949), he proposed that in the analysis of the raw all-or-none data as illustrated by Lush et al. (1948), the heritability could be estimated from a $2 \times N$ contingency table of surviving and dead progeny, where N is the number of sire groups. Thus, in a half-sib analysis, heritability (h^2) can be calculated as

$$h^2 = 4[\text{Chi-square} - (N-1)]/n \qquad (25.1)$$

where n is the weighted mean number of progeny per sire. In the situation where there are no fixed effects, the expected result from this analysis is the same as from the analysis of variance. The Chi-square procedure suggested by Robertson and Lerner has not been used extensively in animal breeding studies (but see Forrest and Bichard, 1974), perhaps because of the computational convenience of the analysis of variance, especially where fixed effects are involved.

Robertson's second and lasting contribution to this topic concerned the nature of the heritability estimate obtained from a conventional analysis of variance on untransformed data. It had already been recognized that while the character as observed may present itself in an all-or-none manner, there was an underlying genetic predisposition to the character which might be assumed to be normally distributed. The notion of an underlying liability or predisposition had been used previously by Wright (1934b) in a genetic context, and was also the basis for the probit transformation. The liability model assumes that all individuals with a liability which exceeds some threshold will exhibit the condition, otherwise they will not. Robertson and Lerner (1949) pointed out that if the threshold model was appropriate, then the heritability estimate (h^2) for an all-or-none trait will be related to the incidence of the condition such that

$$h^2 = h_L^2 z^2 / [p(1-p)], \tag{25.2}$$

where the term h_L^2 is the heritability on the underlying scale, p is the incidence of the condition, and z is the height of the ordinate at the threshold corresponding to the proportion p. In this formula, the denominator is simply the formula for the variance of a binomially distributed character, and reflects the reduction in phenotypic variation in changing from a normally distributed trait with variance of unity to the binomial state. Correspondingly, the factor z^2 represents the approximate reduction in the additive genetic variance in moving to the binomial scale of measurement. It is approximate because it represents a first order or linear approximation to the nonlinear relationship between liability and incidence. Robertson presented a more formal derivation of (25.2) in his Appendix to Dempster and Lerner (1950).

There are two consequences of the relationship described by (25.2). Firstly, the heritability on the observed scale will always be less than on the underlying scale, by the ratio $z^2/p(1-p)$. Secondly, for any given heritability on the underlying scale, heritability on the binomial scale will be symmetrically distributed with a maximum value at an incidence of 50 %. Thus for $p = 0.1$ or 0.9, $h^2/h_L^2 = 0.342$, and for $p = 0.2$, 0.3, 0.4 and 0.5, $h^2/h_L^2 = 0.489$, 0.579, 0.622 and 0.636, respectively.

In their study on mortality in poultry, Robertson and Lerner (1949) specified that these relationships would apply only where there was a single cause of the mortality, and where the notion of a normally distributed underlying liability to this specific disease was appropriate. Despite this warning, it is common to find the heritability on the observed scale transformed to the underlying scale, even if in disease studies the aetiology of the condition is unknown or where there might well be more than one cause of death. There are a number of obvious advantages in working on the underlying scale. For example, when comparing heritability estimates obtained in different populations, we will usually want to know if any differences observed truly reflect unequal scope for improvement, or if they are simply a consequence of different incidences in the populations. This distinction can be made most satisfactorily by comparing estimates on the underlying scale. Further, heritability on the underlying scale provides a more reliable basis for predicting or explaining responses to selection for an all-or-none trait, where the predisposing environmental causes of the condition may vary over time or in different environments.

Dempster and Lerner (1950) were the first to attempt to evaluate empirically the relationship described in (25.2) by using normally distributed data on egg production and recreating all-or-none data sets with different incidences from these by altering the threshold value. At each incidence level, a heritability was calculated from the binomial data, scaled according to (25.2), and compared with the heritability estimate for egg production. There was good agreement between the two sets of estimates over the range of incidences examined.

Essentially the same procedure has been employed in a number of computer simulation studies. In the two best known of these, Van Vleck (1972) and Ollausson and Ronningen (1975) again reported good agreement between the two sets of half-sib heritability estimates, but with a tendency for the scaled all-or-none estimates to be too high at the more extreme incidences studied. In a recent simulation study, Mercer and Hill (1984) examined more extreme incidences and found the discrepancy between the two sets of estimates was more marked. Again the heritability estimates which had been scaled up from the binomial to the underlying scale were too high, especially as the true underlying heritability increased.

We have recently completed a similar computer simulation study but with more replication than in the earlier studies (B.J. McGuirk and R. Thompson, unpublished). A variety of family sizes were used in half-sib analyses, and while only the results for family sizes of 100 will be shown here, the results obtained were quite general and agree with those reported by Mercer and Hill (1984). At each of the three heritability values studied (Table 25.1), the scaled binomial heritability estimates were higher than the true estimates on the underlying scale, especially at extreme incidences. The degree of discrepancy also increased as the true heritability increased. When the two components of heritability, the additive genetic and phenotypic variance, were examined separately, the phenotypic variance (scaled by $p(1-p)$ was close to its true value, while the additive genetic variance (scaled by z^2) followed the same pattern as the heritability. These results are not surprising, and indicate that the assumption of a linear relationship between the genetic variances on the underlying normal and all-or-none scales is less satisfactory at the extremities of the normal distribution, and these discrepancies are magnified as the true underlying genetic variance increases.

Table 25.1

Simulation results showing the relationship between the estimate of heritability scaled from that on the observed data and the true underlying heritability

Underlying heritability	Incidence (%)						Average SE
	50	20	10	5	2.5	1	
	Mean estimate of heritability						
0.10	0.100	0.097	0.097	0.100	0.101	0.115	0.006
0.25	0.247	0.251	0.258	0.267[1]	0.274[1]	0.304[1]	0.007
0.40	0.393	0.410	0.428[1]	0.451[1]	0.474[1]	0.507[1]	0.012

[1]Estimate differs significantly from true underlying value ($P < 0.05$)

The proportion of replicates in our study which yielded negative estimates of heritability was also interesting. The probability of obtaining negative heritability estimates is known to increase as either the true heritability or family size decrease. For any heritability on the underlying scale, the heritability estimate from an analysis on the observed sale is always expected to be lower, declining as the incidence becomes more extreme, and the probability of obtaining a negative heritability estimate will increase accordingly. For example, with a family size of 10 and a true (underlying) heritability of 0.25, the frequency of negative heritability estimates increased from 1.5% at an incidence of 50%, to 21.3% at an incidence of 10%, and to 55.9% at an incidence of 1%. Even after allowing for the heritability estimates expected at these incidences on the all-or-none scale, these frequencies are higher than predicted by Gill and Jensen (1968) for normally distributed data. For those populations with a low incidence of the all-or-none trait, it appears that a negative heritability estimate is produced unless there is one family with at least two affected offspring. For this reason, the incidence of negative estimates is particularly high when family sizes are small and incidences extreme.

Proband analyses

This method of analysis was introduced by Falconer (1965) and enables the heritability of liability to be estimated directly from the data on the assumption that there is an underlying normally distributed liability to the condition. The method requires information on the frequency in the general population, the frequency of affected individuals, or probands, and the corresponding frequencies among their relatives. These incidences are then transformed into their underlying liabilities. The difference between the mean liability of the affected individuals and of the general population is equivalent to a selection differential, while the corresponding liabilities in their relatives can be considered as the response. Heritability can then be calculated from the ratio of response to the selection differential applied, allowing for the degree of genetic resemblance between probands and their relatives.

Described in this way, this form of analysis is easier to visualize for an analysis which is overtly in the form of a regression, as in the relation of performance of offspring to that of their parents, but it is quite general. For example, within the context of a half-sib structure, one sib can be regarded as the propositus, while the siblings are the relatives. For such an analysis, it is possible to calculate a 'proband' half-sib heritability directly from a conventional analysis of variance of 0/1 data (R. Thompson, unpublished). In the form described initially by Falconer (1965), such an analysis would be expected to give the same estimate of the heritability on the underlying scale as would be obtained by scaling the binomial heritability obtained in a conventional analysis of variance of all-or-none data. If, however, allowance is made for the difference in variance among the population of propositi and the population of relatives (Mendell and Elston, 1971), then the proband estimates should be closer to the true estimates, as Mercer and Hill (1984) found in their simulation study.

While proband analyses have been used quite extensively in human genetics, they have been less commonly employed in animal breeding studies, perhaps because it is not clear how to accommodate fixed effects, due to years, herds, etc., with accompanying differences in incidences between levels.

Offspring–parent and mixed model analyses

Until recently, little attention had been given to offspring–parent estimation procedures for all-or-none traits, other than by Falconer (1965) within the general framework of proband analyses. Van Vleck (1972) also considered offspring–parent correlation analyses in his simulation study, using the same incidence in both generations. He found poorer agreement between underlying and scaled binomial heritabilities in the offspring–parent analyses than in the comparable half-sib estimates.

The more recent interest in offspring–parent analyses was prompted by data sets in which incidences in the parent and offspring generations were known to differ. This situation can arise when environmental factors have an important role in influencing an animal's liability to the condition, and these factors can fluctuate over time. Where such differences in incidence exist, the notion of a common population incidence is clearly inadequate.

James and McGuirk (1982) showed that where parental and offspring incidence differ, both influence the heritability for the binary trait, with the incidence in the offspring generation having the larger effect. If the incidence in the parental generation is p, then the observed heritability is equal to the heritability of liability multiplied by $zz_0/p(1-p)$. Here z refers to the ordinate corresponding to the incidence p in the parental generation, while z_0 refers to the ordinate corresponding to the offspring incidence p_0. Parental incidence has less effect on the heritability on the observed scale, as it affects both numerator and denominator. Where offspring and parent incidences are equal, the expression described above reduces to that for half-sib analyses (25.2). As with the formula describing the relationship between half-sib estimates for liability and on the observed scale, the above expression for offspring–parent estimate is again approximate, and James and McGuirk (1982) suggested ways of improving it.

Thompson *et al.* (1985) subsequently presented a method for directly estimating the heritability of liability from incidences in the offspring and parent generations. This is done within the more general context of Generalized Linear Models (GLIM) analyses, which enable the simultaneous estimation of fixed effects. Within each level of a fixed effect, both the total number of progeny and the number of affected progeny are recorded for both affected and unaffected parents, together with the mean liability for both groups of parents, as the independent variate in the regression analysis. It is usual to assume that the underlying distribution of liability is either normal (probit) or logistic in form, and a 'link' function then relates changes on this underlying scale to changes in the observed incidence.

This method of analysis is essentially an extension of proband analysis, and like proband analysis, can also be seen as a single generation selection experiment. Although it contains some approximations, this analysis is particularly useful in that it can be extended to cover categorical traits, where more than one threshold is involved. It also has other applications, including the estimation of repeatability on the underlying scale from the regresssion of subsequent on early performance, and could be used to estimate genetic regressions involving combinations of binary or categorical traits, or either of these with normally distributed traits.

Similar extensions are needed to analysis of variance procedures for estimating heritability, both to account for fixed effects and the existence of more than one threshold. In addition to specifying random and fixed effects, it is also necessary to specify a link function, as with the offspring–parent analyses. Where more than two scores are involved, these may represent an ordered categorical trait, or one of the extremal type, where the states are exclusive. Gilmour *et al.* (1985) or Thompson (1988) should be consulted for a discussion of the statistical methods and computing strategies suggested for such analyses.

In recent years there has been considerable interest in improved procedures for predicting breeding values for categorical traits (see for example Gianola and Foulley, 1983; Meijering and Gianola, 1985; Foulley *et al.*, 1987). This topic is of special relevance in dairy breeding programmes, where the bulls being ranked are represented unequally by daughters in many different herds. A variety of nonlinear statistical procedures have been used on both real and simulated data sets, and the predicted breeding values of the bulls then compared by rank or product moment correlation with prediction methods that assume the data are normally distributed. In the simulation studies, the correlations between the two sets of predicted breeding values have generally been high, greater than 0.95, although the nonlinear methods have generally been superior in terms of expected genetic progress. Perhaps not surprisingly, the superiority of the nonlinear methods is larger when there are few categories, when the average incidence in the categories is extreme, as the differences in incidence among subclasses increases, the more unbalanced the distribution of progeny across subclasses, and the higher the underlying heritability. While the practical advantages of using nonlinear prediction procedures may not always be large, with developments in computing strategies and technology they will become more widely used in industry evaluation and breeding programmes.

Selection responses

The development of selection strategies to alter the incidence of all-or-none traits is the obvious extension to the estimation of the heritability of the trait. As already mentioned, heritability estimates obtained from either proband or GLIM regression analyses would provide evidence of the effectiveness of selection. However, in a deliberate directional selection programme, say for increased incidence, we would normally wish to maximize the selection differential achieved, given the biological constraints of the species involved. For a normally distributed trait the selection differential can be predicted from the proportion of the parental generation selected as parents.

Unfortunately, with an all-or-none character, the selection differential achieved will always be less than this, unless the proportion to be selected (say f) is equal to the proportion affected (p). At the level of the underlying liability, then if f is less than p, animals are being chosen at random whose average selection differential is less than if $p = f$. Where f is greater than p, the selection differential achieved is a weighted average of the selection differentials of the affected and unaffected groups (see Dempster and Lerner, 1950). In either case, the selection differential achieved is less than if the parents could be selected directly on liability itself.

One approach suggested to overcome this difficulty if environmental methods for altering liability are possible, is to modify the incidence in the parental generation so that $p = f$. An alternative would be to focus on indirect selection, using a character correlated with the all-or-none trait. Offspring–parent estimates of genetic correlations between either a pair of binomially distributed traits or a continuous and all-or-none trait can be obtained as genetic regressions within GLIM analyses as described previously. In contrast to the situation for heritability estimates, half-sib estimates of the genetic correlation involving one or two binomial traits should not be biassed. Essentially this is because the correlations are calculated between family means, and while the observations on individual animals are of the 0/1 form, the family means should more closely approximate the normal distribution. Whether this is true even for small family sizes and extreme incidences is not clear, and this is a topic requiring investigation. Under these same conditions, our simulation study on heritability estimates indicate that, in a high proportion of data sets, estimates of genetic correlations will not be possible because the heritability of the all-or-none trait will be negative. The situation will be exacerbated where both traits are binomially distributed.

Acknowledgement

I thank Robin Thompson for his very helpful comments.

Chapter 26
Estimation of Breeding Values

L. DEMPFLE

Institut für Tierwissenschaften, Technische Universität München D-8050
Freising-Weihenstephan, Federal Republic of Germany

In animal production we are often faced with the question of how much an animal will produce next
year (e.g. next lactation) or how much will be produced by the offspring either if the animal is mated
at random to animals of a given population (a quantity related to the breeding value) or if mated to
a specific animal or group of animals. Of special interest is the statistical criterion to be applied in
order to predict the production capabilities of the offspring. The breeding values of the available
animals then can be ranked and from these the best selected so as to maximize the increase in average
production from the present generation to the next, i.e. genetic progress. Quite often the problem is
looked at in a slightly different way. In the simplest case genetic progress can be expressed as

$$\delta G = i\rho_{HI}\sigma_H,$$

where i is the intensity of selection, which depends mainly on the number selected out of those tested;
σ_H is a measure of genetic variation, which usually can not be influenced by the breeder; and ρ_{HI} is
a measure of the correlation between true and estimated breeding value. This correlation obviously
depends on the statistical procedure adopted to predict breeding value, and a high correlation is
desirable. In some practical situations, however, this formula can be misleading.

Some common situations

Many experiments in quantitative genetics are carried out with drosophila. Animals are usually raised
together in a culture bottle and, shortly after hatching, they are collected and scored for bristle number
or other traits. In such situations all candidates have, at least are perceived to have, the same
environment, thus corrections for environmental effects are neither possible nor necessary. On the
other hand, pedigrees are unknown, so information from relatives cannot be utilized for predicting
the breeding value. In this situation ranking the individuals on phenotype and picking the best is as
efficient as any other selection procedure. It is only if we want to predict the performance of the
offspring that we have to apply more genetical and statistical theory, but this will not increase genetic
gain. A similar situation might be found in very extensive sheep production systems.

In livestock breeding we usually know at least part of the pedigree and we often have more than
one record on an animal, so we have unequal information on each animal and the question of the
criterion on which to base the ranking becomes more critical. In addition the animals which are
candidates for selection do not perform in the same environment – they can be located on different
farms or perform in different years – so statistical procedures are needed to adjust the records. Finally,
in modern breeding schemes we might have, even if transitory, a mixture of populations from which

we want to select. This problem is especially important if the superior breed is small in numbers and the inferior breed is very numerous.

Principles used in the estimation of breeding values

The basic methodology used in estimation of breeding values is essentially Bayesian in nature, although that terminology was carefully avoided for a long time.

A breed of cattle can be regarded in an abstract manner as a population of breeding values with given mean and variance (often normality is also assumed, but this is not essential for the argument). Since breeding values are often defined as deviations, the mean can be taken as zero and the variance is (by definition) the additive genetic variance (σ_A^2). If we take an animal without additional information *at random* our best estimate is the population average. This information is our prior estimator. The average squared deviation of this estimator from the true breeding value (error variance) is equal to σ_A^2. If the animal has a performance record (data) then a second estimator is the phenotypic deviation from the population mean. The error variance of this estimator is equal to $\sigma_E^2 = \sigma_P^2 - \sigma_A^2$. Having two independent estimators we combine them linearly in the best way using the reciprocal of the respective variances as weights and arrive at

$$\hat{A} = \left(\frac{\text{prior estimator}}{\sigma_A^2} + \frac{\text{data estimator}}{\sigma_P^2 - \sigma_A^2} \right) \bigg/ \left(\frac{1}{\sigma_A^2} + \frac{1}{\sigma_P^2 - \sigma_A^2} \right)$$

$$= \left(\frac{0}{\sigma_A^2} + \frac{y - \mu}{\sigma_P^2 - \sigma_A^2} \right) \bigg/ \left(\frac{1}{\sigma_A^2} + \frac{1}{\sigma_P^2 - \sigma_A^2} \right)$$

$$= h^2 (y - \mu).$$

This can be applied in a very similar way in sire evaluation. The prior estimator (of zero) has (error) variance of $\sigma_A^2/4$ and the estimator from the data (deviation of daughter average) has error variance of $(\sigma_P^2 - \sigma_A^2/4)/n$, where n is the number of daughters. Defining breeding values as deviations we get

$$\hat{s} = \left(\frac{0}{\sigma_A^2/4} + \frac{\bar{y} - \mu}{(\sigma_P^2 - \sigma_A^2/4)/n} \right) \bigg/ \left(\frac{1}{\sigma_A^2/4} + \frac{1}{(\sigma_P^2 - \sigma_A^2/4)/n} \right)$$

$$= \{ n/[n + (4 - h^2)/h^2] \} (\bar{y} - \mu).$$

This approach with two independent pieces of information has practical application, even if we have only the prior estimator. Suppose there is a test station with capacity for 20 animals and 15 animals are needed for breeding. In this case it is best to select the top 10 animals and take the additional 5 animals at random from the population. Here it must be emphasized that taking these at random is crucial, i.e. we must be able to sample the additional 5 animals from the same population as the first 20.

In a short paper Robertson (1955a) very clearly points out the availability and use of the two independent pieces of information. He derives several well-known prediction formulae by this approach, and clearly shows the premises on which the formulae are based.

Adjusting for environmental effects

With the advent of artificial insemination the animal breeder was faced with the problem of selecting among animals which performed in different farms or whose relatives performed in different farms. It is well-known that herd effects had to be considered. The problem was dealt with by Henderson *et al.* (1954) and by Robertson and Rendel (1954). From a purely statistical viewpoint and under the assumption that the expected difference between the progeny of any two bulls will be the same in all herds, there is a (highly) unbalanced cross-classification with herds and sires but without interactions, which can be analysed by least squares (method of fitting constants). However at that time it was very laborious to obtain the exact solution for this problem because, after absorbing herd effects, there is still a set of equations with the number of unknowns equal to the number of sires, which in typical application is in the range of hundreds to several thousands. Such equations can be solved by iterative methods, and an approximate solution was essentially used which is the 'first iterate' of such a method (Jacobi method). That was, however, only one piece of information (from the data) and thus the *a priori* information had still to be included.

The two step procedure becomes clearer if written formally

$$\mathbf{y} = \mathbf{Xh} + \mathbf{Zs} + \mathbf{e},$$

where $\mathbf{h}$ is a vector of herd effects, $\mathbf{s}$ is a vector of sire effects and $E(\mathbf{y}) = \mathbf{Xh}$; $E(\mathbf{s}) = \mathbf{0}$; $E(\mathbf{e}) = \mathbf{0}$; $\mathrm{var}(\mathbf{s}) = \mathbf{I}\sigma_s^2$; $\mathrm{var}(\mathbf{e}) = \mathbf{I}\sigma_e^2$.

The least squares equations are

$$\begin{pmatrix} \mathbf{X'X} & \mathbf{X'Z} \\ \mathbf{Z'X} & \mathbf{Z'Z} \end{pmatrix} \begin{pmatrix} \hat{\mathbf{h}} \\ \hat{\mathbf{s}} \end{pmatrix} = \begin{pmatrix} \mathbf{X'y} \\ \mathbf{Z'y} \end{pmatrix}.$$

Absorbing the herd effects leads to

$$[\mathbf{Z'Z} - \mathbf{Z'X}(\mathbf{X'X})^{-1}\mathbf{X'Z}]\hat{\mathbf{s}} = [\mathbf{Z'y} - \mathbf{Z'X}(\mathbf{X'X})^{-1}\mathbf{X'y}]$$

$$\mathbf{W}\hat{\mathbf{s}} = \mathbf{r}.$$

As an estimate from the data the following is used (first iterate)

$$\hat{s}_{iD} = r_i / w_{ii}.$$

Combining it with the *a priori* value (of zero) gives

$$\hat{s}_{iC} = \{w_{ii}h^2 / [4 + (w_{ii} - 1)h^2]\}\hat{s}_{iD}$$

since the error variance of $\hat{s}_{iD}$ (treating sires as fixed) is

$$(\sigma_P^2 - \sigma_A^2/4)/w_{ii}.$$

The contemporary comparison was accepted very readily in many countries and was more or less the standard method in Europe till the 1970s. Partly due to its success it became necessary to improve the method, for although it eliminates herd effects completely it ignores the competition, i.e. against daughters of what other sires the comparison is made. Thus the method was then superseded by the Best Linear Unbiased Prediction (BLUP) method (Henderson, 1973), but it should be noted that in the contemporary comparison all the essential points of the BLUP method are present.

Using in the first step the exact least-squares solution (instead of an approximation) we get

$$\hat{\mathbf{s}}_D = \mathbf{W}^- \mathbf{Z'My}; \qquad \mathbf{M} = \mathbf{I} - \mathbf{X}(\mathbf{X'X})^-\mathbf{X'}$$

with $E(\hat{\mathbf{s}}_D) = \mathbf{0}$ and

$$\text{var}(\hat{\mathbf{s}}_D - \mathbf{s}) = [\mathbf{W}^-(\mathbf{Z}'\mathbf{MZ})(\mathbf{Z}'\mathbf{MZ})\mathbf{W}^{-\prime} - \mathbf{W}^-\mathbf{Z}'\mathbf{MZ} - \mathbf{Z}'\mathbf{MZW}^{-\prime} + \mathbf{I}]\sigma_s^2 + \mathbf{W}^-\mathbf{Z}'\mathbf{MZW}^{-\prime}\sigma_e^2.$$

Combining in the second step the two pieces of information (having vector variables and covariance matrices), we get

$$\hat{\mathbf{s}}_C = [\text{var}(\hat{\mathbf{s}}_{ap} - s)^{-1} + \text{var}(\hat{\mathbf{s}}_D - s)^{-1}]^{-1}[\text{var}(\hat{\mathbf{s}}_{ap} - s)^{-1}\hat{\mathbf{s}}_{ap} + \text{var}(\hat{\mathbf{s}}_D - s)^{-1}\hat{\mathbf{s}}_D].$$

If an appropriate g-inverse of $\mathbf{Z}'\mathbf{MZ}$ is used (e.g. $\mathbf{W}^- = (\mathbf{Z}'\mathbf{MZ} + \mathbf{11}')^{-1}$) and the *a priori* estimator $\hat{\mathbf{s}}_{ap}$ is zero we get

$$\hat{\mathbf{s}}_C = [\mathbf{Z}'\mathbf{MZ} + \mathbf{I}\sigma_e^2/\sigma_s^2]^{-1}\mathbf{Z}'\mathbf{M}'\mathbf{y},$$

which is exactly the BLUP estimator for the statistical model used.

It should be noted that in the second step not only the variance but also the covariances between the elements of $\hat{\mathbf{s}}_D$ are taken into account. Thus by two improvements of the contemporary comparison method, namely taking the exact least-squares solution and considering also the covariances between the elements of $\hat{\mathbf{s}}_D$, we arrive at the BLUP estimator.

Methodology currently used

Current methodology for estimating breeding values of continuously varying traits is almost completely based on the BLUP principle. There are several derivations available, three of which will be given below, following Dempfle (1977). For each we have the following model and assumptions

$$\mathbf{y} = \mathbf{Xb} + \mathbf{Zu} + \mathbf{e} = \mathbf{Wt} + \mathbf{e}; \qquad \mathbf{W} = [\mathbf{X} \quad \mathbf{Z}]; \qquad \mathbf{t}' = [\mathbf{b}' \quad \mathbf{u}'];$$

$$E(\mathbf{u}) = \mathbf{0}; \qquad E(\mathbf{e}) = \mathbf{0}; \qquad E(\mathbf{y}) = \mathbf{Xb};$$

$$\text{var}(\mathbf{u}) = \mathbf{G}; \qquad \text{var}(\mathbf{e}) = \mathbf{R}; \qquad \text{cov}(\mathbf{u}, \mathbf{e}') = \mathbf{0}; \qquad \text{var}(\mathbf{y}) = \mathbf{ZGZ}' + \mathbf{R}.$$

The *first derivation* uses Bayes theorem, where we assume in addition that $\mathbf{t} \sim N(\mathbf{r}, \mathbf{m})$; $\mathbf{e} \sim N(\mathbf{0}, \mathbf{R})$; $\mathbf{y} \sim N(\mathbf{Xb}, \mathbf{WMW}' + \mathbf{R})$ with

$$E(\mathbf{t}) = E\begin{pmatrix}\mathbf{b}\\\mathbf{u}\end{pmatrix} = \mathbf{r} = \begin{pmatrix}\mathbf{r}_1\\\mathbf{0}\end{pmatrix} \qquad \text{and} \qquad \text{var}(\mathbf{t}) = \mathbf{M} = \begin{pmatrix}\mathbf{S} & \mathbf{0}\\\mathbf{0} & \mathbf{G}\end{pmatrix}.$$

For $\mathbf{t}$ given, $\mathbf{y} \sim N(\mathbf{Wt}, \mathbf{R})$, and for $\mathbf{r}$ given, $\mathbf{t} \sim N(\mathbf{r}, \mathbf{M})$. The likelihood of $\mathbf{t}$ for $\mathbf{y}$ given is

$$L(\mathbf{t}; \mathbf{y}) = k_1 \exp[-(\mathbf{y} - \mathbf{Wt})'\mathbf{R}^{-1}(\mathbf{y} - \mathbf{Wt})/2],$$

the *a priori* distribution of $\mathbf{t}$ is

$$f(\mathbf{t}) = k_2 \exp[-(\mathbf{t} - \mathbf{r})'\mathbf{M}^{-1}(\mathbf{t} - \mathbf{r})/2],$$

and thus the *a posteriori* distribution of $\mathbf{t}$ is proportional to

$$f(\mathbf{t}, \mathbf{y}) \sim L(\mathbf{t}; \mathbf{y})f(\mathbf{t}) \sim k_3 \exp\{-[(\mathbf{y} - \mathbf{Wt})'\mathbf{R}^{-1}(\mathbf{y} - \mathbf{Wt}) + (\mathbf{t} - \mathbf{r})'\mathbf{M}^{-1}(\mathbf{t} - \mathbf{r})]/2\}.$$

Since this distribution is symmetric and unimodal, the mode, the median and the mean are identical and thus a large class of common loss functions leads to the same estimator. Determining the mode leads to

$$[\mathbf{W'R}^{-1}\mathbf{W} + \mathbf{M}^{-1}]\hat{\mathbf{t}} = \mathbf{W'R}^{-1}\mathbf{y} + \mathbf{M}^{-1}\mathbf{r}.$$

In animal breeding applications the prior information on the fixed effects is usually taken as uninformative, which is expressed as $\mathbf{S} \to \infty$ and thus $\mathbf{S}^{-1} \to \mathbf{0}$. This leads to the Mixed Model Equations

$$\begin{pmatrix} \mathbf{X'R}^{-1}\mathbf{X} & \mathbf{X'R}^{-1}\mathbf{Z} \\ \mathbf{Z'R}^{-1}\mathbf{X} & \mathbf{Z'R}^{-1}\mathbf{Z} + \mathbf{G}^{-1} \end{pmatrix} \begin{pmatrix} \hat{\mathbf{b}} \\ \hat{\mathbf{u}} \end{pmatrix} = \begin{pmatrix} \mathbf{X'R}_y^{-1} \\ \mathbf{Z'R}^{-1}\mathbf{y} \end{pmatrix}.$$

The *second derivation*, which is in the same spirit as Robertson's (1955a) derivation is the combination of two independent estimators:

(i) the BLUE or generalized least-squares estimator of $\mathbf{t}$

$$\hat{\mathbf{t}}_1 = (\mathbf{W'R}^{-1}\mathbf{W})^{-1}\mathbf{W'R}^{-1}\mathbf{y},$$

and
(ii) the *a priori* estimator

$$\hat{\mathbf{t}}_2 = E(\mathbf{t}) = \mathbf{r}.$$

The variance–covariance matrices are

$$\mathrm{var}(\hat{\mathbf{t}}_1) = (\mathbf{W'R}^{-1}\mathbf{W})^{-1} + \mathbf{M} \qquad \text{and} \qquad \mathrm{var}(\hat{\mathbf{t}}_2) = \mathbf{0},$$

and the error variances are

$$\mathrm{var}(\hat{\mathbf{t}}_1 - \mathbf{t}) = (\mathbf{W'R}^{-1}\mathbf{W})^{-1} \qquad \text{and} \qquad \mathrm{var}(\hat{\mathbf{t}}_2 - \mathbf{t}) = \mathbf{M}.$$

Combining the two estimators

$$\hat{\mathbf{t}}_C = [\mathrm{var}(\hat{\mathbf{t}}_1 - \mathbf{t})^{-1} + \mathrm{var}(\hat{\mathbf{t}}_2 - \mathbf{t})^{-1}]^{-1}\{\mathrm{var}(\hat{\mathbf{t}}_1 - \mathbf{t})^{-1}\hat{\mathbf{t}}_1 + \mathrm{var}(\hat{\mathbf{t}}_2 - \mathbf{t})^{-1}\hat{\mathbf{t}}_2\}$$

leads to

$$[\mathbf{W'R}^{-1}\mathbf{W} + \mathbf{M}^{-1}]\hat{\mathbf{t}}_C = \mathbf{W'R}^{-1}\mathbf{y} + \mathbf{M}^{-1}\mathbf{r}.$$

Making the same assumptions about the fixed effects, gives the Mixed Model equations. In application the model is often not of full rank and thus instead of the regular inverse an appropriate generalized inverse must be taken.

The *third derivation*, which gave the method its name, is provided by Henderson (1973), who gives a systematic treatment. There $\mathbf{b}$ is regarded as fixed and it is required that the estimator of $\mathbf{w} = \mathbf{P}_1\mathbf{b} + \mathbf{P}_2\mathbf{u}$ is:

<table>
<tr><td>linear</td><td>$\hat{\mathbf{w}} = \mathbf{l'y}$</td></tr>
<tr><td>unbiased</td><td>$E(\hat{\mathbf{w}}) = E(\mathbf{w})$</td></tr>
<tr><td>best</td><td>$E(\hat{\mathbf{w}} - \mathbf{w})^2 \to \min.$</td></tr>
</table>

These requirements lead to

$$\hat{\mathbf{w}} = \mathbf{P}_1\hat{\mathbf{b}} + \mathbf{P}_2\hat{\mathbf{u}}$$

with

$$\hat{\mathbf{b}} = (\mathbf{X}'\mathbf{V}^{-1}\mathbf{X})^{-1}\mathbf{X}'\mathbf{V}^{-1}\mathbf{y}$$

and

$$\hat{\mathbf{u}} = \mathrm{cov}(\mathbf{u}, \mathbf{y}')\,\mathrm{var}(\mathbf{y})^{-1}(\mathbf{y} - \mathbf{X}\hat{\mathbf{b}}) = \mathbf{G}\mathbf{Z}'\mathbf{V}^{-1}(\mathbf{y} - \mathbf{X}\hat{\mathbf{b}}).$$

It is well known that the BLUP estimator of $\mathbf{b}$ and $\mathbf{u}$ is identical to the solution of the Mixed Model equations. The BLUP equation of $\mathbf{u}$ also shows its relationship to the selection index. In the classical selection index we have $\mathbf{H} = \mathbf{a}'\mathbf{g}$, which is estimated by $\mathbf{I} = \mathbf{b}'[\mathbf{y} - E(\mathbf{y})]$, $\mathbf{a}$ being the economic coefficients; and $\mathbf{b}$ is obtained from $\mathbf{Pb} = \mathbf{Ga}$, hence $\mathbf{b}' = \mathbf{a}'\mathbf{G}'\mathbf{P}^{-1}$. Writing this in the usual terminology we have $\mathbf{b}' = \mathbf{a}'\mathrm{cov}(\mathbf{u}, \mathbf{y}')\,\mathrm{var}(\mathbf{y})^{-1}$.

The main difference between BLUP and the selection index is thus:

(i)   the selection index does not prescribe what to do with $E(\mathbf{y})$;
(ii) the selection index does not prescribe what to do if we select from several populations. In BLUP, selection is then based on $\mathbf{P}_1\hat{\mathbf{b}} + \mathbf{P}_2\hat{\mathbf{u}}$;
(iii) with the selection index only a subset of the data is usually used in order to estimate the breeding value of an animal, e.g. its own record and records of near relatives.

Where are we today?

The principles of estimating breeding values via BLUP are fully utilized in dairy cattle breeding and to a somewhat lesser extent in other livestock species, and due to the availability of supercomputers the applications are becoming more and more ambitious. If we look at the different aspects separately the following trends can be noted.

Fixed effects containing environmental components

Corrections for fixed effects are no longer so laborious and thus many effects are now being fitted. However, in most applications the effect of time is still not satisfactorily taken into account. In a typical dairy farm in Central Europe the cows calve all year round, and it is well known that season has a profound effect (through feeding etc.) on milk yield. Having a herd–year–season effect in the model accounts for that factor, but it is not satisfactory because in small herds the number of contemporaries may be very small (zero in quite a lot of cases), animals are compared only within a herd–year–season classification so a lot of information is lost, and animals calving at the end of one season and at the beginning of the next are not compared with each other even though they have nearly identical environments. One way out of this problem is to abandon the fixed block (herd–year–season) concept altogether and define some kind of covariance function. Work in this direction was carried out by Chauhan and Thompson (1986).

Fixed effects containing genetic components

In classical treatments of breeding values only one population was considered, and breeding values were defined as deviations. In modern breeding work we often deal with a mixture of populations where the expected values are different due to genetic causes. Examples of this are animals born in different years or animals of different breeds or different crosses. For the latter a fixed genetic effect is always needed, which essentially expresses the difference between the subpopulations. Things are different for the genetic trend. There if all observations on which selection was based *and* good estimates for the genetic parameters are available then group effects are unnecessary and the genetic trend can be accounted for by use of the complete numerator relationship matrix.

Random effects containing genetic components

The following equivalent models for an observation, for example in sire evaluation with several herd–year–seasons, can be written:

$$\text{var}(e_1) = V; \qquad \text{var}(A) = V_A; \qquad \text{var}(e_2) = V_{ENV}$$

(i) $y = hys + e_1$,

(ii) $y = hys + A + e_2$,

(iii) $y = hys + A_s/2 + A_d/2 + A_w + e_2$,

(iv) $y = hys + A_s/2 + e_4 = hys + s + e_4$.

In model (i) we would estimate the fixed effects and (if we wish) e_1, which is the sum of all random effects. In model (ii) we partition the phenotypic deviation of each animal into its breeding value and the remainder (environmental effects within herd, dominance deviations and epistatic effects). In model (iii) the breeding value is further subdivided into half the breeding value of the sire and of the dam respectively and into the within full sib family deviation of the breeding value. In model (iv) the dam component and the within-family component are put together with e_2 to comprise e_4.

If the variances are specified correctly such that $\text{var}(y) = V$ we get identical results for the effects estimated, i.e. in model (ii) and (iii) we have $\hat{A} = \hat{A}_s/2 + \hat{A}_d/2 + \hat{A}_w$. In practice only models (ii) and (iv) are frequently used. Model (ii) is referred to as the Animal Model (Quaas and Pollak, 1980), in which each animal with recorded data is represented with its breeding value in the model, and breeding values of other animals of interest are also included. The required variance–covariance matrix of A, the numerator-relationship matrix, and especially its inverse is very easy to construct (Henderson, 1976; Quaas, 1976). In this Animal Model nearly all the relationships between the observations and between the breeding values are taken into account. In model (iv) the point of interest is the sire effect which is in some way a measure of the value of an average gamete of the sire. Thus the model is referred to as the gametic model in contrast to the animal model and usually the relationships which exist among the elements of e_4 (e.g. relationships between dams) are only partly taken into account or ignored altogether.

As a result of the general flexibility of the linear model methodology and with the ever increasing power of modern computers, more and more sophisticated models are being applied in animal breeding. It should not be overlooked, however, that these models are optimals only if the parameters (variances and covariances) are correct and in many applications this can be questioned. Another problem arises due to culling: in animal breeding we are not working with a random sample of observations, but have more observations on the surviving animals; and if the information on which culling is based is not included, bias is really unavoidable. This problem makes working with huge populations somewhat unattractive because they are only partly under the control of the breeder. Smaller, well organized schemes can be much better handled, but there is then the danger of genotype × environment interactions, so if animals are not tested on a range of environments (herds) the genetic gain in the nucleus may only be realized to a small extent in the field.

Another aspect should also be discussed critically. The procedures are getting more and more elaborate, but do we really achieve so much more genetic progress? In operational breeding schemes it is very difficult to determine the realized efficiency. The derived correlation between true and estimated breeding value can be misleading since the correlation is increased when we take more information from relatives (e.g. in the Animal Model), but the correlation between the estimates of relatives can also increase considerably and thus lower the expected genetic gain as pointed out by Hill (1976) and Rawlings (1976).

For further progress in animal breeding we thus need not only sophisticated statistical procedures but, since genotype × environment interaction and culling bias are real, we also have to implement more ideas from the field of experimental design, thereby making us less dependent on statistical procedures and on hard-to-verify assumptions.

Chapter 27
Design of Animal Breeding Programmes

J.W. JAMES

Department of Wool and Animal Science, University of New South Wales, Kensington, Australia 2033

The design of animal breeding programmes can be seen from many perspectives. The first major element is the choice of objective, which is usually a combination of several traits important for production. Next one must consider possible selection criteria and their correlations with overall breeding value, since the rate of gain is proportional to the correlation between selection criterion and objective. The process of record collection and analysis also demands attention, both for its feasibility and its costs. Then possible population structures must be reviewed, and the expected rates of genetic improvement in overall breeding value under various alternative combinations of selection criteria and breeding structures need to be calculated. These rates of genetic gain, together with costs and practicalities, are then used to decide on the recommended breeding programme.

Short-term selection response

Following the work of Dickerson and Hazel (1944) on prediction of rates of selection response, attention has been paid to selection intensity, accuracy of estimating breeding value, and generation length. Rendel and Robertson (1950b) and Robertson and Rendel (1950) provided a framework which has been used very extensively. Their equations take account of the pathways of improvement, from sire to son, sire to daughter, etc., and enable a ready assessment of the contribution of each path to genetic gain. This framework has been used by very many workers, such as Morley (1952) and Ollivier (1974) to balance the conflicting pressures for greater selection intensity, more accurate estimation of breeding value and shorter generation interval to best advantage.

One point not addressed in these approaches is that when generations overlap, any progeny crop will be produced by parents of different ages who differ in their history of previous selection, so that there are subgroups of progeny of different expected breeding value. Selection on phenotype would then be expected to retain more progeny from younger parents with a greater cumulative selection history. Bichard et al. (1973) showed that the age of parents gave information on breeding value which could be used to increase rate of selection response. Further, Hopkins and James (1977) noted that the ages of potential parents likewise provided information on breeding value, which could be used in sequential culling decisions and in a strategy they termed 'parent selection'. Use of this information to define optimum age structures of breeding animals could lead in many cases to significantly improved rates of genetic gain. Hopkins and James (1977) assumed the same selection criteria in males and females, while Crook and James (unpublished) extended the analysis to a range of selection criteria in both sexes, and found that if breeding values of males are much more accurately estimated than those of females, the advantage of parent selection is much reduced. An application of these methods to pig selection was given by Hagenbuch and Hill (1978).

Hopkins and James (1977) suggested that in principle it is better to use both own age and parental ages in making selection decisions, but noted that in a short time the number of subclasses to be dealt with would become unmanageable. Pursuing this idea further, selection histories, not of age groups, but of individuals, could be used in estimating breeding values. This leads to the use of Best Linear Unbiased Prediction (BLUP), (Henderson, 1975), as the basis of selection, and it is well known that BLUP can account for genetic trends in estimating breeding values. Looked at in this way, parent selection is seen as using the predicted mean of the age group in which an animal was born to adjust for genetic trend, and BLUP is expected to be superior because estimates of genetic changes that actually occur rather than predicted are used, and because all relatives provide information. Parent selection can, however, be a very simple procedure, with decisions on when animals are to be culled being made at one time. For example, an ear tag could be attached to every animal at a given stage, indicating the date at which that animal was due to be culled. It has not been established under what conditions the rate of progress using BLUP would be sufficiently greater than that achievable through parent selection to warrant the greater expense involved in its use.

In a seminal paper, Robertson (1957b) drew attention to the importance of allocating testing resources to optimize the balance between accuracy of estimation of breeding value and intensity of selection in the context of progeny testing, and later developed the theory further (Robertson, 1960b). These questions were pursued in the context of family selection (Rendel, 1959; Nordskog, 1959), in planning performance testing schemes (Smith, 1960), in the deployment of resources in testing males and females to optimize response (Smith, 1969), and a more comprehensive analysis was undertaken for Merino sheep breeding plans (Jackson et al. 1986). James (1979) extended the method to the case where prior information on breeding value was available and suggested that unequal amounts of information should be obtained on candidates for selection in some situations. Van Raden et al. (1984) have shown that in two-stage selection with differing amounts of information available on candidates after the first stage, it may be better to retain some animals of lower estimated breeding value but lower precision for final selection at the second stage. This is because greater residual variation in breeding value can compensate for the lower mean after first-stage selection. It would not apply if animals were mated after the first culling (Fernando and Gianola, 1986; James, 1987).

In practice, finite resources will be available for obtaining estimates of breeding value, and a very wide range of allocations are possible. For instance, sex, age, pedigree and data on performance already obtained could be used as criteria for allocating the remaining resources for testing. While it seems likely that a fairly simple system could be found which would be close to optimum in any particular case, I am not aware of any systematic analysis of this kind having been completed. If an optimum allocation of costs were found, and the economic value of genetic gain were known, it would be possible to find the total outlay on testing which would maximize benefits from the breeding programme. This is a highly desirable position for a breeder to be in, though there are many difficulties in the way of reaching it.

Population structure

Animal breeding programmes are typically hierarchical in structure, with genetic improvement being made in a 'nucleus' of animals, and improvement transferred by migration of stock, generally through at least one multiplier level, to commercial stock. This flow of genes has important consequences, since genetic change in the whole population is determined by a small élite. If this nucleus makes rapid progress, the hierarchy is an efficient way of disseminating gains, and its nature was first clearly established by Robertson (Robertson and Asker 1951a; Robertson, 1953a). The numerical description in the second of these papers showed clearly how influence over genetic change was concentrated in few animals, while the first established the simple concept of an improvement lag of two generations. That is, if one level in a hierarchy used sires of average breeding value from the level above, the two levels would change at the same rate, but the lower would have the same mean breeding value as the higher had two generations previously. These ideas were developed with respect to traditional livestock breeds, but the same concepts apply to integrated systems such as those run by large breeding companies in the poultry and pig industries.

It was not until 1971 that a detailed theoretical analysis of genetic lag was published, which included the effects of selection practised at lower levels on lags (Bichard, 1971). It is an important concept, surprising to most practical breeders, that if they are not at the top of the hierarchy their selection efforts affect the distance their stock lag behind the upper level, but not the rate of genetic change in their population. An alternative method for calculating lags was presented by Guy and Smith (1981).

At about this time, interest in Australia and New Zealand was directed to a different structure, usually formalized as a cooperative group breeding scheme. This structure is not tied to group breeding schemes, however, so it is better referred to as an open-nucleus system and its characteristic is that gene flow is not unidirectional. Usually all males used for breeding in the whole population are bred in the nucleus, but some females used for breeding in the nucleus have been bred in the rest of the population, the base. The rationale is that despite the lag of the base behind the nucleus, the best base females are genetically superior to some nucleus females who would be used for breeding if the nucleus were closed. A number of such systems were established. A quantitative treatment of their design was given by Jackson and Turner (1972) and some equations relevant to these systems developed by Elsen and Mocquot (1974). James (1977) gave an extensive theoretical analysis with equations for rates of gain and lags, showing that when selection could be intense in males, but not in females, open-nucleus systems could be designed to give 10% to 15% greater rates of gain than closed-nucleus systems. Hopkins and James (1978) gave a modified theoretical approach, which allowed for the effects of parental age pointed out by Bichard et al. (1973); and Mueller (1984) considered a range of selection criteria for the two sexes and nucleus and base, showing that if selection criteria in the nucleus are much more accurate than those in the base, the advantage of an open nucleus can be much reduced. Mueller and James (1984) reviewed developments in open-nucleus theory, including equations for a subdivided base.

A consideration of some importance in open-nucleus systems (as indeed in other systems) is the possibility of genotype×environment interactions. The circumstances in which it is important and the way it can be assessed have been investigated (Del-Bosque-Gonzales and Kinghorn, 1987). If there are two environments, two closed-nucleus systems are better if the genetic correlation between performance in both is less than 0.8, while if every flock has its own environment, one open-nucleus system is best if the average genetic correlation is greater than 0.55.

Almost all theory of open-nucleus systems has been developed for two-layer systems. However, the largest of all such systems (the Australian Merino Society) has a three-layer structure, and while it is not hard to write down equations applicable to a three-layer system, their use in optimization requires considerable work because of the numerous parameters involved. Perhaps this is why, though some specific calculations have been made with respect to particular practical problems, no detailed analysis of the three-layer problem has been made. The main reason for a three-layer system is for multiplication as in traditional livestock breeds, but a good design should allow a faster rate of genetic gain in a three-layer than in a two-layer system. Such extra gains would need to be balanced against the longer lags suffered by the bottom layer. At the limit, multi-layer open-nucleus systems would become assortative mating programmes, with use of parental and progeny performance to estimate breeding value, and analysis of such programmes should give an upper limit to the progress possible with open-nucleus systems.

Inbreeding

Breeding programmes have usually been designed to maximize immediate responses to selection, though it has commonly been recognized that some constraint on inbreeding is necessary, both to avoid inbreeding depression and to limit loss of genetic variation. Rates of inbreeding in populations with discrete generations had long been known, so it is somewhat surprising that the first published equation for rates of inbreeding in populations with overlapping generations in an animal breeding context was that of Turner and Young (1969), though Alan Robertson had a simple derivation of this result in his filing cabinet. The rate of inbreeding per year was given as $1/8mL^2 + 1/8fL^2$, where m and f are the numbers of *new* sires and dams used each year, and L is the generation length. A more rigorous treatment of this problem was given by Hill (1972) who showed how complications such as nonrandom variation of family size could be included. Johnson (1977a,b) showed how short-term inbreeding and genetic drift can be dealt with year by year when generations overlap, rather than simply in terms of asymptotic rates. Procedures for assessing genetic drift and inbreeding in small populations with complicated structures were discussed by Chevalet and de Rochambeau (1985) with particular emphasis on aspects of the process which are not adequately described by inbreeding coefficients, such as the number of ancestors which continue to be represented.

Using a different method to find rates of genetic drift and inbreeding in an open-nucleus system, James (1978) showed that in most cases the effective size of a population would be about twice that of a closed nucleus of the same size. These results were extended to allow for different pathways in both nucleus and base. Of course, in a traditional hierarchy or any other structure with gene flow in one direction only, the rate of inbreeding depends on the nucleus alone, at least in the long term (Togashi *et al.*, 1987).

A simple derivation of inbreeding rates in populations with overlapping generations which can be extended to cover open-nucleus and other structures has been given by Hill (1979). The key idea is to identify different groups of breeding animals within which individuals have equal probabilities of contributing genes to future generations. These may be age groups, groups contributing to the different paths of improvement, or animals of a specific sex within the nucleus or base of an open-nucleus system, for example. If n_i is the number of animals in the ith such group, and f_i is the expected contribution of genes from this group, the rate of inbreeding is

$$\Delta F = \tfrac{1}{2} \sum_i (f_i^2 / n_i)$$

where $\sum_i$ denotes summation over i. This readily yields such well-known results as $1/8M + 1/8F$ for unequal numbers of sires and dams as well as the equations of Turner and Young (1969) and James (1978).

These methods enable the breeder to predict rates of inbreeding, but ignore the fact, apparently first pointed out by Morley (1954), that selection itself affects inbreeding because selected animals are more likely to be related than randomly chosen animals. Robertson (1961c) gave an analysis which showed the importance of heritability and selection intensity as factors increasing the rate of inbreeding. The approximations involved led to overestimation of these effects (Jones, 1969) but no improved prediction procedure appears to be available, though Robertson's formula is easily adapted to deal with several paths of improvement. The absence of an improved theory is a problem for the accurate evaluation of programmes based on new reproductive technology such as schemes using multiple ovulation and embryo transfer (Nicholas and Smith, 1983). These depend on family selection which increases inbreeding rates (Robertson, 1961c). While the ready availability of computing power nowadays means that simulation can be used to assess expected inbreeding under various designs, it would clearly be better to have a (preferably fairly simple) reasonably accurate approximate formula.

The flow of improvement

Breeding programme plans have mostly been based on calculations of asymptotic response rates in populations with overlapping generations. It takes some time for these asymptotic rates to be achieved, however, and in the shorter term gains do not accrue at a constant rate. In addition, it may be several years after the initiation of a breeding programme involving significant investment before genetic improvement of productive ability appears in commercial animals. Whether a breeding programme is undertaken by a private company or by a government authority, some form of investment appraisal seems necessary. This can be done by predicting flows of costs and returns and discounting them to present value. This discounted cash flow procedure was first applied to animal breeding plans by Poutous and Vissac (1962), though they assumed asymptotic rates of gain applied from the start. Hinks (1972) calculated the flow of benefits from a dairy cattle breeding scheme, showing the fluctuations to be expected. Hill (1971a) gave a detailed treatment of the use of discounting in appraising large-scale breeding plans, and this paper may be seen as the origin of a series of papers applying discounting methods in the evaluation of breeding programmes in many countries. A recent unusual application is the suggestion of Smith (1985) that, because of the uncertainty of long-term goals, there would be an advantage to selecting several stocks with different objectives in the hope that these would cater for changing future needs. How to choose the range of objectives remains a problem.

These studies of short-term responses were greatly assisted by the development of a theory of gene flow which predicted the expected breeding value for any class of animals in any particular year (Hill, 1974a; Elsen and Mocquot, 1974). In addition, this analysis could be used to derive classical theory as the limiting state of the dynamic system. The theory has been developed to include division of the population into subunits (Elsen, 1980), to allow differential selection from parental age subgroups discussed above (Hopkins and James, 1979), and to include complex breeding structures (Danell *et al.*, 1976; Togashi *et al.*, 1986).

The problem motivating the development of gene flow matrix theory was the discounting of uneven early genetic gains. It seems, however, that in many cases the differences are small between these accurate analyses and those based on asymptotic rates with a delay between initiation of the programme and appearance of first benefits, especially given the uncertainty associated with estimating future benefits. But without the exact analysis the validity of the approximation is unknown, and the gene flow theory has another important contribution to make. This is the insight it gives into the pattern of change to be expected during a programme, which is valuable for advance planning and for the interpretation of observed changes. The theory has also been the basis for other developments such as Johnson's (1977a) work on inbreeding.

Long-term response

The studies just discussed have been concerned with genetic changes in the early years of selection and the balancing of the costs of the scheme against its benefits. Another important aspect is the balance between short- and long-term improvement. In an important paper, Robertson (1960c) pointed out, among many other things, that many practices which increase immediate response also reduce the limits attainable, the use of family information as an aid to selection being one example. He also showed that limits are highest when half the candidates are selected as parents in each generation, though this would often mean greatly reduced gains in the early stages. If linkage has a significant effect on limits, more than half should be selected (Hill and Robertson, 1966).

In practice, however, breeders cannot afford to sacrifice significant amounts of short-term response in order to increase gains at a distant future time. For one thing, the future is uncertain, as recognized in Smith's (1985) proposal to select several lines for different goals. But even if goals remain unchanged, competition and other pressures would force an emphasis on immediate gains. Although Robertson (1970c) considered maximizing total gains over a fixed time period and found an elegant solution, I doubt whether a breeder would retain a fixed time horizon as it came closer and closer. An alternative way of balancing short- and long-term gains is to adopt the discounting procedure used in investment appraisal to find the net present value of responses for programmes which differ in their balance between immediate and future gains. This approach was used by James (1972) to find the optimum selection intensity in a breeding programme, and also to evaluate how large an investment is warranted. The discounting of gains in this way may be subject to fewer uncertainties than in benefit-cost analyses, which appear to be dependent on more unverifiable assumptions.

A specific example of the balancing of immediate and future gains is the proposal by Dempfle (1974) that selection of dairy bulls should be done within sires in order to decrease the rate of inbreeding, with a consequent increase in the selection limit. Because of the intense and accurate selection applied to bull sires, there would be little genetic variation between such sire families, and little would be lost in the short term by not using it. Although the balance between long- and short-term gains remains an important topic, it seems not to have attracted much attention, though it is one of the criteria considered by Goddard (1987) in designing dairy bull breeding programmes.

Conclusion

In this review I have concentrated on those aspects of the design of animal breeding programmes to which Alan Robertson made major contributions, and have given less attention to some very important problems such as the choice of objective or the design of crossbreeding programmes. I have also not referred to his work on choice of selection criteria such as the contemporary comparison method or prediction of breeding value (Robertson, 1955a). This is not because these topics are less important, but because the ones I have stressed are those in which his influence has been predominant.

The main focus of animal breeding theorists in recent years has been on estimating breeding values (BLUP and associated methods) with an apparent assumption that when this can be done accurately the breeder's problems are solved. While this would not, I think, actually be asserted by anyone, it does appear to be a reasonable inference to be drawn from practice, given that breeding value estimation is pursued in great detail while important problems in planning breeding operations are left untouched. There are, of course, exceptions such as the very interesting work of Kinghorn (1986) on methods of combining selection and mating plans using costs and expected returns in an overall framework. Nevertheless, such work is an exception from the mainstream. Significantly, Alan Robertson has devoted attention both to improved estimation of breeding values and to improved breeding plans. As in so many other ways, his lead here is worth following.

Chapter 28
Animal Breeding Practice

A. John WEBB

Cotswold Pig Development Company Limited, Rothwell, Lincoln LN7 6BJ, England

The practice of animal breeding is based on the science of quantitative genetics. The early description of the nature of genetic variation led to a set of simple principles for selection which could be applied in farm animals. Since then the development of accurate methods for estimating breeding value, together with the understanding of population structure have paved the way for the large scale genetic improvement programmes of today. This chapter reviews the penetration of quantitative genetics into practice from the viewpoint of an international pig breeding company, highlighting some of the latest developments.

Principles of practical breeding

The majority of traits of economic importance in pig production are assumed to be controlled by many genes of small effect. The two principal tools for improvement are selection, which exploits additive genetic variance and leads to permanent and cumulative changes, and crossbreeding, which makes use of nonadditive (mainly dominance) variance but must be regenerated at each mating. Growth and carcase traits show moderate heritabilities but low heterosis, whereas reproductive traits show low heritabilities accompanied by high heterosis. In the past, selection has therefore concentrated on lean growth traits, while crossbreeding has been used to boost sow productivity.

To select several different traits simultaneously, an aggregate breeding value (T) is defined as a linear function of genetic values weighted by their relative economic importance. To date, a classical selection index (I) has been used to predict T, weighting a series of observations according to their genetic association with traits in the objective. The accuracy of selection, expressed as the correlation between I and $\bar{T}$ (r_{TI}), would then be around 0.5 for lean growth. With no evidence of imminent selection limits, population size has been determined by the balance between inbreeding and the cost of measurement, typically between 100 and 200 sows per line.

Breeding objectives

For 20 years pig populations have undergone index selection to improve growth rate, feed efficiency and lean content. During the 1970s, unselected control herds maintained by the UK Meat and Livestock Commission (MLC) showed annual rates of **genetic** improvement in excess of 2% for both lean tissue growth rate (LTGR) and lean tissue food **conversion** (LTFC) (Mitchell *et al.*, 1982). The present value of the return of an annual investment of £2 million was estimated to be of the order of £100 million per year.

Following this success, some pig populations are now approaching optimum fatness. While past genetic improvement in food conversion has been obtained mainly by substituting lean for fat, further improvement will be achieved by growing the lean more quickly. The slowing of genetic improvement in food conversion will provide a stronger incentive to select for litter size, in spite of its low heritability of around 0.1 (Haley *et al.*, 1988). In addition, attention is now turning towards possible genetic aspects of meat quality, especially adverse quantity–quality relationships. Pig breeding is therefore now entering its 'second era', with a shift of objectives away from fat and towards overall efficiency and product quality.

Choice of breeds

Immigration or breed substitution gives an immediate genetic improvement which can clearly save many generations of selection. Due to quarantine and cost constraints, between-country comparisons based on live animals often involve small samples leading to low precision. Nevertheless, international distribution of live pigs from a common Norwegian Landrace control population has been successfully used to rank breeds in seven European countries (King *et al.*, 1975). The advent of frozen semen now allows international comparisons of crossbred progeny based on adequate numbers of sire families (Sutherland *et al.*, 1984). Although indirect comparisons can be made between countries which have already exchanged genotypes, the opportunity to establish a frozen semen 'reference' population has yet to be taken up (Sutherland *et al.*, 1985).

In Europe current interest is centred on the Duroc and Meishan breeds. The Duroc, with its higher levels of intramuscular fat, may confer an improvement in juiciness of meat (Wood *et al.*, 1987). Crossbred dams containing Duroc may show additional heterosis in litter productivity (Buiting and Merks, 1986).

Compared with Large White, the Meishan rears 3 to 4 extra piglets per litter, reaches sexual maturity 3 months earlier, but is slow growing and fat. Western research so far suggests that the advantage in litter size arises from an improvement in embryo survival, due to earlier development and greater uniformity of the embryos (Bazer and Thatcher, 1986). There is speculation that this could be the effect of a single gene, as in the Booroola Merino. In 1987 a sample of 32 Meishan were imported to Edinburgh from the People's Republic of China for physiological and genetic research.

Crossbreeding

Crossbreeding has mainly been used to improve performance for less heritable traits such as litter size and survival from birth to weaning. Two distinct types of crossing system exist: 'discontinuous' systems in which for example lines A and B are repeatedly mated together to give an F_1 A×B cross in each generation, and 'continuous' systems where A and B males are mated to crossbred females in rotation (A×(B×(A×B)). Since maximum heterosis is obtained when both dam and offspring are crossbred, the discontinuous systems have been preferred in Europe. Until recently, the heterosis advantages for crossbred boars in libido, semen quality and conception rate averaging from 5 to 50% for F_1 crosses (Buchanan, 1987) have been largely ignored in Europe in favour of purebred terminal sires.

Synthetic lines have been poorly accepted due to low levels of heterosis, high inbreeding, and the absence of additional genetic variation (e.g. Robertson and Lerner, 1949). An exception has been the Pietrain/Hampshire synthetic, combining the extreme muscling of the Pietrain with the stress resistance of the Hampshire. An attempt to develop an open synthetic line with continuous competitive immigration from any source was unsuccessful due to lack of information on the means of the immigrant populations (Webb and King, 1976).

Limited studies show apparently large maternal additive and heterosis effects on litter size for crosses of Meishan with European breeds (Table 28.1). At present, the advantages of up to two extra piglets weaned for females containing 50 or 25% Meishan appears largely offset by poorer lean growth (Bidanel and Legault, 1986). The option exists either to select for lean growth in purebred

Meishan or in a Meishan/European synthetic. One theoretical study suggests that the synthetic approach would be most efficient in the short term, but that selection in the pure Meishan is the best long term strategy (Bidanel, 1988).

Table 28.1
Estimated crossbreeding parameters for litter size in
Meishan and Large White (Bidanel, 1988)

Effect	Born alive	Weaned
Additive genetic[1]		
direct	−0.74	0.12
maternal	3.44	2.97
Heterosis		
direct	0.88	1.24
maternal	2.51	2.35
Recombination loss	0.59	0.07

[1]Difference: Meishan minus Large White.

Specialized sire and dam lines

It was quickly realized that selection for different objectives in specialized sire and dam lines could accelerate genetic improvement (Smith, 1964). Since litter size is determined genetically largely by the mother, selection could be confined to lean growth in the sire line, while the dam line would be selected both for lean growth and litter size. For the past 20 years, the benefit from selection for litter size in a pig dam line would have been small due to its lower heritability and economic value, and negligible genetic correlation with lean growth. However, as optimum fat levels now approach, the relative economic value of litter size will increase. In addition, use of relatives' records can increase the accuracy (r_{TI}) of selection for litter size by around 50% (Avalos and Smith, 1987). The extra economic improvement from selection on litter size in a dam line can therefore be expected to increase to around 15%, rising to over 20% in the event of an adverse genetic correlation with lean growth (Webb and Bampton, 1988).

Within-breed selection

The first method of selection in the 1950s was the 'progeny test', in which breeding value (T) was predicted from measurements (I) on progeny. This was quickly superseded by the 'performane test', with direct recording of the candidate for selection, in order to shorten generation interval and increase selection intensity. The performance test involves recording growth rate, ultrasonic backfat depth and often individual food consumption over a fixed period, usually roughly 30 to 90 kg live weight. The success of this system stems from its simplicity and low cost. Addition of individual food consumption or records on full-sibs to measurements of growth rate and backfat on the candidate improves accuracy (r_{TI}) by relatively small amounts (5 to 15%).

Electronic feeding stations are now being developed to record individual food intake for pigs fed *ad lib.* in groups. As well as the size and frequency of individual meals, automatic weighing platforms may shortly allow individual growth curves to be plotted. Two-dimensional ultrasonic technology, developed for human medicine, allows more detailed *in vivo* carcass measurement at reasonable cost. For within-population selection on lean growth, the new BLUP methodology, which utilizes all available information on relatives, offers relatively small increases in accuracy, maximum benefits deriving from sequential culling for a trait of low heritability (Wray, 1988). Nevertheless, BLUP does offer an ongoing measure of the rate of genetic change, giving visible evidence of a return on investment.

Population structure

The success of pig improvement owes much to the pyramid structure for distributing genetic change. Selection takes place in a relatively small nucleus where genetic improvement accumulates. The improvement passes first to multipliers, who multiply and cross purebred lines to produce dams and sires of pigs for slaughter, and then to commercial farms (Figure 28.1(a)).

While nucleus herds of less than 200 sows were adequate for selection on lean growth, the between-family element of BLUP selection for litter size requires a much larger nucleus population (e.g. Toro *et al.*, 1988). For example, Cotswold has established two dam line nucleus populations of 1000 sows each by combining the nucleus and purebred multiplication tiers of the breeding pyramid (Figure 28.1(a) and (b)). Reduction from four to three tiers then also has the benefit of lowering the genetic time lag for improvement to pass from nucleus to commercial level. For health reasons, each of the 1000 sow populations is distributed over four to five separate units with gene movement by artificial insemination (AI), in a new structure termed a 'Group Nucleus' (Figure 28.2).

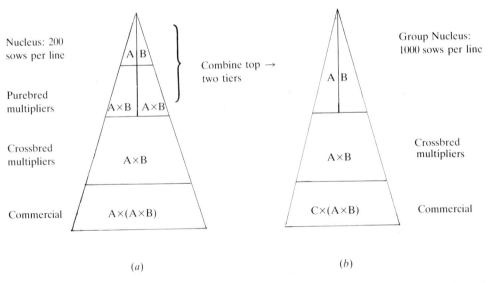

(a)   (b)

Figure 28.1. Integration of top two tiers of breeding pyramid to allow family selection for litter size. (a) Traditional pyramid: 4 tiers. (b) Group Nucleus pyramid: 3 tiers.

In the Group Nucleus, the 15 boars per line are mated across units, and replaced five times per year giving over 75 sire families annually. BLUP predictions of genetic merit are calculated at birth, at the end of performance test, and after each farrowing to allow sequential culling. Thus the full power of BLUP has been to increase the flexibility of structure, allowing selection among herds and generations.

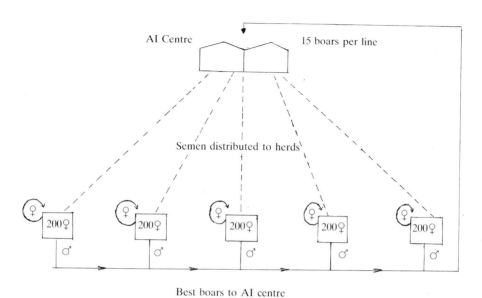

Figure 28.2. Principle of the Group Nucleus: a group of 200-sow nucleus herds form a 1000-sow breeding population

Selection regime×production system interaction

Current performance testing programmes are giving rise to interactions of two kinds. First, genetic correlations between performance in testing station and commercial farm environments are proving lower than expected, averaging 0.25 for growth rate and 0.40 for backfat (Merks, 1988). These may arise from sire×feeding level interactions, or from the absence of competition during individual feeding at test stations. The second interaction is the observed reduction in voluntary food intake from programmes giving a high emphasis to efficiency rather than rate of lean growth on *ad lib.* feeding (e.g. Webb and Curran, 1986). For example, voluntary food intake declined at roughly 0.4% per annum in the MLC scheme (Mitchell *et al.*, 1982).

The role of intake has been investigated by selecting for five generations on an index giving a high weighting to efficiency of lean growth. While selection on twice daily to appetite feeding caused a reduction both in intake and growth rate on full *ad lib.* feeding, selection on restricted feeding caused a correlated increase in both traits under *ad lib.* feeding (Table 28.2). Theoretical predictions of two distinct pathways for selection therefore appear to be borne out (Figure 28.3). Selection for LTGR on *ad lib.* feeding appears to lead to a primary increase in rate of protein deposition. Selection for LTFC appears to lead to a primary reduction in rate of fat deposition via a lowering of voluntary food intake. The optimum selection strategy depends on the genetic relationships of LTFC and LTGR with their component traits across feeding regimes (Figure 28.3). These are now being investigated in a selection study at Edinburgh and Wye College involving four objectives: LTGR on *ad lib.*, LTGR on restricted, LTFC on *ad lib.*, and voluntary food intake on *ad lib.* feeding (Cameron *et al.*, 1988).

In the meantime, the optimum performance testing regime is likely to involve *ad lib.* group feeding, with recording of individual food intake. This will allow genetic variation in intake and competitive ability, and still measure any genetic differences in maintenance (e.g. protein turnover).

Table 28.2
Comparison of selected and unselected control lines on *ad lib.* and scale feeding following 5 generations of index selection either on twice daily to-appetite or on scale feeding (McPhee 1981, 1985)

Feed	Ad lib.		Scale	
Line	Selected	Control	Selected	Control
Selection on appetite feeding				
Food conversion ratio	2.67	2.78	2.56	2.66
Backfat (mm)	18.7	21.8	18.5	21.1
Daily liveweight gain (kg)	0.78	0.81	0.76	0.74
Daily food intake (kg)	2.07	2.24	1.95	1.97
Selection on scale feeding				
Food conversion ratio	2.71	2.99	2.42	2.62
Backfat (mm)	14.8	17.9	12.9	15.3
Daily liveweight gain (kg)	0.89	0.76	0.75	0.69
Daily food intake (kg)	2.42	2.27	1.81	1.79

Components of selection objective		Results of selection
TIME		Lean per day △ (increased)
	LTGR	Food per day △ (increased)
		Fat per day — (static)
LEAN		
		Fat per day ▽ (decreased)
	LTFC	Food per day ▽ (decreased)
FOOD		Lean per day — (static)

Figure 28.3. Illustration of two possible pathways to genetic improvement following selection for lean tissue growth rate (LTGR) or lean tissue food conversion (LTFC) on *ad lib.* feeding to a constant weight.

Although not ideal, the new electronic feeding stations provide a practical solution to the measurement of individual food intake under *ad lib.* group feeding. However, if the genetic correlation between test and farm performance remains below 0.8, the increased accuracy from the test environment may be more than offset by selection regime×production system interaction, giving a faster rate of progress from selection in the commercial environment (Brascamp *et al.*, 1985).

Single genes

Single genes with economically important effects are rare. In pigs, the exception is the halothane gene, which has additive beneficial effects on lean yield and largely recessive harmful effects on stress susceptibility and meat quality (Simpson and Webb, 1988). The resulting heterozygote (Nn) advantage can then be exploited by mating a homozygous normal (NN) dam to a nn or the Nn terminal sire (Figure 28.4). In the longer term, the gene may be undesirable due to the high cost of maintaining nn lines, and a possible deterioration in the meat quality of the Nn with increasing lean growth. It may soon therefore be necessary to eliminate the gene altogether, possibly using linkage relationships with neighbouring blood type genes (Archibald and Imlah, 1985) or restriction fragment length polymorphisms (A.L. Archibald, personal communication).

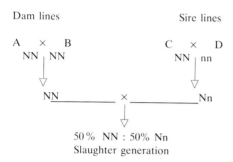

Figure 28.4. Mating scheme to exploit lean advantage of halothane heterozygote (Nn), while avoiding stress susceptibility (nn).

In pigs, a single gene affects liability to K88 *Escherichia coli* infection (Walters and Sellwood, 1988). There is evidence that the SLA complex may influence both liability to rhinitis and reproductive performance, with implications for general disease resistance (Rothschild, 1985). Colour inheritance, determined by a series of genes, is important in the UK due to the use of females containing Saddleback and Duroc on outdoor units.

Laboratory animal models

How successful are selection studies in laboratory animals in modelling the responses of their agricultural counterparts? Selection experiments in mice have shown broadly similar patterns of response to pigs from selection on reproduction and growth traits. However, while the growing mouse uses some 90% of its daily food intake for maintenance, the growing pig uses less than 50%, leaving a large surplus for fat deposition. The consequence is illustrated by lines selected for high and low voluntary food intake (Hill and Bishop, 1986). The high voluntary food intake lines showed a reduction in fatness, accompanied by an increase in maintenance requirements, probably due to a more rapid rate of protein turnover. In pigs it is likely that selection for high intake would be accompanied by an increase in fatness rather than a decrease. Thus, although laboratory animals may offer a good model for genetic mechanisms, they may not accurately reflect the genetic relationships among production traits. Genetic research on farm animals is therefore essential for the future.

New developments

The technology of animal breeding is developing rapidly on three frontiers: biotechnology, biometrics and 'bioelectronics'. Advances in molecular biology will of course eventually allow the transfer of genes between species, together with the full control of their expression. In the shorter term biotechnology is more likely to allow the manipulation of reproduction in the form of cloning, *in vitro* maturation of oocytes, and sex determination. DNA technology will offer Restriction Fragment Length Polymorphism (RFLP) markers and genetic finger-printing, which could in turn pave the way for animal breeders' rights in the same way as plant varieties.

For biometrics, new mixed model techniques (e.g. BLUP and REML) will offer more efficient methods of breeding value and parameter estimation, utilizing all available information from many relatives, traits and environments, and harnessing even greater computing power. Bioelectronics will bring more accurate methods of recording individual food intake, live weight changes, and body composition. Already, X-ray tomography can display and mathematically describe rashers of bacon in the live pig. New methods will be available for estimating the value of each carcase to the processor, and rewarding the individual producer accordingly.

Conclusion

The science of quantitative genetics has achieved a high degree of penetration into practical pig improvement. The speed of technology transfer has been exceptional, for example with new theory on selection for litter size together with mixed-model statistical methodology immediately impacting on practice. In future, quantitative genetics will provide the vehicle for the practical exploitation of the new biotechnology. The success of scientific genetic improvement in commerce is a tribute to Professor Alan Robertson and his many students, and their ability to combine research and practice.

Chapter 29
Incorporation of New Reproductive Technology in Genetic Improvement Programmes

F.W. NICHOLAS

Department of Animal Husbandry, University of Sydney, NSW 2006, Australia

For many decades, reproductive biologists have continued to challenge the ingenuity of animal geneticists by developing new reproductive technologies that are directly relevant to animal improvement programmes. The aim of this review is to show how animal geneticists have responded to these challenges, and to summarize current knowledge of the most effective ways to incorporate new reproductive technologies into animal improvement programmes. In the course of this review, Alan Robertson's significant contribution to this field of knowledge will become evident. Since his major interest in this area has been in dairy cattle breeding, I shall concentrate primarily on that industry.

Artificial insemination (AI)

In the 1930s and 1940s, there was a rapid increase in the commercial use of AI in many countries, especially in dairy cattle, and by the mid 1940s geneticists had begun to discuss its implications. In 1944, Dickerson and Hazel briefly mentioned the potential of AI in progeny testing, but did not attempt any formal analysis of its benefits.

Two years later, Alan Robertson and Jim Rendel commenced a thorough examination of the genetic implications of AI in dairy cattle improvement. They soon realized that AI provided a marvellous opportunity to exploit the quite substantial differences that they had shown to exist between the average yield of daughters of different sires used in commercial herds. This could best be achieved if bulls were selected on a progeny test based on their daughters' performance relative to the performance of their daughters' contemporaries in a wide range of commercial herds. If a cycle could be established in which selected bulls were then used to breed the next generation of young bulls, which in turn would be progeny-tested in the same way as their fathers had been, then continued genetic progress would result.

This realization gave rise to the first scientifically-based proposal for the use of AI in an animal improvement programme (Robertson and Rendel, 1950), and to the development of the contemporary comparison method of progeny testing (Robertson, 1953b; MacArthur, 1954; Robertson and Rendel, 1954).

Robertson and Rendel's proposal for the use of AI in dairy cattle improvement was examined and extended by many other workers, particularly in relation to the determination of optimum designs. Examples of such studies include those by Specht and McGilliard (1960), Van Vleck (1964) and Skjervold and Langholz (1964). In one form or another, the basic Robertson and Rendel proposal was eventually applied in most dairy cattle populations throughout the world. Most of these programmes are still in operation today.

The potential dangers of AI

From the beginning of their work, Robertson and Rendel were aware of the dangers of the widespread use of AI in large-scale progeny-testing schemes. They were concerned about the potential of AI for spreading deleterious recessive genes through the overuse of particular sires, and, at a more general level, they could see the undesirable consequences of having a whole breed consisting of one interbreeding population, especially if the breed was reliant upon and involved in a single breeding programme that was under the control of just a few individuals (Rendel and Robertson, 1950c). They were also aware of the need for the rate of inbreeding to be kept at an acceptably low level because of the undesirable consequences of inbreeding depression and depletion of genetic variance (Rendel and Robertson, 1951).

At the same time, it was clearly appreciated that AI offered the means to avoid these problems, by offering a far wider choice of unrelated bulls than was ever available to individual breeders under natural mating. To maintain diversity, Robertson and Rendel pictured a breed being composed of partly isolated units each associated with an AI centre running its own progeny-testing programme. This type of breed structure has developed to a certain extent, but in more recent times there has been a strong tendency for AI units to become more global in their choice of bulls and cows. Given this tendency, it may be opportune to reconsider the suggestion that was often made by Robertson and Rendel, namely that if a bull is to be used widely through AI, he should first be mated to twenty or so of his own daughters, in order to provide some indication of the deleterious genes that he carries.

A general method for examining the effect of reproductive technologies

In addition to its direct practical consequence in dairy cattle breeding, Robertson and Rendel's (1950) paper was also important because it showed how Dickerson and Hazel's (1944) formula for response to selection could be used to investigate the effect of incorporating any new reproductive technology into any animal improvement programme. This matter was first mentioned very briefly in a companion paper by Rendel and Robertson (1950b).

Dickerson and Hazel had pointed out that response to selection per unit time equals the average genetic superiority of those animals selected as parents, I, divided by the average age of parents when their offspring are born (called the generation interval, L). They had also shown that average genetic superiority can be calculated as

$$I = ir\sigma_A$$

where i is the intensity of selection, which is the average phenotypic superiority (in standard deviations) of selected parents; r is the accuracy of selection, which is the correlation between true breeding value and the criteria on which selection is based; and σ_A is the standard deviation of breeding values.

Rendel and Robertson's (1950b) major contribution was to show how selection could be envisaged as operating along four pathways (sires to sons, sires to daughters, dams to sons and dams to daughters), with the contribution of each pathway to the rate of improvement depending on the values of $\sum I$ and $\sum L$ pertaining to that pathway. Since each pathway provides one-quarter of the genes to the next generation (assuming a 1:1 sex ratio in offspring), the average genetic superiority across the four pathways is $\sum I/4$, and the average generation interval is $\sum L/4$, which gives the annual response to selection as

$$\Delta G = \sum I / \sum L$$

which is the well-known formula first presented by Rendel and Robertson (1950b).

In order to express genetic superiority in terms of a readily available reference point, they expressed I in each pathway as a proportion of the average performance of the population at the commencement of the breeding programme. This meant that annual response to selection could then be expressed as a proportion (or, more commonly, as a percentage) of the population mean at the commencement of selection.

This practise of expressing response as a percentage of average performance of the trait undergoing selection is now very widespread. Unfortunately, it has caused considerable confusion, because it has given rise to the common misconception that response in each year of a breeding programme is a fixed percentage of the mean in the previous year of the programme, i.e. that annual response to selection increases geometrically! This confusion can be avoided by remembering that the percentage improvement is always with respect to the mean performance before the breeding programme commenced.

Noting that σ_A is a constant factor common to I in all four pathways, it is evident that the annual response to selection depends on the values of intensity of selection, accuracy of selection and generation interval. Rendel and Robertson (1950b) realized that the effect of any reproductive technology could be evaluated in terms of its effect on these three factors in each of the four pathways. For example, by evaluating the effect of AI on these factors, Robertson and Rendel (1950) showed that with progeny testing in a population of 2000 cows, the maximum possible annual response to selection is of the order of 1.5%; and with 10000 cows, the maximum possible response is around 2.0%.

Among the cases of reproductive technology cited (but not investigated) by Rendel and Robertson (1950b) were the use of multiple ovulation and embryo transfer and sex determination of embryos. They pointed out that both of these technologies could be used to increase the intensity of selection of dams, and hence to increase the rate of improvement.

In the last four decades, there have been many studies undertaken using the approach first presented in these two Robertson–Rendel 1950 papers.

One of the more interesting applications has been in evaluating the effect of AI on honeybee improvement (Moran, 1984). Because of the haplodiploid nature of this species, the sire-to-son pathway does not exist. But the same basic approach was still able to be used to great effect.

We shall now briefly summarize the results of studies that have investigated the effect of other reproductive technologies in domestic mammals.

Multiple ovulation and embryo transfer (MOET)

One of the first formal genetic investigations of MOET was Land and Hill's (1975) study which showed that for traits measurable on both sexes prior to puberty in cattle, MOET could double the rate of improvement. These results were confirmed and extended by Smith (1984), who showed, for example, that if MOET could provide a three-fold increase in female reproductive rate from 0.67 to 2.0 replacements per year (thereby enabling the proportion of females selected to change from 1/1 to 1/3), then annual response in weight gain could be increased from 1.4% (without MOET) to 2.6% (with MOET), while annual response in percentage lean could be increased from 0.5% (without MOET) to 1.0% (with MOET).

While these results were immediately applicable to beef cattle, they were not directly relevant to dairy cattle, in which the traits of major interest are sex-limited and cannot be measured prior to puberty. For the latter situation, it was soon shown that if MOET were used as an integral part of a large-scale traditional progeny testing scheme, there would be only a small increase in the rate of improvement, mainly because selection of dams to breed sons was already quite intense even without MOET (Bradford and Kennedy, 1980; McDaniel and Cassell, 1981). An earlier study had shown, however, that the effect of MOET could be quite substantial in a progeny testing scheme in a small (500-cow) herd, because MOET resulted in a relatively larger increase in intensity of selection of females (Nicholas, 1977). The potential uses of MOET in a small dairy herd were extended by Nicholas (1979) and Nicholas and Smith (1983). In essence, these studies showed that if females can produce 'litters' of offspring, then the rate of improvement is even greater if progeny testing is discarded, and both male and female replacements are instead selected on whatever information is available at puberty (pedigree selection) or at a slightly later age (sib selection). While the accuracy of both these forms of selection is less than the accuracy of progeny testing, the consequent decrease in generation interval is so great that the overall rate of improvement is substantially increased.

To pig and poultry breeders, this 'discovery' was already common knowledge – they had learnt this lesson at least two decades earlier, and had been applying it in practice to great effect for many years. But this conclusion presented quite a challenge to dairy breeders, who by now felt very much at home with the Robertson/Rendel type of progeny testing scheme.

The results of Nicholas and Smith indicated that in a population of, say, only 500 recorded cows, MOET could double the response to selection, thereby achieving rates of improvement substantially greater than the maximum possible annual response in large-scale (state or national) progeny-testing schemes. Further calculations have confirmed these conclusions. For example, Woolliams and Smith (1988) have shown that if MOET enables 4 male and 4 female progeny to be obtained from each of 64 donors by 34 to 36 months of age (so-called 'adult' MOET), then in a population with only 512 recorded cows, the possible annual response is around 2.4%. In contrast, the maximum possible annual response without MOET is around 1.1% in a herd of the same size (Nicholas, 1979), and around 2.0% in a conventional national progeny-testing programme involving thousands of cows (Robertson and Rendel, 1950). If embryos can be obtained from donors at 12 to 14 months of age ('juvenile' MOET), and/or if marker traits (such as blood urea nitrogen level after a short fast) turn out to be useful, then even greater annual responses, up to 5.7%, are possible from MOET nucleus breeding programmes (Woolliams and Smith, 1988).

It is interesting to note in passing that in one of his earliest reviews of livestock improvement, Robertson (1954b) predicted that 'in an efficient progeny-testing scheme, the effect of [MOET] would not be great'. However, he then went on to say that 'in the absence of progeny testing, [MOET] would have more effect and might almost double the rate of improvement in small herds'.

From the outset of consideration of how best to use MOET in dairy improvement, it was realized that it was not necessary for a MOET nucleus to completely replace well-established large-scale progeny-testing schemes: it could be more practical (and still very effective), at least in the immediate future, to run a MOET nucleus alongside a progeny testing scheme, using the nucleus as a major source of young bulls for conventional progeny testing (Nicholas, 1979; Nicholas and Smith, 1983; Colleau, 1985, 1986; Christensen and Liboriussen, 1986; Goddard and Nicholas, 1987). A full account of this and other aspects of the use of MOET in dairy improvement is given by Ruane (1988).

Similar studies of the effect of MOET have now been conducted in other species, most notably sheep (Smith, 1986). As summarized by Smith (1988a), the general conclusion is that MOET will have a significant effect on rate of improvement in any species whose reproductive rate is naturally low. Thus by increasing female reproductive rate of cattle and sheep up to the levels that occur naturally in pigs and poultry, MOET provides rates of improvement in cattle and sheep that are very similar to those achieved naturally in pigs and poultry.

The longer-term role of MOET nucleus herds

One of the greatest virtues of a MOET nucleus is that it involves only a very small proportion of the total population of any breed; a far smaller proportion than that involved with an AI unit. It seems, therefore, that MOET offers new possibilities for the maintenance of genetic diversity within a breed, by providing the means for a number of independent units to each go their own way in attempting to breed more profitable livestock. As pointed out by Rendel and Robertson (1950c), although inevitably there will be a higher rate of inbreeding within each of a number of smaller units than in one interbreeding population, the undesirable effects of this inbreeding can be eliminated at the commercial level simply by ensuring that all commercial animals are produced by crossing between units.

On a related topic, Smith (1988b) has recently highlighted the paradox that the relatively sophisticated technology involved in MOET may offer the greatest hope for improving livestock populations in developing countries. Although the requisite resources are scarce in these countries, the relatively small scale of MOET nucleus herds enables these scarce resources to be concentrated and hence to be used more effectively, rather than being dispersed to much less effect in extensive field recording schemes.

Sexing of semen and embryos

Despite occasional claims to the contrary, it appears that semen sexing is not yet a practical possibility. There are, of course, many reports of slightly altered sex ratios resulting from subjecting semen to various treatments, but the ability to completely separate the two types of sperm has so far eluded scientists. However, since there obviously is a difference between X-bearing and Y-bearing sperm, it must only be a matter of time before someone discovers how to either separate the two sperm types, or eliminate one of the types from a semen sample.

More progress has been achieved in sexing embryos. In most cases, this involves sampling a few cells from a young embryo, testing these cells for the presence of a Y chromosome, and then if the embryo proves to be the desired sex, transferring the remaining portion into a recipient. Initially, cytogenetic methods were used to determine the embryo's sex, but these are now being replaced by immunological tests for the presence of a product of the Y chromosome, or by the use of DNA probes specific for some part of the Y chromosome.

Because of these developments, both actual and potential, geneticists have used the Robertson/Rendel approach to investigate the effect that these technologies could have on improvement programmes.

For a conventional large-scale progeny-testing scheme, Van Vleck (1981) showed that the ability to sex semen would enable a substantial increase in female selection intensity, which could result in an increase of around 25% in overall rate of improvement. In a MOET nucleus, sexing of embryos before transfer would enable the same rate of improvement to be achieved with fewer transfers, but would result in an increase in the rate of inbreeding (Nicholas and Smith, 1983). Alternatively, if the number of transfers remained the same, then sexing of embryos would result in a decrease in the rate of inbreeding.

Of course, there are other ways in which these technologies could substantially enhance the efficiency of animal production, e.g. in single-sex beef cattle systems (Taylor et al. 1986), but they are outside the subject area of this review.

Embryo splitting and cloning

Splitting embryos into two identicals is now a relatively common practice, but the regular production of larger numbers of identicals from a single embryo appears to be less practicable. While this technology is bound to improve in the future, it seems that geneticists can safely confine themselves to considering the implications of no more than eight identicals from an embryo. In fact, it turns out that in a MOET nucleus, splitting of embryos will produce relatively small increases in response to selection, and carries with it the substantial disadvantage that the rate of inbreeding is increased by a factor equal to the number of identicals into which each embryo is split. If the total number of transfers possible in a MOET nucleus is fixed, as will usually be the case in practice, then splitting embryos into any number of identicals appears to be of strictly limited value (Woolliams, 1989). Once again, however, there are other applications of this technology that may well have quite substantial effects on other areas of animal production.

Inbreeding and genetic drift

In studying the effect of MOET and/or any of the other associated technologies on livestock improvement, it has become customary to calculate the rate of inbreeding from the standard adaptation of Wright's (1931) formula, namely $\Delta F = 1/(2N_eL)$ per year, where N_e is effective population size (calculated from $N_e = (4sdL)/(s + d)$, with s males and d females enter the breeding population each year), and L is the average generation interval. For example, with 64 new females (donors) and 8 new males entering the population each year, and with a generation interval of 3.67 years, this formula gives the rate of inbreeding in the adult MOET scheme mentioned above as being 0.0013 per year, which is similar to rates that occur in national progeny testing programmes. Because they have shorter generation intervals, juvenile MOET schemes have substantially higher rates of inbreeding, up to four or five times the rates in comparable adult schemes (Nicholas and Smith, 1983).

Most genetic studies on reproductive technologies have gone no further than calculating ΔF, but it is appropriate to suggest that reproductive technologies should also be evaluated in terms of their effect on genetic drift. In his first published paper, Robertson signalled his lifelong interest in the problem of genetic drift by mentioning the chance loss of genes from cattle populations (Robertson and Asker, 1949). Largely as a result of his later work, together with the contributions of colleagues, we now have quite a clear understanding of the implications of genetic drift in practical breeding programmes. Furthermore, we now also have several relatively simple measures by which the practical implications of genetic drift can be assessed (see, for example, Hill, 1980).

One of the most important practical effects of genetic drift is variation in response. No matter how well-designed a selection programme is in terms of expected rate of improvement, the only way to be confident that the expected response will actually be achieved in practice is to ensure that the effective population size is not too small.

A simple measure that provides a very useful guideline is the coefficient of variation of response, CVR. After making various simplifying assumptions that are discussed by Hill (1980) and Nicholas (1980), a reasonable approximation is $CVR = (2L)^{1/2}/[G(N_et)^{1/2}]$, where G is the average of the product of intensity of selection and accuracy of selection for each of the four pathways of improvement, t is the number of years over which the programme is to be evaluated, and the other terms are as defined above.

Of course, the use of this approximation leaves open the question of what magnitude of CVR is acceptable. Opinions will differ, but most geneticists would probably be unwilling to stake their reputation (not to mention their chance of long-term employment) on a selection programme with a CVR of greater than 20%. On the other hand, they could be quite confident of achieving something near the expected result if the CVR was less than 5%.

Both the rate of inbreeding and the coefficient of variation of response are functions of effective population size, N_e. Unfortunately, the formula by which N_e is calculated (as shown above) strictly applies only when the s males and d females are chosen at random: it does not take account of the tendency for directional selection to concentrate on only the best families. Consequently, it underestimates the true values of ΔF and CVR in a selection programme.

What is needed is a more accurate measure of effective population size that takes account of the effects of directional selection, and that takes account of the increasingly common situation whereby replacements are chosen on the basis of selection indices or BLUP procedures incorporating data from several types of relatives. One of the earliest attempts to provide a better formula was made by Robertson (1961c), but the resultant expressions were of limited applicability and because of this, they have not been widely used. Other attempts have been made more recently, but a satisfactory and yet simple formula is still lacking.

One advantage of considering *CVR* as well as the rate of inbreeding when evaluating reproductive technologies, is that the effective population size generally needs to be larger to achieve a satisfactory *CVR* than to achieve an acceptable rate of inbreeding. In the adult MOET programme cited earlier, for example, a simple rearrangement of the formula for rate of inbreeding shows that an effective population size of 27 is required to keep the rate of inbreeding at around 0.005, which is commonly regarded as acceptable. In contrast, rearrangement of the formula for *CVR*, and calculation of *G* from the response and generation interval shown in Woolliams and Smith's (1988) Table 2, leads to the conclusion that in order for this same programme to have a *CVR* of, say, 10% after ten years, the effective population size needs to be around 100. If geneticists feel that a *CVR* of no greater than 5% after ten years is needed to ensure their continued employment, then an effective population size of around 400 is required.

By noting that the mating ratio in the above programme is 1 male : 8 females, and that all breeders are replaced each year, these requirements can be translated into the number of donors required each year. In order to satisfy the inbreeding criterion, only 17 donors per year are required; whereas to have a *CVR* of 10% after ten years, around 60 donors are required per year; and almost 250 donors are needed each year to achieve a *CVR* of 5% after ten years.

The more stringent requirement imposed by *CVR* will help to alleviate the current difficulty experienced in underestimating the true rate of inbreeding in selection programmes. It does not, however, remove the need for a better expression for N_e.

Conclusions

The basic approach to assessing the effect of new reproductive technology on animal improvement programmes was developed by Robertson and Rendel in 1950, and their proposal for the use of AI in progeny-testing schemes has been applied in dairy populations throughout the world. More recently, their approach has been used to evaluate both MOET and associated technologies. Further work remains to be done before we have a full understanding of the genetic and economic implications of these technologies. Although this work will employ procedures imported from other disciplines (systems analysis, discounted cash flow, etc.), it is a fitting tribute to Alan Robertson and Jim Rendel that the basic approach introduced by them so successfully in 1950 will still form the backbone of future studies.

Acknowledgement

Thanks are due to Chris Moran for comments on a draft of this paper, and to Gareth Evans for providing information on the current state of reproductive technology.

D GENETIC MANIPULATION AND ANIMAL BREEDING

Chapter 30
The Application of Molecular Manipulation to Livestock Improvement

R.B. LAND

AFRC Institute of Physiology and Genetics Research, Edinburgh Research Station, Roslin, Midlothian EH25 9PS, Scotland

Genetic change, whether it underlies animal improvement or evolution, depends upon mutation, recombination and sexual reproduction as the source of genetic variation. Alan Robertson quickly recognized that the advent of recombinant DNA technology gave the opportunity to control mutation and hence that it might be possible to introduce particular desired characteristics. He first addressed the application of this technology to animal breeding at Madrid, and subsequently at New Dehli and Lincoln, Nebraska (Robertson, 1982b, 1984, 1986a).

Writing in 1982 amidst the excitement of the early results of gene transfer in the mouse, but before the 'giant mouse' paper in *Nature* (Palmiter, Brinster *et al.*, 1982), he presented a series of maxims for the use of genetic engineering in animal improvement. Perhaps the most perceptive of these was that 'we may find that the methods (for gene transfer) are available before we know what to do with them'. 'Catching' genes (as he went on to call the identification of genes for transfer) is now the major issue facing the application of molecular biology to animal breeding. The genetic basis of transgenics and the technology itself are described by Bishop and by Church (in this volume), so 'catching' genes forms the core of the current chapter.

Single genes are unlikely to be useful

Gene transfer allows the transfer of single genes. What evidence is there that single genes are likely to be useful? Firstly, if one looks across the principal domestic species and excludes genes which affect traits such as coat colour and horns, there are few examples of single genes which have useful effects. The exceptions are well known: the Booroola and Thoka genes which affect fecundity in the sheep, the double muscling gene in cattle and the 'halothane' gene in pigs which increase the yield of lean meat, and the dwarfing gene in poultry. Single genes with major favourable effects on production traits can be quickly fixed by selection, but the examples above all have disadvantageous as well as advantageous effects. Litter size can be too high with the Booroola gene, the doubling muscling gene increases calving difficulties, the 'halothane' gene depresses fecundity and meat quality, and dwarfing is only useful in the exceptional circumstances of reducing feed consumption in broiler mother strains selected for high rates of growth and hence adult size. Secondly, if one looks at biochemical control theory it can be seen that single gene transfers are unlikely to be useful. Kacser and Burns (1979) developed this in the context of understanding biochemical systems, but it is equally relevant to gene transfer in circumstances where the gene concerned codes for an enzyme. They consider metabolic processes in terms of fluxes and use the concept of sensitivity coefficient to describe the effect of changing the level of any one enzyme in the overall rate of flux through the system. The important characteristic of the sensitivity coefficients is that for any system the sum of the coefficients is one.

Any change in the level of one enzyme will reduce its coefficient and increase others, no matter whether the change is made artificially or genetically by gene transfer, so there will be diminishing returns to increasing the level of any enzyme.

With little evidence for the importance of single gene segregation in existing systems it is therefore important to look for exceptions, and indeed Robertson (1982b, 1984, 1986a) does so, although not formally.

Strategies for gene 'catching'

Analogies with man

Alan Robertson was quick to point out the difference between application of molecular biological technology to farm animals and to humans.

> The doctor is usually concerned with a quite specific defect which he wishes to cure – to restore normality. The animal breeder begins with the normal individual and wishes to improve some aspects of its performance. The doctor knows in many cases that one defect is due to a single identifiable gene. The animal breeder does not – we shall see later that one critical problem is to arrive at this level of understanding. (Robertson, 1982b)

> Superficially it might seem that the improvement of the normal (animal breeding) is simply the mirror of the restoration of the defective (medicine). The difference however is fundamental and biochemical control theory indicates why. Although increases in the activity of individual steps in a metabolic or physiological pathway do little to increase the activity of the pathway as a whole, the absence of any step in an unbranched pathway would block it. A recessive defective gene could have such an effect and many of the human genetic diseases are caused by such alleles. They not only have large effects and are therefore identifiable, as recessives they persist at low frequencies once established in large populations. By contrast, in livestock any recessive with a major favourable effect would either be selected rapidly to fixation or remain sufficiently rare as to be undetected.

Higher control systems

> Most biological processes are under the influence of feedback controls operating at many levels – physiological, hormonal and genetic. A mere increase in the number of genes coding for a particular product, even though the problem of regulation has been solved, need not lead to a proportional increase of the rate of accretion of the final product. Might it not be more sensible to make changes at the control level – many of the hormones involved in animal growth, lactation and reproduction are polypeptides whose sequences are specified by genes. (Robertson, 1982b)

This prediction was to be vindicated within months with the demonstration that the human metallothionein-rat growth hormone construct dramatically increased the growth rate and adult body size of mice. As Robertson put it in 1984, 'the peptide hormones, already used so brilliantly by Palmiter, Brinster et al. (1982) are an obvious target'.

As with the anologies between human medicine and animal breeding, the arguments bear development. The growth hormone construct of Palmiter, Brinster et al. (1982) affected growth because the metallothionein promoter directed expression to the liver and other tissues and hence escaped the normal feedback control of the expression of the gene in the pituitary gland. The mouse experiment demonstrates the opportunity for transgenic modification of higher control systems. When metallothionein-growth hormone genes were transferred to pigs the animals exhibited increased feed efficiency and reduced backfat. However the animals did not grow faster and the chronic, deregulated expression of growth hormone from the metallothionein promoter caused a number of adverse effects including arthritis, susceptibility to stress and reduced fecundity.

Rather than override the control system another approach would be to modulate the system and 'set it at a higher level'. The immune modulation of feedback systems indicates that this would be likely to be successful. In the case of reproduction, for example, the immunosuppression of the ovarian feedback hormones which normally restrict the secretion of gonadotrophins from the pituitary gland leads to an increase in the ovulation rate and litter size of sheep without any increase in the variance. Feedback control is retained but at a different level.

In the case of growth hormone a comparable approach would be to decrease the activity of the feedback system which normally inhibits secretion. This could be achieved if somatostatin (the inhibitory hypothalamic peptide concerned) were less active or the sequences responsible for its regulation less sensitive. We are some way from the ability to identify and modify control sequences but the advantages of being able to do so are clear. A more subtle intermediate approach might be to link the growth hormone gene to a regulator which would induce its expression in desired circumstances, for example at a specific phase of growth or during lactation.

To return to reproduction, genetic manipulation to reduce the rate of synthesis of feedback steroids through alteration of the enzyme systems concerned or to reduce the efficiency of the protein hormones (inhibin), would both be expected to modify the balance between stimulation and feedback and so increase the number of eggs shed. The attainment of twinning in cattle might succumb more rapidly to a molecular approach than to conventional selection.

Structural proteins

The rate of production of proteins is not related simply to the number of copies of the structural gene. However, where the production trait is a protein, the physical characteristics of that protein might be altered by changes to its chemical characteristics. The properties of caseins in milk, for example, could be changed to make them more suitable for cheese making or to give a more appropriate base product for other uses in the food industry. Some caseins stabilize emulsions and this might be better achieved by changes to the balance between the hydrophilic and hydrophobic parts of the molecule.

Meat, white of egg and wool might also be changed if we knew the nature of the changes which would make them more desirable. The rate of accretion of protein in meat, however, is likely to be influenced by the enzymes responsible for synthesis and degradation rather than the structure of the protein itself (see Bulfield, this volume). Robertson (1986a) draws attention to the possibility that a decrease of the proteolytic enzymes responsible for the breakdown of muscle proteins might be economically desirable.

Multiple constructs

Some critical pathways may contain few enzymes; more than one structural gene might be transferred – either simultaneously or separately and then 'combined' by crossing. An elegant example of where this might be useful is given by the work of Ward *et al.* (1986). The rate of synthesis of wool is limited by the availability of the essential amino acid cysteine. Serine is available but the enzyme systems are not present to enable the sheep to convert it to cysteine. Appropriate enzymes (serine transacetylase and O-acetylserine sulphydrylase) are present in *Escherichia coli*. The success of their transfer to sheep will be awaited with great interest, not only with regard to wool production but more importantly to verify the practicality of the principle itself.

Biomedical proteins

A final example of the way in which genes may be caught takes us outside the range of traditional animal products. Rather than find 'uses' for existing animal proteins it is now possible to start with the proteins required and consider how best to harvest them. An extreme example is where proteins such as Factor VIII required to treat haemophilia can be obtained at present only from human blood. With the requirements of post-translational processing this protein is not amenable to production in prokaryote systems and the principle that the secretory powers of the mammary gland might be harnessed in this way has now been demonstrated (Simons *et al.*, 1987). They have gone on to pioneer this approach using DNA sequences encoding Factor IX and α-1-antitrypsin linked to the sheep ß-lactoglobulin promoter.

Early results with the ß-lactoglobulin gene itself were in themselves exciting. Even though normal mice do not have this gene, it was expressed specifically in the mammary gland and specifically during lactation in transgenic mice. Pilot fusion genes have been incorporated into sheep and express the protein at meaningful concentrations. It is now a question of identifying appropriate control sequences to facilitate useful levels of expression and of establishing the downstream processing and acceptance of the technology. Once established it adds a new dimension to the relationship between livestock and mankind.

Prospects

Robertson recognized from the beginning that the identification of useful genes would be much more difficult than their transfer and this has proved correct. In considering the subject he drew greatly on his experiences in population genetics in which he was so influential. For the future, success is likely to be dependent upon the combination of a range of disciplines. Further understanding of molecular genetics in combination with the appropriate biometry will facilitate the design, transfer and evaluation of genes.

The combination of physiological genetics with traditional physiology and biochemistry, could be the key to the identification of genes. One approach is to try to identify the genes which are responsible for existing variation, either within or between populations, and the expression 'gene catching' implies that the search is for existing alleles. This is logical but *it misses the point that molecular biology offers the opportunity to design genes with specific, desired, effects*. The interactions between molecular biology and physiology are therefore two-fold. They provide the potential to catch existing genes. More importantly physiological understanding of how animals actually work has the potential to indicate the type of physiological events likely to have a favourable effect while new knowledge of molecular biology indicates how best genes might be designed to induce such an effect. Like much in science the interactions are likely to be two-way. Physiological knowledge may indicate the route to transgenes; the study of the effects of transgenes will contribute to physiological knowledge.

Greater rates of incorporation are likely to be developed through systems based on embryonal stem cells and nuclear transfer. Retroviral based vectors may have a role, particularly in poultry where the germinal disc is opaque and fertilization polyspermic. These topics are outwith this review but even with present rates of incorporation the resources needed to evaluate transgenes are likely to be substantially greater than those needed to create them. It is not just a question of detecting favourable effects, indeed this is likely to be the simplest task, it is also important that systems should detect unfavourable effects. Given a response to selection of 2% per generation and say, five generations to create and evaluate a transgene, an appropriate minimum effect of a potentially useful transgene on economic merit would be 10%. Smith *et al.* (1987) point out that for a trait such as growth with a coefficient of variation of 10%, 35 animals might be a reasonable number to test. However 5% deterioration in a trait such as reproduction could well more than outweigh the advantage, and with a coefficient of variation of 25%, reproductive performance on a sample 800 animals would be required.

The investment needed to establish transgenics will therefore be considerable and beyond that normally employed in breeding programmes. Fortunately the technology itself not only needs the investment but also provides the means by which it might be obtained. Transgenes can be used to identify the origin of superior stock and there is no longer any reason why animal breeders should not be accorded the commercial protection of their investment so long enjoyed by the plant breeder. Perhaps a corollary of this protection should be a comparable responsibility to test and demonstrate the superiority of the strain. An important factor in the evaluation of transgenes is that while advantages may be apparent in the hemizygote, disadvantages may not be manifest until the transgene is homozygous. Homozygous as well as hemizygous stock would therefore have to be tested.

Robertson, always alert to discovery, recognized the enormous excitement and potential of molecular biology. Not only did he point out the opportunities, he also outlined some of the problems to be resolved. He will continue to challenge students of animal breeding well into the future. To close with two examples, control theory and human diseases show how single alleles are much more likely to have major effects if they 'remove' a protein than if they add one. The success of research to turn genes off could therefore have a major impact on the extent to which molecular biology is applied to enhance the characteristics of livestock. With regard to the choice of trait, we have seen that the investment required is considerable. Equally we know that traits such as growth can be readily changed by conventional selection.

Acknowledgements

The editors wish to thank John Clark for helpful comments on the late Roger Land's draft manuscript.

Chapter 31
Quantitative Variation and the Control Analysis of Enzyme Systems

H. KACSER

Department of Genetics, University of Edinburgh, West Mains Road, Edinburgh EH9 3JN, Scotland

Anybody interested in quantitative variation must at some time face the question of its biochemical basis. The methodology of classical population studies does not require an answer to that question and such studies rarely address it. Genetic variation is judged by its outcome. 'By their fruits ye shall know them' (Matthew) seems to be an adequate motto for this approach. In the last 50 years or so, however, an increasing body of work in biochemistry and its offspring molecular biology, has been concerned with the nature of the gene, the mechanism of gene expression and the 'control' of the gene product, usually a protein. Additionally, enzymology has accumulated a large body of information on the mechanism of catalysis by these proteins, the variety of different enzymes and their organization in metabolism. The elucidation of these problems has been the outstanding achievement in these fields. The principal methodology has been *in vitro* experimentation rather than the study of whole organisms. '*In vitro veritas*' (Sidney) seems to be the guiding principle which informs this approach.

Although both approaches are ostensibly concerned with the same question – the role which the genes play – their practitioners largely do not talk to each other and they use different languages. There is a gap, not only in mutual understanding, but also in applying the results in one field to the problems in the other. This gap has been brought into sharp focus by one of the most interesting recent advances, namely genetic engineering.

If it is desired to 'improve' a phenotype by using this technology – be it increased antibiotic production in a micro-organism or reduced fat content in a pig – the question as to which genes to manipulate has to be addressed. The molecular biologist may be able to identify, clone and insert a chosen gene into a chosen organism and may even succeed in realizing its expression in the appropriate tissue, resulting in the production of an altered gene product or, by an increase in gene dose, a higher concentration of the same product. While such techniques are essential, by themselves they are insufficient to achieve the aim of altering the phenotype in a specified way. Even if the gene product is a well-characterized enzyme with a specified role in the metabolic map, the net effect on the phenotype is not predictable. The highly interactive nature of metabolism makes it impossible to assert a simple enzyme–phenotype relationship. The art of genetic manipulation, with all its deep understanding of the processes of gene expression, is uninformative of its effects beyond the protein. The idea of manipulating 'the rate limiting enzyme' is largely an illusion, and the choice of gene (or genes) to manipulate remains the problem.

Similarly, the quantitative geneticist cannot identify candidate genes for the improvement programme. Even a selected line, presumably containing all the 'positive' alleles and epistatically favourable combinations, does not provide information on where to look for suitable sites (see, however, Bulfield, this volume). Where 'genes with large effects' can be located to relatively small regions of a chromosome, (for example in drosophila (Shrimpton and Robertson, 1988b)), these regions are still much larger than single genes. In any case, it is an entirely open question whether further substitutions or manipulations at these regions (loci) will be effective. Furthermore, it is perfectly possible that substitutions at other loci at which there happened to be no variation in the population, would be more effective in improving the desired phenotype.

The above considerations may sound like a counsel of despair but they are really intended to point to the nature of the gap and give directions to a search to close it. In what follows a method, called control analysis, will be described which makes theoretical and experimental contributions to this ecumenical task.

Gene products and phenotypes

In trying to establish the link between the genotype and the phenotype it is obvious that the biochemical and physiological processes underlying all phenotypic manifestations are 'controlled', or at least influenced, by the activities of the gene products of which catalytically active enzymes are the best studied. We now have a clear picture of the elaborate organization of metabolism where different enzymes are kinetically coupled to each other in pathways and cycles, and coupled further via a variety of effectors arising in one part of the system and acting in another. The rate of each individual transformation step – the functional unit of metabolism – can be described in a standard way by the genetically determined parameters of the enzyme (its concentration, turnover number and affinities for the various metabolites on which it is acting, etc.) and by the variables of the system (the concentrations of the intermediate metabolites and the flux through the step). Such formulations (the rate equations), one for each transformation step, are, in general, nonlinear functions of the variables and therefore a complete description of the system would involve many thousands of simultaneous nonlinear equations. These are not solvable analytically and are impractical to treat computationally even in the unlikely case that sufficient quantitative information is available for all the enzymes in the organism.

This dilemma can be sidestepped by considering only the effects of *variation* in the enzymes on some output of the system. In essence it is a version of the genetic methodology applied to the biochemical processes *in vivo*. When as geneticists we are considering an allelic substitution, as biochemists we are considering a change in the rate of an elementary transformation step. One can impose a change in activity to one of the enzymes (for example by the use of an inhibitor) or use a naturally occurring enzyme change (caused by the substitution of an allele) and measure the resulting change on some chosen flux or metabolite (character), thereby relating a change in a 'local' rate to the net 'global' effect *in vivo*.

This approach is demonstrated by the molecular analysis of the phenomenon of dominance (Kacser and Burns, 1981). A large number of mutants were known in man, fungi and bacteria whose effect was the virtual elimination in the mutant of any catalytic activity of the enzyme specified by the wild-type allele. When the heterozygotes (or heterokaryons) were examined, the extracted enzyme activity was normally found to be the mean of the two homozygotes (additivity), i.e. about 50% of wild-type, in contrast to the almost complete recessivity of the mutant when the enzyme was embedded in the rest of metabolism and its effects were measured *in vivo*. Careful measurements showed that the rate sustained *in vivo*, the system flux, and the steady-state concentrations of the metabolites were virtually identical for heterozygote and wild-type homozygote. If the change in enzyme activity in the heterozygote is denoted by ΔE and the change in the system flux by ΔJ, the ratio of 'effect'/'cause', expressed as fractional changes, is for these cases:

$$(\Delta J / J)/(\Delta E / E) \approx 0.$$

This apparent paradox of additivity at the (extracted) enzyme activity level but (almost complete) dominance at the organismal (character) level was resolved when the analysis showed that this is the expected outcome in a multi-enzyme system when alleles with greatly differing activities of their enzyme products are considered (Kacser and Burns, 1981). The essence of the phenomenon is that there is a highly nonlinear (hyperbolic) relationship between activity of any one of the enzymes of a pathway and the system flux sustained by all of them (Figure 31.1). Thus, a substantial reduction in activity (homozygotes in near-null mutants) gives a very low system flux while the heterozygote (approximately

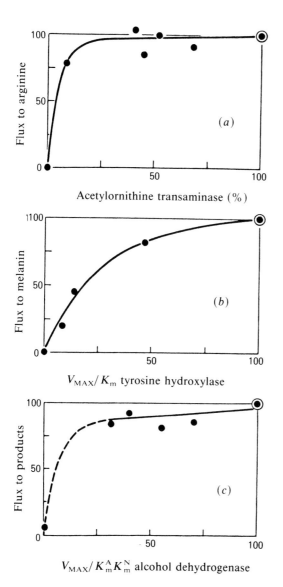

Figure 31.1. Three examples of flux–enzyme relationships *in vivo*: (*a*) Neurospora crassa: one enzyme in the arginine pathway (from Flint *et al.*, 1981); (*b*) Mouse: one enzyme in the melanin pathway (from Kacser and Burns, 1981); (*c*) Drosphila: the first enzyme in the ethanol catabolism pathway (from Middleton and Kacser, 1983).

50% activity) has a flux which is still on the plateau of the curve, often not distinguishable from the wild-type flux. Dominance was therefore seen to be the automatic result of the interactions inherent in the structure of metabolic systems. No 'dominance modifiers', allegedly accumulated by natural selection, are therefore required to explain the observed phenomena. The treatment was extended (Keightley and Kacser, 1987) to deal with the dominance of pleiotropically related characters. It was shown that the common observation, that the same dominance relationship was observable when the effects of allelic substitutions were measured on different characters, was the most likely expectation based on the kinetic structure of metabolism.

These examples serve to introduce a more rigorous analysis of multi-enzyme systems (Kacser and Burns, 1973) which lead to the development of methods and concepts to treat enzyme variation *in vivo*. (For reviews see e.g. Westerhoff *et al.*, 1984; Kacser and Porteous, 1987; Kacser, 1987.)

Control coefficients

In view of the nonlinear relationship between enzyme activity and net metabolic flux observed (Figure 31.1) and expected on theoretical grounds (see below and Kacser and Burns, 1979), the formulation $(\Delta J/J)/(\Delta E/E)$ is an unsatisfactory general measure of *in vivo* effects of enzyme variation because the ratio depends on the size of ΔE. In control analysis we therefore consider infinitesimal changes. As $\Delta E \to 0$ we obtain

$$\frac{\partial J_j}{J_j} \Big/ \frac{\partial E_i}{E_i} = \frac{\partial J_i}{\partial E_i} \times \frac{E_i}{J_j} = \text{slope} \times \text{scaling factor}$$

or

$$C_{E_i}^{J} = (\partial \ln J_j)/(\partial \ln E_i).$$

This ratio, called the control coefficient of a system flux, J_j, with respect to an enzyme activity, E_i, is a dimensionless number whose magnitude specifies the 'importance' of that particular enzyme in controlling the flux in that particular system under one particular set of conditions and one particular set of enzyme activities. The condition of the system is usually assumed to be the steady state. If the magnitude of the control coefficient is small, say 0.01, (as it often is) it means that this enzyme, and hence locus, has very little control and increases of its activity, either specific activity or concentration, is going to have little impact on the dependent flux and any character related to it. Such loci are therefore unlikely to be useful candidates for genetic manipulation, whereas large coefficients, if found, identify loci that are potential targets. It should, however, be emphasized that the value of the coefficients changes not only with value of the enzyme activity to which it refers (see Figure 31.1) but can also change with changes in other enzymes in the system (genetic background) and in environmental and physiological conditions. The control coefficient is therefore itself a systemic property, showing epistatic, pleiotropic and environmental influences. Much of biochemistry, based largely on *in vitro* experimentation, uses such metaphors as 'key-enzyme', 'rate-limiting-step', 'pace-makers' etc. and fails to appreciate the fundamentally different behaviour of enzymes when they are acting in concert with other enzymes. Similarly, the literature on the 'control of gene expression' stops short at the point where the functional properties of the gene products enter the functional description of the flow of metabolites and their concentrations in the physiology of the organisms. Some progress has been made in using cloned genes to study metabolism (Walsh and Koshland, 1985; Brindle, 1988).

The concept and experimental estimates of the control coefficient is central to this approach. The flux specified need not be the pathway in which the enzyme is acting. Every flux in the system will be affected by changes in one enzyme, some of course more than others, specified by the respective control coefficients: $C_{E_k}^{J_k}$, $C_{E_i}^{J_l}$, $C_{E_i}^{J_m}$, ..., *Mutatus mutandis*, any one flux will be affected by changes in every enzyme in the system, and again, some enzymes will have greater effect than others.

There are, of course, similar coefficients for substrates, defined by

$$\frac{\partial S_j}{S_j} \Big/ \frac{\partial E_i}{E_i} = \frac{\partial \ln S_j}{\partial \ln E_i} = C_{E_i}^{S_j},$$

and designated the concentration control coefficients. Again, there are as many concentration control coefficients for one metabolite as there are enzymes in the system, and any one enzyme will specify as many concentration control coefficients as there are metabolites in the system.

One of the clearest conclusions to emerge from these studies is that control is distributed among many steps in metabolism (Groen *et al.*, 1982; Salter *et al.*, 1986). We should therefore no longer ask 'Where is the control', but 'What is the quantitative distribution of control?' (For experimental evidence, see references in Kacser and Porteous, 1987.) Such a distribution has immediate practical consequences in the 'improvement strategy' by genetic manipulation. Insertion of a number of genes simultaneously is more likely to be the method of choice since no single enzyme may have a large enough coefficient (e.g. Flint *et al.*, 1981). Identification of potential targets by the methods of control analysis is therefore a first step in the strategy, but there are constraints on the values of the coefficients which must be considered.

The summation theorems

The underlying assumption of the treatment of metabolic systems is that the enzymes are independently acting catalysts, one for each chemical transformation. Such an independent array of catalysts explains the existence of the 'metabolic map' which implies that there is coupling between the enzymes via the substrates/products of adjacent steps. Although the enzymes are independent of each other, the *rates* sustained by them are not. The effects of changing one enzyme activity will therefore spread throughout the metabolic system, via changes in metabolite concentrations to all the other enzymes which have not changed. Since the control coefficients measure these effects, it is clear that there is a coupling between their magnitudes. The constraints are displayed in the Summation Theorems:

$$\sum_{i=1}^{n} C_{E_i}^{J} = 1 \qquad \text{Summation Theorem of Fluxes}$$

and

$$\sum_{i=1}^{n} C_{E_i}^{S_j} = 0 \qquad \text{Summation Theorem of Metabolites.}$$

The Flux Summation Theorem states that the sum of the coefficients of one particular flux with respect to all the enzymes in the system equals unity. Leaving the existence of negative control coefficients aside for the moment, it is clear that the 'average coefficient' is $1/n$ where n is the number of enzymes. Since this number is very large, say 5000, this average coefficient is likely to be very small. We may assume that there will be a skewed distribution of values so that most of them will be virtually zero while a few, perhaps 20, will have measurable values. The 'zero class' of enzymes are likely to be those which are 'distant' (in the kinetic sense) from the point where the flux is measured. A zero coefficient with respect to one flux/character does not, of course, imply that the same enzyme variation cannot have a significant effect on another flux, one that is 'nearer' to the action of the enzyme. This is the reason for the practical observation that an allele, shown to affect character 1, may have 'no effect' on character 2. This makes Mendelian segregation possible without massive pleiotropies being observed at all levels of observation. Conversely, pleiotropic or correlated responses are possible and indeed not an uncommon feature of genetic variation (Wright, 1934a; Keightley and Kacser, 1987).

The argument so far has assumed there are only positive coefficients, as would be the case in an unbranched pathway. A realistic model of metabolism, however, will have a number of branch points where a substrate is the beginning of two competing pathway branches. Variation in the enzymes of one branch will have a negative effect on the flux of the competing branch.

For divided pathways the summation relationships are slightly more complicated (Kacser, 1983a; Fell and Sauro, 1985) and involve measurement of the relative fluxes in two branches.

$$\longrightarrow (J_T) \xrightarrow{E_T} \quad S \underset{E_2 \, (J_2)}{\overset{E_1 \, (J_1)}{\diagup \diagdown}}$$

In such a branched system, negative flux control coefficients arise: Since the fluxes are constrained by the relationship $J_T = J_1 + J_2$, an increase in E_1 increases J_1 i.e. $C_{E_1}^{J_1}$ is positive but reduces the flux J_2, i.e. $C_{E_1}^{J_2}$ is negative and $C_{E_1}^{J_T}$ is positive. Again, the summation property applies. Taking any *one* flux, say J_1,

$$C_{E_1}^{J_1} + C_{E_2}^{J_1} + C_{E_3}^{J_1} = 1.$$

Since $C_{E_2}^{J_1}$ is negative, the sum of the other two (positive) coefficients must exceed unity which leaves it open for one of them to exceed unity (Kacser, 1983a). No verbal device of labelling such steps (rate-limiting etc.) would adequately describe this condition, and only the assignment of the actual magnitude of the coefficient would be necessary and sufficient quantitation of the control exercised by the enzyme.

Negative coefficients are, in practice, much rarer than a casual glance at the metabolic map may suggest because, quite frequently, a branch is the beginning of a loop which 'returns' positively into the main flux.

Selection

Although control analysis and the resulting theorems apply to small (infinitesimal) changes, the movement of metabolites and fluxes are often reasonably monotonic so that an extrapolation to larger, finite, changes can usually be made. The important statement of the summation theorem is that *at any constellation* of enzyme activities, $\sum C = 1$. If, therefore, a large change has been effected (by allelic substitution or by the increase in the gene dose) the new sum will again equal one but the distribution will have changed. The large change in one enzyme will alter its control coefficient and, as an automatic consequence, all other coefficients will change (although their activities have remained the same), so there is a built-in epistasis in such systems. Unlike classical population treatments, this epistatic component is not arbitrary but precisely defined by the structure and parameter values of the system. The consequences to our view of selection are important.

Let us assume we have multiple alleles at several loci in a pathway contributing variation to the flux, but that under the initial conditions this flux is a very small component of the fitness. The alleles will therefore be virtually 'neutral'. Let us now assume that the artificial selection regime is changed to maximize this flux. The alleles will now cease to be 'neutral' and we may think of the consequences as a series of stages, although in fact all processes will proceed simultaneously. The change in allelic frequencies will now proceed by 'first' increasing the alleles of the locus with the highest (average) control coefficient, since changes at that locus will give the greatest response. Two consequences will follow. The average control coefficient of the 'selected' locus will decline as the selection proceeds, and the control coefficients at the other loci (which had lower coefficients) will increase. What had been a small component of the variance of the flux at these loci has now become greater and 'in the second round' they have become more selectable. This 'bootstrap' effect is due to the changing distribution of control coefficients as selection proceeds which therefore would continuously change the selection coefficients. Assumptions of constant selection coefficients, usual in population models, are therefore not the expectations given by the constraints of the summation theorem.

Hartl and his colleagues (Dean *et al.* 1986, 1988; Dykhuizen *et al.*, 1987) have applied the general approach of control analysis to their experiments on selection of *Escherichia coli* in chemostat cultures. Variation of two enzymes in the pathway of lactose utilization was obtained by either genetic substitutions or physiological means. This enabled them to apportion control to the permease, catalysing the transport of lactose into the cells, and to the β-galactosidase, the first step of hydrolysis in the utilization of the sugar. They showed that the induction of the operon (which increases both enzymes coordinately) has its principal effect through increase in activity of permease which has a control coefficient of 0.55, while the β-galactosidase, with a control coefficient of about 0.02, contributes almost nothing to the increase in growth rate. At high levels of induction both enzymes have 'lost' almost all control and enzymes catalysing other steps of sugar diffusion and catabolism jointly control growth.

Considerations of these results and evidence from other studies on *in vivo* control coefficients and polymorphisms of enzyme variants, resulted in the suggestion (Hartl *et al.*, 1985) that there was a general tendency for many genes to evolve towards selective (near) neutrality (see also the ensuing debate: Burton and Place, 1986; Hartl *et al.*, 1986; Dean *et al.*, 1988). While this may well be a mechanism in cases where directional selection acts on some output which is a major component of fitness, there is also a nonselective mechanism tending towards near-neutrality. The constraint of the Summation Theorem implies that, even in the absence of strong selection, the average control coefficient of the enzymes involved in the output will be small simply because of the number of genes (and hence enzymes) involved. Small variations around the 'wild-type', the most frequent class of variants, will therefore contribute little individually to the variance in fitness.

The Summation Theorem for metabolite pools, $\sum_{i=1}^{n} C_{E_i}^{S} = 0$, shows other interesting properties. In some cases the concentration of a pool can be regarded as a character or some simple function of a character. The recessivity of a pool-character for near-null mutants has exactly the same explanation as that for fluxes (Kacser and Burns, 1981).

The summation to zero (as opposed to the summation to unity for fluxes) implies that, in a population in mutation–natural selection equilibrium subjected to strong directional selection for a particular pool, the tendency to maximize this pool would automatically result in *not* maximizing the fluxes associated with the variation. There is therefore variation in such a population with the potential to be selected for both increased or decreased flux. If such selection experiments were carried out, the results could then be interpreted by assuming that the population had been under selection for a flux optimum, while the proper explanation for the observation is simply the opposing pleiotropic consequences of variation on certain pools and fluxes (see also Beaumont and Clarke, 1988, Int. Congr. Genet. Toronto). The 'rejection of extremes by natural selection' which is often given as an 'explanation', can now be seen as the outcome of certain complex interactive events in metabolism.

In this brief sketch of the application and implication of control analysis to quantitative genetics, much of the algebraic treatment and theoretical foundations have been omitted. In particular no mention has been made of the relationship of the global control coefficients to the properties of the individual enzymes which, jointly, must give rise to the behaviour of the system represented by the so-called elasticity coefficients. Theoretical treatment of these gives rise to the second set of theorems, the connectivity theorems of flux and of metabolites. Additional relationships to deal with conserved metabolites (Hofmeyr *et al.*, 1986), cycles (Fell and Sauro, 1985; Sorribas and Bartrons, 1986) and branched pathways (Kacser, 1983a; Sauro *et al.*, 1987) have been established.

The main purpose of this survey has been to point to the impact of a radically new approach to the old and almost abandoned problems of the biochemical and physiological basis of genetic variation. In our vigorous pursuit of the technology of genetic manipulation, we are in danger of ignoring the questions which the techniques set out to answer. The analytical approach – to look at ever smaller pieces of the organism in ever greater detail – must be supplemented by considering the whole. Control analysis is a method – of thought and of experimentation – which, by gentle probing, discovers how each part responds when it is embedded in the matrix of all other parts. It thereby bridges the gap which has developed between the study of the gene and the study of the whole organism.

Chapter 32
The Biochemical Control of Quantitative Traits

Grahame BULFIELD

AFRC Institute of Animal Physiology and Genetics Research, Edinburgh Research Station, Roslin, Midlothian EH25 9PS, Scotland

I first came to think about the problem of the biochemical control of quantitative traits in October 1963, after I had heard Alan Robertson speak at a meeting at the British Oil and Cake Mill (BOCM)'s experimental farm at Barlby in Yorkshire. In response to a now-forgotten question, he replied that for animal breeding to progress past the point of selecting for single characteristics like growth using statistical techniques, we had to understand the biochemistry and physiology of processes such as fat deposition and muscle growth. I have to admit that 25 years have undoubtedly transposed his reply into my own terms, but that must have been the essence of his answer as I have clearly had this conundrum in mind all these years. More recently Alan Robertson has been interested in two areas: the nature of the switch of ingested nutrients to either protein or fat deposition, and the application of molecular genetic techniques to livestock improvement. It is the interface of these areas: biochemistry, molecular genetics and animal breeding (where he identified the problems so clearly) that is the topic of this review.

Problems

The genetic manipulation of laboratory animals is now routine (for a review see Palmiter and Brinster, 1986). Application of these techniques to farm animals, although logistically complicated and expensive, has been achieved for cattle, pigs and sheep (Brem et al., 1986; Simons et al., 1988) and will soon be possible for poultry (Bulfield, 1985; Perry, 1988). The ability to insert single cloned genes into farm animals, have them expressed in ectopic sites and designed so that they respond to specific regulatory signals, is a powerful new tool in the hands of the animal breeders. The major problem preventing the widespread application of this technique is our almost complete lack of knowledge of the 'trait-genes' controlling the major important commercial characteristics such as fertility, muscle growth, lactation, disease resistance and egg production. We do not know the numbers of these trait-genes, their nature, chromosomal locations or primary biochemical products. This ignorance is a major and serious obstacle preventing the application of genetic manipulation to animal improvement.

Conversely, we know the products of *only* around 1–2 000 mammalian genes; these are mainly the classical enzymes of intermediary metabolism, the protein hormones and hormone receptors. The vast majority of the approximately 100 000 expressed mammalian genes are unknown to us; each cell type translates around 2000 proteins (Duncan and McConkey, 1982) of which about half appear to be cell specific (Klose, 1982).

We must therefore face a problem of some magnitude before we will be able to obtain an understanding of the molecular and biochemical control of commercially important traits.

Solutions

With the present paucity of knowledge on the biochemical control of commercially important characteristics, there is no single strategy capable of identifying trait-genes simply and with certainty.

I have therefore divided potential solutions to this problem into three approaches: First, the analysis of genes and gene-products whose role has been suggested from the work of physiologists and biochemists. Second, 'dissecting' a commercial trait into its physiological components with the aim of eventually locating the products of trait-genes through their actions – the 'top-down' approach. Third, finding a way of displaying and assessing the expression of a large number of genes or gene products at random (as already mentioned over 90% of genes are unknown to us) – the 'bottom-up' approach. Normal biochemical or physiological analyses are not on their own satisfactory to make any of these approaches successful as we have no way of knowing, for example, which factor in a physiological system is the important one (Kacser, this volume). We need some way to modulate commercial characteristics genetically and look for response in expression of putative trait-genes.

There are two ways available at present of doing this; the first is to use genetically engineered 'transgenic' animals themselves and the second is to use animals divergently selected for commercially important traits.

The first strategy, the use of transgenic animals, can be exemplified by the classic paper of Palmiter, Brinster et al., (1982) where 'giant' mice had been produced. These animals, when single-celled embryos, had several hundred copies of a cloned growth-hormone gene injected into one of their pronuclei (Bishop and Al-Shawi, this volume); the resulting animals produced up to one thousand times the normal levels of circulating growth hormone and grew to about twice the size of their normal littermates. Therefore, in principle, any cloned gene can be inserted into transgenic animals and its expression correctly regulated in order to assess its effect on a commercially important trait like growth. Fortunately many of the genes we are interested in seem to be highly conserved, for example luteinizing-hormone, releasing hormone (one amino acid different between man and chicken) and growth factors and their receptors (Heldin and Westermark, 1984). This conservation makes the analysis of these transgenes in mice a valuable approach. It is of course a time-consuming and lengthy analysis and would only be performed with genes for which physiologists and biochemists have produced strong evidence that they are implicated in the control of a trait. For the other two approaches, the 'top-down' from the trait to the gene or, the 'bottom-up' from the gene to the trait, it is not possible to make use of transgenic technology until a gene has been identified.

We are therefore left with the second strategy, to use animals divergently selected for commercially important traits. Fortunately we have at our disposal valuable genetic resources in strains of laboratory and farm animals which have been divergently selected for those traits. For example, there are lines of mice and chickens available that have been selected for growth (Falconer, 1973; see review by Bulfield, 1980) or some component of it such as appetite, fat content or muscle development (Sharp et al., 1984; Bulfield et al., 1988; McCarthy and Roberts, this volume). These strains can be used to attempt to assess which physiological and biochemical parameters have been affected by the selection process. That many of the strains are replicated from the same or similar base populations accounts for the problem of chance events; i.e. genetic drift, occurring during the selection process and giving a false indication of the results of selection (Falconer, 1973). What remains is to identify the loci controlling these traits and this is a daunting task as demonstrated by the inability, despite a wealth of research, to identify the primary lesions in such gross mutants as the *obese* and *dwarf* genes in the mouse (reviewed by Bulfield, 1980).

Results

Mice

As an illustration of the results obtained I shall concentrate on growth as a commercially important trait; most work has in any case been done on growth or some component of it. The work up to 1980 in the area has been reviewed previously (Bulfield, 1980) and can be summarized as follows. First a substantial number of physiological differences have been found between animals selected for growth (weight-for-age) and controls at a fixed age. Most of these differences, however, disappear when a fixed weight is used as the basis for comparison and Falconer (1973) concluded that selection appears to have altered developmental age in relation to chronological age. Second, although there were some changes in activities of enzyme levels in relevant metabolic pathways in selected animals, no logical pattern was found except for a minor alteration in some aspects of lipid metabolism (see later). There were also some puzzling reductions in *both* protein synthesis *and* protein degradation in Small versus Large mice (Priestley and Robertson, 1973; see later).

The most dramatic result was the finding that the *Hbb*[s] allele of the ß-haemoglobin locus was fixed in six replicated large lines of the Q strain (Falconer, 1973; Garnett and Falconer, 1975; Garnett, 1976). No effect on body weight from the *Hbb*[s] allele could be seen in a segregating population, suggesting that the trait-gene was closely linked to the *Hbb* locus and was not *Hbb* itself. The *H-2* haplotypes have also been found to be nonrandomly associated between the Q lines (Simpson *et al.*, 1982). The *H-2*[q] haplotype was nearly at fixation in the six Large lines and *H-2*[b] haplotype was fixed in the majority of the Small lines. This suggests that *H-2*, or a gene closely linked to it, is a trait-gene for growth. This could now be directly tested in transgenic mice as the genes of the *H-2* haplotypes have been cloned.

The above expriments were all performed on strains selected for weight-for-age; more recently attempts have been made to select animals for components of growth or to initiate selection from a genetically defined base population. The replicated G lines were created by Alan Robertson and Bill Hill (Sharp *et al.*, 1984) and were selected divergently for one of three traits: appetite, fat-content or lean-mass. The F (fat-content) lines in particular have been the subject of detailed biochemical analysis. The Fat and Lean F lines differ 2.5-fold in gonadal fat pad weight at 5 weeks of age and nearly 5-fold at 10 weeks of age (the age of selection). The Fat animals have significant alterations in *de novo* hepatic and, especially, gonadal fat pad lipogenesis at both 5 and 10 weeks of age. The enzyme activities in this specific pathway (ATP citrate lyase, acetylCoA carboxylase and fatty acid synthetase) are elevated up to 3-fold as is the flux from ^{14}C acetate; the animals also have 2-fold higher levels of circulating insulin (E. Asante, I. Hastings, W.G. Hill, G. Bulfield, unpublished observations). The enzymes and fluxes in peripheral pathways such as glycolysis, pentose phosphate and the TCA cycle are only marginally elevated. These results indicate that the metabolic alterations in the Fat line mice are specific to the *de novo* lipogenesis pathway and mechanisms causing this can now be investigated in detail with the aim of identifying the genes and their products.

A further set of replicated selection lines (the X lines) for 6-week body weight has recently been developed from a cross between the C57BL/6J and DBA/2J inbred strains of mice (G. Bulfield and C. Sinnett-Smith, unpublished). These two strains have allelic differences in over 200 genes (B.A. Taylor, personal communication), differ in the concentration of about 100 protein spots on two-dimensional (2D) gels (Klose, 1982) and also differ substantially in DNA restriction fragment length polymorphisms (RFLPs) and at DNA hypervariable loci (Jeffreys *et al.*, 1987). Not only are the progenitor strains of the X-lines genetically well-characterized, but there are also 26 BXD recombinant inbred (RI) strains which have been produced by crossbreeding the progenitor strains and inbreeding the F_2s (Taylor, 1978). These BXD strains are well-characterized and have been typed for approximately 200 allelic differences so that the 16 Morgan mouse genome is genetically marked on average at around ±4 cM intervals.

The combination of a large number of known allelic differences between the progenitor strains and the existence of the 26 BXD RI strains, will make the X-lines valuable for identifying and analyzing trait-genes or their products (or closely linked marker genes). Alleles found fixed or at high frequency in all six replicated lines, for either high or low six-week body weight, will be candidate trait-genes. By analyzing the strain distribution pattern of such an allele across the BXD RI strains its locus can be mapped (Taylor, 1978; for an example see Jeffreys et al., 1987). In particular, allelic differences in gene products identified amongst the X-lines using 2D-gel electrophoresis can be analyzed by reverse genetics. A protein present at average concentration can be isolated from between 5 to 10 gels in sufficient quantities to raise an antibody and isolate its cDNA from an expressed gene library. Therefore, in principle, the gene can be cloned, its amino acid sequence determined and, potentially, a function assigned to it. In this way it is possible to study putative trait-genes by recombinant DNA techniques and to analyze their function in transgenic mice.

Chickens

In our Institute we have chosen to analyze initially three lines of chickens in detail: a broiler line selected for weight-for-age and muscle development for 36 generations, an offshoot of this line where selection was relaxed after 18 generations, and a line not selected for growth rate. These lines differ up to 4-fold in body weight and 8-fold in muscle weight at 7 weeks of age (Bulfield et al., 1988). Rapidly growing broiler chickens have a much higher rate of muscle growth than unselected slow growing animals. In an extensive study of muscle growth from hatch to maturity we have shown that the DNA to protein ratio remains constant irrespective of genetical differences in muscle growth or age. This implies that nuclear division is continuing in proportion to increase in muscle mass posthatch. As the nuclei in the myfibrils do not divide this has focused our attention on cell division of the satellite cells lying alongside the myofibrils and their capacity to 'donate' nuclei to the rapidly expanding fibres (A. Tinch, unpublished observations).

It has been found that at least one indicator of nuclear division, ornithine decarboxylase (ODC) activity, is dramatically altered in rapidly growing animals. ODC is the first step in polyamine synthesis and is a truly remarkable enzyme in having a half-life of only 10 minutes; it is inducible by a wide range of hormones and growth factors in a wide variety of tissues. Furthermore, ODC synthesis is induced during rapid cell division, in response to known mitogens, and in rapid growth of tissues, embryos and tumours and transformed cell lines. Rapidly growing broiler chickens have 20-fold more ODC in their breast muscle than unselected animals at the time of most rapid growth (Bulfield et al., 1988). Mouse embryos from the rapidly growing selection lines also have elevated levels of ODC (3 to 5-fold) during the period of rapid cell proliferation and this is associated with differences in the ODC gene itself (A. Gray, A. Tait, R. Johnson and G. Bulfield, unpublished observations). As the ODC gene has been cloned, the molecular basis of these differences can now be investigated.

It has also been shown by both in vivo and in vitro methods that rapidly growing broiler chickens have much lower protein degradation rates than unselected slow-growing animals. Therefore protein degradation may be important in genetic variation in protein deposition, since increased protein growth appears to be achieved by decreased protein degradation rather than increased synthesis. The biochemical and molecular control of protein degradation, and in particular the role of individual proteinases has been investigated (J.C. McKay, unpublished observations). It was noted earlier that differences in protein degradation rates were also found in the Q-strain mice (Priestley and M.S M. Robertson, 1973).

Circulating levels of growth hormone and insulin-like growth factor (IGF-1) have also been determined for these strains. There are no differences in plasma IGF-1 levels and GH levels are higher in slow- than fast-growing animals both between and within strains (C. Goddard, personal communication). The mechanism of action of IGF-1 in muscle cells and the role of the IGF-1 receptor, therefore, may be more important than the quantitative amount of hormone produced in determining phenotypes divergent for growth. Both chicken genes have now been cloned to further this analysis (D. Fawcett, unpublished observations).

It is clear that the search for trait-genes controlling growth (or some component of it) is going to be a long and arduous process. The use of genetically divergent lines of animals differing significantly in growth characteristics will be important tools in this programme. Already attempts to 'dissect-down' from the trait to the gene in mice and chickens have identified:

(i)   the *de novo* lipogenesis pathway as controlling differences in fat content;
(ii) the role of protein degradation in producing differences in muscle mass;
(iii) the role of muscle nuclear division, satellite cells, and ODC in muscle growth; and
(iv) increases in circulating levels of GH and IGF-1 are not involved in changes in growth rate, but receptor and target-cell controls may be important.

The potential of 2D-gels to display several thousand potential cellular trait-genes products and the use of specially designed selection lines has been discussed; the value of this 'bottom-up' approach from the gene/gene-product to the trait has yet to be determined.

Conclusions

Genetically engineered, transgenic, laboratory and farm animals containing specific cloned genes can now be routinely produced. To make full use of this transgenic technology in animal breeding it is essential to develop strategies to identify the genes controlling commercially important traits. I have outlined in this review the approaches currently in use. Unfortunately it is not easy to develop a strategy that will lead us unerringly to the identification and cloning of trait-genes. The gains that can now be made, however, from a complete understanding of the relationship between genotype and phenotype and their modification by transgenic technology make this one of the most important areas for animal science in the next 20 years.

Chapter 33
Physiological Predictors of Genetic Merit

James C. McKAY

Genetics and Animal Breeding Consultant, Ravenscraig, Roslin, Midlothian EH25 9RD, Scotland

Alan Robertson established the theoretical framework for combining predictors of genetic merit (Robertson and Rendel, 1950; Robertson, 1959c; Latter and Robertson, 1960). He illustrated problems in the use of physiological predictors (Neimann-Sorensen and Robertson, 1961; Jamieson and Robertson, 1967) and has been consistent in urging that new physiology, biochemistry and molecular biology should become an integral part of animal breeding practice (Robertson, 1970a, 1982b,c, 1984, 1986a).

A physiological predictor is any characteristic of an animal's metabolism which can be used to estimate the genetic merit of that individual or a relative for a production trait of interest. Levels of hormones, enzymes or metabolites may be measured during normal development or in response to an external stimulus. The physiological connection between the predictor and the production need not be direct or obvious. Value as a predictor does not imply that the enzyme or hormone's gene is a major gene affecting the trait or is linked to such a major gene (the uses of major and marker genes are discussed in separate chapters by Piper and Shrimpton and by Smith and McMillan).

Livestock breeders seek to maximize the rate of response to selection and physiological predictors are sought to improve the accuracy of selection, reduce the generation interval or increase selection differentials. The expected response to direct selection on a production trait (trait 1) is:

$$R_1 = ih_1\sigma_{A1} = i\sigma_{A1}{}^2/\sigma_{P1},$$

(where i = selection intensity, h^2 = heritability, $\sigma_A{}^2$ = additive genetic variance and $\sigma_P{}^2$ = phenotypic variance). The response in trait 1 to selection on a physiological predictor (trait 2) is:

$$R_1 = ir_A h_2\sigma_{A1} = i\,\mathrm{cov}_A/\sigma_{P2},$$

(where r_A = genetic correlation and $\mathrm{cov}_A = r_A\sigma_{A1}\sigma_{A2}$ = genetic covariance), and the relative efficiency is

$$R_1(\text{indirect})/R_1(\text{direct}) = (\mathrm{cov}_A\sigma_{P1})/(\sigma_{A1}{}^2\,\sigma_{P2}) = r_A h_2/h_1.$$

Thus the efficacy of a physiological predictor depends on its phenotypic variance and genetic covariance with the trait to be improved. When available, data from direct and predictor traits would be combined in a selection index.

Physiological predictors of genetic merit have greatest potential for improving the accuracy of selection and rate of response where the expression of a trait is limited to one sex at maturity (e.g. milk production), the trait has a low heritability (e.g. litter size), or the trait is difficult to measure in an individual before selection (e.g. disease resistance or meat quality). In such cases there are obvious benefits from predictors which can be measured accurately at low cost and in juveniles of both sexes.

Theoretical developments have reinforced the view that physiological predictors could be powerful tools (Sales and Hill, 1976b; Hill, 1985) especially when combined with new technologies in animal breeding such as multiple ovulation and embryo transfer (Woolliams and Smith, 1988). There are serious risks in using such predictors where genetic parameters are poorly estimated (Hill, 1985), in extrapolating across species or different strains of one species (Lee and Land, 1985) and in predicting the long-term effects of selection (Sejrsen and Løvendahl, 1986). These dangers mean that physiological predictors can only be integrated into breeding programmes when we have a thorough knowledge of the genetic and phenotypic relationships between the predictors and all production traits within the selected population. Such exacting conditions mean that there are few examples of successful application. By reviewing the current status of research on a number of traits, different aspects of the use of physiological predictors will be highlighted.

Growth traits

Growth rate and feed efficiency are not especially suited to the application of predictors because they are simple to measure on all individuals and have relatively high heritabilities. However predictors of body composition and meat quality could be valuable and are being sought in a number of species.

Müller (1986) has shown that selection for decreased activity of a number of enzymes of lipogenesis can decrease fatness in the pig. She argued that selecting to restrict the potential for production of the essential co-factor NADPH in biopsies of subcutaneous deposits would limit the rate of deposition of fat. This predictor can be successfully combined with other measurements of fatness such as ultrasonic scanning. In broiler chickens, plasma triglyceride levels have been used as predictors of fat deposition (Whitehead and Griffin, 1985) and plasma glucose levels show similar potential (Leclerc et al., 1987). These predictors use an important difference in physiology between poultry and mammalian species. In pigs and sheep the majority of fat synthesis occurs at the site of deposition whereas in poultry the liver is the main site and triglycerides in the plasma transfer it to the site of deposition. The excretion of N^t-methylhistidine is being developed as an indicator of skeletal muscle mass and protein degradation rates in poultry (Jones et al., 1986). N^t-methylhistidine is released during the degradation of skeletal muscle myosin and is excreted unchanged in poultry whereas it is re-utilized or metabolized in other species. In none of these examples can the physiology be applied across species and its application to different genotypes within species is uncertain.

Milk production

This is a trait where much physiology is known and for which physiological predictors could have great impact, yet none are used in current breeding programmes (Christensen and Liboriussen, 1986). Blood group and serum enzyme level variation associated with differences in milk yield have been reported (Neimann-Sørensen and Robertson, 1961; Graf, 1984). Such variants are generally the result of segregation at a single locus and favourable alleles would be efficiently fixed by selection unless their effects are trivially small.

The search for physiological predictors has focused on hormones because of their roles in the control of growth and lactation and metabolite levels, especially in fasting young stock. Fasting is used to simulate the metabolic stress and energy deficit of lactation. Several hormones have been found to differ between lactating cows of different breeding value (reviewed by Bauman et al., 1985). More usefully, several aspects of thyroid hormone function in young bulls are related to their daughters' milk production (Sejrsen et al., 1984), but concern about other long-term effects of selection for high levels of thyroid hormones has meant that breeders are cautious about using these as selection criteria. Changes in other hormones or metabolites such as urea and free fatty acids in fasted young cattle also show promise as predictors (Barnes *et al.*, 1985; Woolliams and Smith, 1988).

To use an index of physiological measurements in a selection programme requires confidence that the parameter estimates are accurate and that the physiological changes will not produce an unbalanced response. Thus, although the potential gains are considerable, there may be little impact in this century (Christensen and Liboriussen, 1986).

Reproductive traits

Walkley and Smith (1980) have argued that physiological predictors should have most effect in breeding programmes for improved female reproduction. However there has been no success to date despite a considerable research effort and varied strategies. For example, prolific breeds of pigs and sheep have been studied in many laboratories in a search for the physiological basis of large differences in litter size (e.g. Sellier and Legault, 1986; Bindon *et al.*, 1985). In the case of sheep breeds single genes account for much of the variation in ovulation rate. Experience in many species (including man) in tracing the primary change of a mutation of large effect has not been encouraging. Breed comparisons may focus on differences which are irrelevant or misleading. Either strategy requires a profound understanding of the control of metabolism affecting the trait and must continuously assimilate new knowledge and techniques.

Testis size in young males proved effective as a predictor of their daughters' reproductive potential in the mouse. However sheep responded unpredictably to such selection or selection for response to leutenizing hormone releasing hormone (Lee and Land, 1985). These results again emphasize that genetic parameters have to be estimated accurately in the population of interest, and direct and correlated responses cannot necessarily be predicted across species or across strains of the same species.

Disease resistance

Resistance to specific diseases has been improved, especially in poultry, by selecting for particular B-haplotypes (the major histocompatibility complex) or other single gene resistance markers (reviewed by Hartmann, 1985). The physiology of these host-pathogen interactions transfers well from experimental inbred lines to commercial strains.

Predictors of general disease resistance would be of great interest in all species. One trait widely studied has been the immune response to novel antigens. Improved response has in some cases produced resistance to disease challenges (e.g. Martin *et al.*, 1986). Other approaches to improved animal health (e.g. the use of vaccines and antibiotics) are now widely used in animal production. The flexibility of these approaches may be preferred to seeking any residual variation in disease resistance except in species such as poultry where the cost of treating individuals is prohibitive.

Prospects

While opportunities do exist for the use of physiological predictors of genetic merit, the research required is substantial, there are risks and the time scale for gains is long. All the traits discussed can be affected by direct physiological intervention. Reproductive performance in the large farm species is now manipulated by the administration of drugs and routine surgical procedures. Growth, body composition, feed efficiency and milk yield can be improved by synthetic hormones and drugs such as beta-agonists.

Such treatments are equivalent in their effects to many generations of efficient selection. They are available now, are reversible and will continue to be improved by further research. Livestock production will continue to adopt new treatments unless they are unacceptable on the grounds of animal welfare or consumer reaction.

Animal breeding should not be in competition with these physiological approaches; the findings should be used to redefine the livestock genotypes required for the future and integrated into selection programmes in the form of physiological predictors.

Chapter 34
Use of Identified Genes in Animal Breeding

Charles SMITH and Ian McMILLAN

Centre for Genetic Improvement of Livestock, Department of Animal and Poultry Science, University of Guelph, Guelph, Ontario N1G 2W1, Canada

Alan Robertson's research work and papers provided an important link between genetics and livestock improvement. He continued and extended the work of Fisher, Wright and Lush on the development and use of quantitative genetics to animal breeding. In the topic of this chapter, the use of identified genes in animal breeding, his 1961 paper was seminal in showing how known genes could be included in selection index theory (Neimann-Sorenson and Robertson, 1961). This paper has been the basis for most of the subsequent developments. Another important area was his experimental research to determine the number of genes affecting quantitative traits and the distribution of the size of their effects (McMillan and Robertson, 1974; Piper, 1972; Shrimpton, 1982; Shrimpton and Robertson, 1988a,b). More recently (Robertson, 1970a, 1984, 1986a) he initiated the discussion about the use of molecular genetics in its application to livestock improvement.

Identified genes

The use of identified genes in animal breeding has been limited, and they have had a marginal role in the genetic improvement of livestock. Colour and other simply inherited traits were used as convenient trademarks in the formation of many breeds. They served as markers to maintain breed purity and kept stocks genetically isolated. More recently blood groups and protein polymorphisms have been used to check on parentage and to monitor error rates in mating practice, for example in fish (Hershberger and Iwamoto, 1985). Another useful role has been in the understanding and control of simply inherited genetic disorders. Some use has been made of polymorphism to estimate genetic distances between breeds, as a guide to historical origin in planning breed conservation and possibly to predict heterosis from crossbreeding. Good examples of the practical use of individual genes are the early and late feathering genes for sexing in some poultry stocks, egg shell colour genes to fit different markets, and marker genes introduced into fish subgroups to identify them when the groups are mixed.

Major genes

Unlike plant breeders, animal geneticists have not practised systematic screening of large populations for extreme individuals which might carry a major gene, one having a large effect (say more than two standard deviations between homozygotes) on performance traits (Roberts and Smith, 1982). However, several genes with large effects on individual traits have been found in extreme stocks developed by breeders. Examples of genes currently in use or being considered in practical breeding systems are given in Table 34.1. Incorporation of these into breeding-production systems is not always straightforward. Information is needed on their effects on all economic traits in order to assess their overall economic merit in the breeding system. Often the major genes have some deleterious effects, such as calving difficulty in double muscled animals (Hanset and Michaux, 1985); or else the changes may be too large for the current production system to cope with, as is the case for the high fertility Booroola gene in sheep (Piper and Bindon, 1985; Davis and Hinch, 1985). Introgression and fixation of a major gene in nucleus breeding stock take time and selection effort, which are at the expense of genetic responses from conventional selection methods, for example in improvement of leanness in pigs using the halothane gene (Webb *et al.*, 1982; Smith, 1982). There are major genes for disease resistance in poultry (B21 for Marek's disease, Gavora and Spencer, 1983), in pigs (K88 *Escherichia coli* strain resistance, Sellwood *et al.*, 1975) and perhaps in sheep (Haemonchus worm resistance, L.R. Piper, personal communication) but their use in practice has been limited, perhaps due to genetic changes in the disease agent over time.

Table 34.1
Examples of useful major genes affecting livestock performance

Species	Gene	Main advantage	Disadvantage

Poultry
  Broilers
	Sex-linked
	  dwarf gene
		Lower feed needs of
		  broiler mothers
			Slightly lower growth
			  rate

Egg-layers	Blood group
	  B21
		Resistance to Marek's
		  disease
			None

Pigs	Halothane
	  gene
		Higher lean meat
		  yield
			Stress mortality,
			  lower meat quality

	K-88	Resistant to *E. coli*	Uncertain
Sheep	Booroola gene (and others)	Higher ovulation rate	Litter size too large

Cattle	Double-
	  muscling gene
		Higher lean meat
		  yield
			Calving difficulty

	Milk protein
	  genes
	  (α, ß, κ caseins,
	  ß-lactoglobulin)
		Higher cheese yield	Uncertain

Many of the genes which have been detected have very large effects: for example the difference in lean weight between normal and homozygous double-muscled animals at the same liveweight is some six standard deviation units, in litter size between the Booroola homozygotes about three standard deviation units, and for leanness between halothane gene homozygotes about one standard deviation. There may be many genes with intermediate effects which have not yet been detected.

Intermediate genes

Little is known about the numbers of genes affecting economic traits in farm animals or about the distribution of the size of effects. Following the detection of polygenes for bristle number in *Drosophila melanogaster* by Thoday (1961) much effort was spent by Alan Robertson and co-workers in Edinburgh in trying to locate and measure polygenes for bristle number (Piper, 1972; McMillan and Robertson, 1974; Shrimpton, 1981; Shrimpton and Robertson, 1988a, b). Usually several chromosome segments with detectable effects on bristle number were found, with a minimum of 5–10 such segments for both sternopleural and abdominal bristles. Such work has not been attempted in farm animals because of lack of sufficient marker loci and the high costs involved. With many more marker loci becoming available, a range of quantitative methods available in human genetics and plant genetics (Roberts and Smith, 1982; Famula, 1986; Hill and Knott, 1988; Sing et al., 1988) is being extended to farm animals. Good examples of locating useful quantitative trait loci (QTL) in plants are given by Tanksley et al. (1982) and Kahler and Wehrhahn (1986). Reviews of methods of identifying genes and on the effects of genes are given by Mayo and by Piper, respectively, in this volume.

Selection automatically employs all loci with useful effects so identification is not necessary for their exploitation. If, however, the phenotypes or genotypes at a locus can be identified directly, then the information on the individual (and its relatives) for the locus can be used in selection to increase response rates. The information can be used on its own, or more efficiently combined with information on other traits in a selection index. This was shown originally by Neimann-Sorenson and Robertson (1961), and the method was adapted by Smith (1967), Soller (1978) and Smith and Simpson (1986) and extended to mixed model methods by Famula (1986) and Kennedy and Sorensen (1988). The value of information on QTL will depend on the proportion of the additive genetic variance associated with identified QTL, and also on the heritability of the trait, as shown by Smith and Simpson (1986). If the heritability of the trait is high then the scope for increasing the accuracy and response to selection is limited. The main benefits of including QTL in selection will be when the heritability is low and when a high proportion of the additive genetic variance is due to identified QTL. However this is the situation where the QTL effects will be small (in standard deviation units) and hard to detect. The information on the QTL must also be regressed to take account of sampling errors in the parameter estimates, so reliable estimates of effects on *all* economic traits are required (Smith and Webb, 1981). So far the methods do not seem to have been much used in practice.

Marker Assisted Selection (MAS)

The polymorphisms detected by conventional or molecular genetics methods are unlikely to be the quantitative trait loci (QTL) themselves, but they may be linked to QTL. With limited numbers of detected polymorphisms the power of linkage information in locating QTL is low (Robertson, 1973c). As the number of polymorphisms increases the more likely, and the closer, the linkages that can be obtained. Using restriction fragment length polymorphism (RFLP), (Botstein et al., 1980), mini-satellite finger printing (Jeffreys et al., 1985), and variable number tandem repeats (VNTR) (Nakamura et al., 1987; Georges et al., 1988), large numbers of polymorphisms will be detected in the next decade, thereby covering the whole human genome. Considerable progress has been made in mapping the human genome with over 500 polymorphic marker loci, some of which (VNTR, Nakamura et al., 1987) have high levels of heterozygosity. A 10 cM map is now available, covering 95% of the genome, and a 5 cM map is likely in 1–2 years (Donis-Keller et al., 1987). There is much homology in linkage groups among species (Womack and Moll, 1986), so that much of the linkage information will be transferrable across the livestock species.

These marker polymorphisms can be used in livestock improvement in marker assisted selection (MAS) (Geldermann, 1975; Geldermann et al., 1985; Soller and Beckmann 1982). Some use can be made of linkage disequilibrium between QTL and marker loci (Gahne and Juneja, 1985) but with decreasing returns in successive generations. Soller and Beckmann (1982) and Kashi et al. (1986) have studied the possible value of MAS in dairy cattle, and estimated that extra genetic response rates of 25–50% may be possible, and the methodology of MAS was considered further by Stam (1986) and Smith and Simpson (1986). MAS is improved when there are markers on either side of the QTL, giving so-called marker brackets or haplotypes (a series of alleles at different loci on a segment of the chromosome). A number of problems arise in applying the methods: the linkage phase must be identified within families; only a proportion of parents are heterozygous (at both the marker loci and QTL); there are cross-overs (detected and undetected); haplotypes are not unique but may overlap among families; QTL effects and linkage distances are estimated; and there may be nonadditive effects among QTL loci. New quantitative genetic methods are needed to handle the information and apply it usefully (Van Arendonk, 1989). Many of the problems are reduced, however, as more polymorphisms are detected, and unique tight linkages become available.

Transgenes

In livestock there has been so far a lack of identified genes with useful effects worth transferring. This is seen as a limitation in developing useful transgenes in practice (Bulfield, 1985). If there were genes with useful effects their DNA promoter and coding sequences would have to be found. This may be done by locating the gene on the chromosome using linked markers and then getting progressively closer by 'chromosome walking' using linked DNA probes and finally cloning the gene (Collins and Weissman, 1984; Koenig et al, 1987). If the protein product of the gene is known, then it may be possible to work back from the amino acid sequence of the protein to the mRNA and DNA coding sequences (Imakawa et al, 1987). Use of genetic information from other species will be useful, since there is a considerable homology in the genome across species with similar DNA sequences for the same genes. Thus DNA probes from laboratory animals and from man might be used in farm livestock (Womack and Moll, 1986).

Kacser (1983b) suggested that modifying genes in a metabolic pathway will have little effect on the flux through the pathway, or in changed output, because generally no factor is limiting (Kacser and Burns, 1979). An exception might be for changes that restrict or block a pathway, suppressing the flux, which might be useful if the flux were for an unwanted product or feedback factor controlling a pathway. The same result might be achieved with antisense DNA. Natural transgenes (transgenes with their normal regulatory sequences) will be regulated by normal feedback mechanisms and their expression limited. Fusion or hybrid transgenes (transgenes with exogenous regulatory sequences) could, however, be separated from normal feedback control (Palmiter and Brinster, 1986) and hence have more effect. Regulatory sequences which could be controlled externally would be especially useful. A more extreme but feasible change is the development of transgenic stocks with new products (Lathe et al., 1986). For example, sheep with the human blood clotting Factor IX (Simons et al., 1988) have been developed.

The role of molecular biology in animal improvement has been considered by Robertson (1986a) and the use of transgenes discussed by Smith et al. (1987). Transgenes should be made in stocks of high genetic level for economic merit to avoid genetic lag from the use of poorer stocks. Each transgene founder is unique in site and copy number, and so in its effects, and needs to be evaluated separately. The founder transgenic individual and its transgenic progeny will be hemizygous (TO) for the transgene. It must be backcrossed into elite nucleus stocks for 3–4 generations to reduce the founder effect and the associated inbreeding, and then intercrossed to produce homozygous (TT) individuals. If the transgene survives the initial screening for expression, effects on performance and transmission to progeny, its effects need to be tested on larger numbers for all economic traits, first in available hemizygotes and then in the homozygous (TT) form in which it will be ultimately used. It may turn out that the laboratory effort in making transgenes may be less than the effort of testing the transgenic stocks in practice. And this testing effort will be additional to the normal selection improvement effort which must be continued until useful reliable transgenic stocks have been developed.

Once a transgene has been identified as useful, it can be treated in selection and in dissemination as a normal gene. If the effect is large it can be introgressed (3–4 generations), intercrossed (1 generation) and homozygotes (TT) selected as for a major gene. To be worthwhile, its effect on economic merit would have to be at least 5–10% of the mean, to offset the loss in conventional selection response over the same time. If there are several transgenes with small effects they can be combined in a transgenic gene pool, giving enhanced genetic variation, and selected for economic merit by conventional selection methods.

Creation of transgenic stocks is not without risk, especially because the transgene may not be stable and there may be unforeseen consequences in the future such as defective progeny and possible litigation for losses involved. However, transgenes offer very large, though unpredictable, opportunities for producing new and modified products and for enhancing genetic responses in conventional improvement systems.

Conclusion

The use of identified genes is likely to play a much more important role in future animal improvement than in the past. With large numbers of polymorphic loci becoming available, marker assisted selection methods will become more useful and efficient. For QTL genes with larger effects attempts will be made to identify markers even closer to the QTL (by chromosome walking) and finally to identify the gene itself. This may then lead to an active role of gene manipulation or enhancement by molecular genetic methods, rather than merely the passive role use in selection. An alternative to identifying QTL may be to locate genes coding for enzymes and proteins in relevant metabolic pathways, and to control or manipulate them directly (Palmiter and Brinster, 1986). These methods appear very powerful and could lead to faster genetic improvement of farm livestock, and to new products from farm animals (Lathe et al., 1986). The use of identified genes in animal breeding may indeed be starting a new era.

Acknowledgement

We are grateful to Alan Robertson for his inspiration and counsel in our research work throughout our careers.

Chapter 35
Genetic Engineering in Animals

R.B. CHURCH

Departments of Medical Biochemistry and Biological Sciences, Faculty of Medicine, University of Calgary, Calgary, Alberta, Canada

Embryo manipulation and gene transfer offer the animal geneticist the flexibility in genome manipulation previously restricted to the domain of bacterial and plant geneticists. Transgenic animals which contain foreign genes in their germ line can now be produced (see Palmiter and Brinster, 1986, and Jaenisch, 1988, for reviews). Such transgenic animals have provided new approaches to the study of developmental genetics, and can be used as models for human disease and to make important biological products. Additional potential applications of this technique are to develop domestic animals with improved production efficiencies, disease resistance and changes in body composition. Of equal impact to the animal geneticist are the new technologies of embryo manipulation which allow the development and multiplication of selected or genetically engineered animals. Willadsen's (1986) success with nuclear transfer in sheep and cattle approaches the classical definition of 'cloning'. Such technologies coupled with embryonic stem cell propagation and transformation make genetically engineered animals a reality for the animal geneticist and breeder. Alan Robertson (1954b, 1958a, 1961a, 1970a, 1984) has stimulated the crossfertilization of molecular and quantitative geneticists for many years. He (Robertson, 1986a) posed criteria for evaluation of these technologies in his usual concise and probing manner. Robertson viewed the production of transgenic animals as 'mutagenic events', and had thought of the impact of altered genotype in terms of phenotypic analysis. The application of mixed model (BLUP) analysis of genetic evaluation for quantitative traits does not, as yet, deal with the problems created by molecular and embryo manipulation (Smith et al., 1987). Accurate mixed model analysis for offspring got by embryo transfer, homozygotic twins from split embryos, genotypically identical clones from nuclear transfer, and new polymorphisms in transgenic animals are difficult without an appropriate data base from recipient females to provide some estimate of maternal influence on commonly measured traits. The biological implications of such embryo manipulations have been reviewed by Polge (1985) and Willadsen (1982).

Reproductive techniques

Manipulation of reproduction has had a major impact on animal breeding programmes for years through the use of artificial insemination and embryo transfer. The dairy industry adopted artificial insemination in the 1950s, and Robertson was quick to recognize the importance of progeny testing through artificial insemination (AI). Demand for continental beef breeds sparked the development of embryo transfer, for it offered increased numbers of offspring from limited female gene pools in the 1970s. Embryo transfer is now commonly used to produce sires for progeny testing from planned matings of proven élite females and progeny tested superior sires. Nonsurgical embryo recovery, freezing and transfer, as developed at Alberta Livestock Transplants, allowed the industry to move from the laboratory to the farm (Seidel, 1984; Church et al., 1985). Breeders can produce increased numbers of offspring from specific planned matings of superior females without the fear of decreased reproductive performance due to surgical damage to the donor. Embryo freezing has increased the flexibility of embryo transfer programmes and transport by cutting costs and the logistics of international transport. The embryo transfer industry now produces thousands of pregnancies each year world-wide and has been the subject of a number of reviews (Church and Shea, 1977; Betteridge, 1977; Seidel, 1984; Church et al., 1985).

Additional techniques which enhance the opportunity for genetic improvement include in vitro fertilization (see Brackett, 1983, for review). In the bovine this provides access to the large numbers of oocytes contained in the ovaries of genetically élite females without the customary superovulation cycles which are plagued by inconsistent results. Successful in vitro fertilization of oocytes would circumvent this problem by increasing the supply of embryos available for manipulation or gene transfer (Xu and Greve, 1988). The success of in vitro maturation and fertilization (Anderson, 1985; Bousquet et al., 1988, and Goto et al., 1988), will result in the production of large numbers of full-sib offspring. The problems of reduced genetic variation in the breeding population which could result is discussed later.

Embryo transfer can also be used successfully to assess whether or not abnormalities in animals are of genetic or environmental origin. Females who have produced abnormal offspring are superovulated and bred to a suspect sire. The embryos are transferred to unrelated recipient females managed under uniform environmental conditions. Pregnancies can be terminated in late first trimester and the foetus examined to determine if the originally observed abnormality has an environmental or genetic basis (Fisher et al., 1984). This approach has been used on numerous occasions by breeders who are interested in the cause of abnormal offspring born in seedstock herds.

Embryo manipulation

The focus of mammalian embryo research has recently shifted from embryo recovery, storage and transfer to micromanipulation of the embryo itself. This is spurred, in part, by the lack of culture conditions which allow an embryo to proceed through all stages of pre-implantation development in vitro. Micromanipulation techniques allow production of identical animals through blastomere separation, which has been used to produce up to four genetically identical animals from a single 8-cell embryo (Willadsen, 1985). Simple bisection of morula or blastocysts has been used to produce identical twins in a number of species (Ozil, 1983; Church et al., 1985; Willadsen, 1985). In cattle, morula splitting is used to obtain higher pregnancy rates (about 110%) than for normal embryo transfer and to produce monozygotic twins (Baker and Shea, 1985). Some breeders use genotypically identical twin bulls in natural service in their herd. This permits double the number of cows to be served by a 'sire' or to have a genetically identical bull in reserve should injury or disease prevent use of his monozygotic twin, without resorting to artificial insemination.

Animal breeders may be dismayed that colour patterns are not identical in all monozygotic twins produced. The extent of colour patterns is a result of gene–environment interactions not just genotypic identity.

The ultimate goal of embryo manipulation is to clone the genetic élite. Willadsen (1986) has been successful in producing identical lambs and calves by the fusion of blastomere nuclei with oocytes.

Willadsen has improved reproductive efficiency of the cow by cloning embryos through nuclear transplantation. One 'cloned' embryo can result in numerous pregnancies, all genetically identical. Willadsen has produced several hundred cloned animals by this method at Alta Genetics Inc. (S.M. Willadsen, 1988, unpublished). Smith et al. (1987) have enunciated the concepts which are needed for development of new genetic strategies involving these 'cloned' animals.

Embryo sexing is an aspect of embryo manipulation which is of interest to animal breeders. It has been possible for some time to biopsy embryos and establish the sex by karyotypic analysis (Betteridge, 1977). However, this technique is time consuming and uncertain. Investigations utilizing

H-Y antigens and other antibodies have not been very useful (Anderson, 1985). Taking the lead from human Y-specific probes, Popescu *et al.* (1988) have developed a bovine DNA probe which has been localized by *in situ* hybridization to the short arm of the Y chromosome. Through the use of sensitive biotinylated Y specific DNA probes hybridized *in situ*, rapid sex determination of intact blastomeres can be achieved. This permits routine embryo sexing of microsurgically isolated blastomeres or nuclei from embryos. Embryo sexing before nuclear transfer 'cloning' would increase the efficiency of the process. Similarly, knowing the sex of embryonic cell lines or transgenic chaemera partners eliminates the chance of male–female chaemeras.

A powerful embryological technique is to produce chaemeras by the aggregation of blastomeres from a number of embryos or the injection of a single blastomere into the blastocyst cavity of an embryo (Brinster, 1974). Chaemeric animals have a cellular contribution from each set of parents and may or may not have germ-line involvement from a particular genome. Interspecific chaemeras offer unique opportunities for the study of cell differentiation and interaction during development. It is possible to produce chaemeras from embryos of a single species or to transgress species. The most spectacular chaemeras are the sheep–goat chaemeras ('geeps') which show phenotypic characteristics of both parental genomes (de Fehilley *et al.*, 1984). Such interspecies chaemeras are designed to neutralize incompatability between the foetus and the recipient mother's uterus by placing the heterologous inner cell mass inside a homologous trophectoderm.

Chaemeras are useful in studies of gene rescue. For example, bovine chaemeras made from morulas obtained from 'double-muscled' and normal embryos have resulted in tetraparental cattle. The double-muscling gene(s) is recessive in the Charolais. The objective of the experiment was to establish whether the 'double-muscle' gene product(s) was diffused and affected normal cells in the same tissue, or if 'normal' gene products were diffused to suppress 'double-muscle' cells. To date, such tetraparental animals have not shown any of the double muscled phenotype, even though the mutant genotype is present in substantial numbers of muscle cells (R.B. Church and R. Bricker, 1988, unpublished). The total contribution of each parental cell type to the chaemeric offspring may vary with age. The development of tetraparental cattle serves as a model for the use of embryonic stem cell-blastocyst chaemeras as a vehicle for gene transfer in domestic animals.

Embryonic stem cells (EK or ES cells) are cultured embryonic lines which, after re-introduction into a host embryo, contribute to normal embryogenesis. This gives rise to chaemeric or tetraparental animals whose tissues are a mosaic of host embryo and ES genotype. A proportion of chaemeras pass the ES genotype to progeny, perpetuating any genetic alteration in ES cells that might follow a variety of *in vitro* manipulations. ES lines have so far been restricted to mice (Evans and Kaufman, 1981). Several groups are attempting their isolation from porcine, ovine and bovine embryos.

The benefits of such technology to animal genetics fall into two broad categories, reflecting the respective applications of the system to basic and applied problems. In the former category are the usefulness of ES cells for *in vitro* differentiation, mutation and transformation studies which elucidate basic aspects of animal embryology. Using directed gene insertion techniques it is possible to generate animal models of human genetic disease and to investigate the developmental and evolutionary significance of specific genes. The ability of ES cells to differentiate either *in vitro* or through their passage as tumours provides an opportunity to assess recombinant fusion gene constructs in specific chromosomal contexts for their levels of expression and control, without recourse to expensive and lengthy animal experiments. This is a particularly important advantage in large species such as cattle. Finally, because of their pluripotential character, ES cells are an obvious source of nucleii for nuclear transfer experiments.

Attempts at the introduction of novel genes into the genome in domestic species have been confined to pronuclei injection. This technique has proven highly inefficient in cattle (Church, 1987). The main advantages of the ES system are the ability to replicate site of incorporation effects by repeated blastocoel injection using subclonal transgenic ES lines, and the opportunity to assess expression by inducing differentiation either *in vitro* or *in vivo*. These advantages have been demonstrated in mice using muscle specific constructs introduced into ES cells (E. Shani, 1988, unpublished). Murine ES chaemera production occurs with an efficiency of 50% (live born chaemeras/ embryos transferred) provided that healthy, euploid male ES lines are used; roughly 50% of these will result in germ line incorporation. If this level of efficiency can be duplicated in cattle it provides an efficient system for producing transgenic animals. McWhir *et al.* (1988) have isolated and characterized a pluripotent porcine ES cell line.

Use of a male stem cell line in mice leads to a disproportionate number of chaemeric males and to a population of germ line chaemeras which are exclusively male. This occurs because XX embryos are either converted to phenotypic males by the ES contribution or are sterile. The XX/YY germ-line chaemeras are of particular interest because XX cells cannot form functional germ cells and viable sperm will be of ES genotype only. A group of such animals derived from a single ES clone are themselves 'germ-line clones' which for sire proving purposes are genetically identical.

Based on mouse data (Bradley *et al.*, 1984) chaemeric male offspring still represent about 70% of chaemeras or 35% of liveborn animals. These animals can be karyotyped shortly after birth and divided into 40XX/YY (germ-line clones) and 40XY (germ line mosaic) groups, about half of each. The former group (15–20%) of potential germ line clones could be produced at the rate of 12–16 per 100 females assuming birth rates of 80%. With twin transfers this rises to perhaps 20, and, with embryo sexing, to about 30 germ-line clones per 100 females bred. These figures are, of course, highly speculative, and among ES lines it should be remembered that the efficiencies of chaemera formation are highly variable.

These observations offer novel advantages to animal breeders who are often thwarted by undesirable genetic correlations in their attempts at multitrait selection. More problematic still is that desirable phenotypes in a herd sire may be undesirable in female animals (i.e. ideal phenotype and genotype do not correspond). Consider the use of a breed or line of animals selected for small mature size, hardiness, libido and longevity as a source of host embryos for ES injection. In such a scenario, ES lines would be derived from male embryos of quite different animals which have been selected for feed lot traits such as high-growth rate and efficient use of high-energy feed. The advantage over simple transfer of 'male-line' embryos into 'female line' recipients is that the 'male line' ES cell lines can be evaluated. ES cell lines might be progeny tested by the same statistical technique used in current sire-proving programmes. The process, however, would be much more accurate because all germ-line, ES, clones could be considered as a single animal in genetic analysis. Systems analysis work suggests that economic opportunities and management dramatically affect the 'optimal' type of beef animal selected (McWhir and Wilton, 1987). Testing of ES cell line progeny under various environmental and management regimes is an efficient and economical process.

Transgenic animals

Geneticists with interests in molecular techniques have been intrigued by the possibility of genetic engineering. However, until conditions for handling mammalian pre-implantation embryos were available, little progress could be made. One of the first attempts at alteration of an animal genome with exogenous DNA was carried out by Stirling Munro (unpublished, 1968) with chickens. He injected DNA isolated from bantam fowl, which are coloured and have a different claw structure, into the ovaries and testis of White Leghorn chickens. He obtained some offspring which had patches of dark feathers and the bantam type claw. Molecular technology was not available to definitively assess the relative success of this gene transfer.

Introduction of foreign DNA (genes) into the germ-line is called transgenesis. A transgenic animal might be defined as one whose genetic composition has been altered to include selected genes from other animals or species by methods other than those used in traditional animal breeding. The most common method of producing transgenic animals is to inject DNA sequences into the fertilized oocyte (Palmiter, Brinster *et al.*, 1982). There are some limitations to microinjection since very little is known about the actual process of genome integration or the factors involved in the control of foreign gene expression (Jaenisch, 1988; Bishop and Al-Shawi, this volume). Many animals do not express genes incorporated into their genome if the regulatory sequences or insertion site are not correct (Palmiter and Brinster, 1986). The isolation of desired gene sequences and the regulatory DNA sequences required for controlled expression in a genetically engineered animal is not a trivial task. The production of a fusion gene involving metallothionein regulatory sequences coupled to the rat growth hormone gene injected as a pBR322 construct into mice by Palmiter, Brinster *et al.* (1982) initiated considerable speculation about the possibility of introducing desirable genes into animals. Subsequent production of transgenic rabbits, pigs and sheep (Hammer *et al.*, 1985) and cattle (Church, 1987) have been accomplished but with varying degrees of success with regard to expression.

Hammer *et al.* (1985) reported that of 1032 sheep ova injected only 1 of 73 newborn sheep incorporated the fusion gene which was injected. The resulting offspring were not physiologically normal. Simons *et al.* (1988) reported that 1.2% of sheep embryos injected were transgenic for secretion of peptides into their milk. Superovulation, natural mating and subsequent flushing yielded 1161 fertilized bovine oocytes for microinjection with three fusion constructs. Only 7 out of 126 calves born were transgenic on the basis of incorporation of the fusion gene into their genome. Only one shows evidence of expression of the bovine AFP-like-human beta interferon fusion gene in its serum, a first step in introducing disease resistance into an animal (Church, 1987).

Transgenic animals can also be viewed as a production system for useful peptides which require post-transcriptional modification prior to secretion. Transgenic animals with fusion genes for a medically important protein such as tPA or Factor IX, coupled to regulation sequences for milk protein, secrete the protein into their milk (Gordon *et al.*, 1987; Simons *et al.*, 1988). The high value peptide secreted into the milk of these transgenic animals can be isolated for therapeutic use. Once a successful transgenic animal has been produced, the breeding of sufficient numbers of that genotype to have an impact on livestock productivity and profitability requires the successful application of embryo transfer, manipulation and nuclear transfer techniques.

The relative lack of success in producing transgenic livestock in comparison with mice is probably not a function of species differences. The differences are more likely to be associated with the lack of knowledge of molecular development, inadequate culture conditions and inappropriate DNA constructs for livestock species. The mutagenic effects of DNA incorporation into the genome of mice are relatively high. Similar mutagenic effects on foetal development in 'transgenic' cattle embryos is suggested by the low pregnancy rate noted for injected embryos after transfer in comparison to noninjected embryo transfer (Church, 1987). The general lack of knowledge of the molecular biology of physiology and reproductive fitness in domestic animals makes the selection of promoter and structural sequences for gene fusion construct very difficult.

Discussion

The animal geneticist has a range of new vistas opened up by new gene and embryo manipulation technologies. Transgenic animal technology can generate precise models for study of development and human disease as well as introducing new genetic variation into animal genomes. Embryo manipulation provides a mechanism for propagation of unique genetic material and for expansion of élite genetic resources. However, work with large domestic animals requires a considerable investment in the number of animals required. The logistics of recovering sufficient newly fertilized bovine embryos without *in vitro* fertilization, their microinjection or transfusion, culture in the laboratory to select those which are developing normally, and their subsequent transfer to recipients, is a major management task. Successful development of technology to allow genetic engineering and embryo manipulation of domestic animals has great importance to animal breeding strategies. Inbreeding will become a major concern. Such programmes will be long term and in the case of cattle, limited by the number of embryos available and long generation time. The potential gains in livestock production include improved efficiencies of reproductive performance, enhanced growth, disease resistance and changes in milk and wool production and composition. The introduction of growth-hormone-related genes may increase growth but an understanding of the molecular physiology of growth to insure such animals are capable of a normal reproductively fit lifespan is needed. Probably the least useful transgenic livestock will feature growth hormone gene constructs. More subtle approaches to repartitioning of metabolism, receptor modification or lymphokine control may well prove more beneficial. The field of animal genetic engineering and embryo manipulation, including 'cloning', will have significant impact on animal breeding strategies. Alan Robertson probably had ideas for these strategies under way.

Acknowledgements

The support and cooperation of Alta Genetics Inc. and the Natural Sciences and Engineering Research Council is acknowledged. Appreciation is also extended to J. McWhir and S. Willadsen for comments. My thanks to Alan Robertson who stimulated my interest in combining animal breeding prospectives with molecular biology while a graduate student at the Institute of Animal Genetics in Edinburgh. This paper is dedicated to the memory of a long-time friend and colleague – Roger Land.

Chapter 36
Gene Expression in Transgenic Animals

John O. BISHOP and Raya AL-SHAWI

Department of Genetics, University of Edinburgh, West Mains Road, Edinburgh EH9 3JN, Scotland

A transgenic animal is one that carries foreign DNA in some or all of its cells. When some but not all cells carry the foreign DNA the animal is a transgenic mosaic. Such an animal may carry foreign DNA in somatic cells only, in some somatic and some germ-line cells, or presumably in germ-line cells only, although for procedural reasons this has not been reported. Transgenic mosaics sometimes result when the foreign DNA is introduced by the embryo microinjection route, and always when it is introduced by retrovirus infection or by manipulating embryonal stem cells. Usually the foreign DNA is integrated into one or a few chromosomal sites, and when present in germ-line cells it is inherited in the normal Mendelian way. In many cases the interpretation of results obtained with transgenic animals is simplified if transgenic 'lines' are first established by breeding, and if the foreign DNA carried by each line is integrated at a single chromosomal site. Such lines usually breed true and can provide unlimited numbers of individual transgenic animals. This allows comprehensive studies to be carried out with enough individuals so that individual (environmental) variation and where appropriate variation due to genetic background effects can be taken into account. Most transgenic mammals have been mice, but transgenic rabbits, sheep and pigs have also been produced (Hammer et al., 1985; Simons et al., 1988). Some general reviews are by Gordon and Ruddle (1985), Palmiter and Brinster (1986), Scangos and Bieberich (1987) and Jaenisch (1988).

Ways to create transgenic animals

Most transgenic animals have been produced by directly microinjecting DNA into one of the pronuclei of a fertilized egg (Gordon et al., 1980; Brinster et al., 1981; Gordon and Ruddle, 1981; Costantini and Lacy, 1981; E.F. Wagner et al., 1981; T.E. Wagner et al., 1981). Usually the male pronucleus is more prominently visible. Sooner or later the embryos are introduced into female recipients which have been made physiologically receptive, and implantation and gestation proceed normally. In mice, typically, 3% of the fertilized eggs injected and 20% of live births develop into transgenic adults. Some of the factors that affect the efficiency of production of transgenic mice by this method have been discussed by Brinster et al. (1985). The proportion of mosaics among the initial (G_0) transgenic animals is difficult to estimate accurately, but most transmit the foreign DNA to at least some of their offspring. Most transgenic animals produced in this way contain foreign DNA inserted at a single chromosomal site, although insertion at more than one site sometimes occurs and is detected by the segregation of sites in the G_1 and subsequent generations. The number of copies of foreign DNA that become integrated at a chromosomal site varies enormously, from one to hundreds, and cannot as yet be controlled.

There are several known alternative routes, none of which has been used very much yet except in special applications. Retrovirus infection is one such route (Jaenisch, 1976, 1977). Following the infection of a cell by a retrovirus, DNA copies of the retroviral genome are inserted into the chromosomes. The number of integration sites per cell can be roughly controlled by controlling the multiplicity of infection. Tandem arrays are not produced and copies with defined ends are inserted into the chromosomal DNA singly. Retroviruses seem to have preferred sites of insertion, but there is a very large number of such sites in the genome of a mammal (Shih *et al.*, 1988). Retrovirus infection of embryos is an effective way of producing insertional mutations (Gridley *et al.*, 1987). Retroviruses have been used to carry foreign genes into cells by the use of specially constructed retrovirus vectors (Cepko *et al.*, 1984; Wagner *et al.*, 1985). In general retrovirus vectors each have two manifestations. First, they can be easily manipulated in a proviral DNA form, which is routinely propagated as part of a bacterial plasmid. Second, to prepare virus particles the plasmid carrying the modified provirus is transfected into a special cell line which provides the proteins necessary for virus maturation without itself producing virus. The contiguous length of foreign DNA that can be introduced in this way is limited to a few kilobases by the packaging limitations of retroviruses, but it seems probable that this limitation will be overcome. A more serious problem in many applications is the likely influence of the provirus LTR sequences on gene expression. Nevertheless, promising results have been reported (Soriano *et al.*, 1986; Stewart *et al.*, 1987).

Another route takes advantage of the ability of pluripotent embryonal stem cell lines (ES cells) to develop and differentiate normally when introduced into the inner cell mass of a normal blastocyst (Bradley *et al.*, 1984). During their *in vitro* cultivation, DNA can be introduced into the ES cells by infection with a retrovirus or by transfection. In certain cases selection may be applied *in vitro* to eliminate nontransgenic cells. For example, the inactivation of a gene (HGPRT) has been selected for (Kuehn *et al.*, 1987). Indirect selection may be employed if a gene is coupled to or co-transfected with a selectable marker such as the neomycin-phosphotransferase gene (Gossler *et al.*, 1986). Recently, transgenic mice have been produced which contain a component derived from ES cells in which resident genes have been altered by homologous recombination with foreign DNA (Thompson *et al.*, 1989).

The foreign DNA is cloned and manipulated

Foreign DNA for microinjection has invariably been cloned and propagated in bacterial hosts, and has very often been manipulated to alter its natural sequence. The reasons for this are obvious. The cloned DNA can be characterized, often to the point that its complete nucleotide sequence has been determined, it can be identified by means of special probes, and its downstream products, both RNA and protein, are limited in number and usually detectable. The impetus to much transgenic work, in fact, is the desire to study *in vivo* a gene system which has already been extensively studied *in vitro*.

In some areas of investigation transgenic animals can provide information of a quality which is otherwise unobtainable. This will be the case wherever cellular interactions are part of the process under study, or when the process is potentiated by events which occur during normal development and differentiation. Thus the development of immune potential is an area in which the utility of transgenic animals is very clear (Storb, 1987); the etiology of cancer is another (Groner *et al.*, 1987). Perhaps a less clear application is to mechanisms of tissue-specific gene expression. Many immortal cell lines show residual expression of tissue-specific genes, and foreign DNA can be introduced into cells by transfection much more easily than transgenic animals can be made. However there is evidence that foreign genes may be expressed in some cell lines in ways which are different from their normal expresssion. In at least some of these cases, the same gene more closely mimics its normal expression pattern when present as a foreign gene in a transgenic animal (Hammer *et al.*, 1987; Pinkert *et al.*, 1987; Dente *et al.*, 1988; Stewart *et al.*, 1988; Brinster *et al.*, 1988).

Cloned genes have often been modified before being introduced into transgenic animals. By deleting parts of a gene or its flanking regions, those sequences required for the correct expression of the gene may be identified. An extension of this approach is to alter specific nucleotides in regions identified as being of interest. The coding region of one gene may be detached from its own promoter and attached to the promoter of another so as to direct its expression to a tissue in which it is not normally expressed. The way in which the domain structure of a protein tends to relate to the linear arrangement of nucleotides in the corresponding gene means that the synthesis of hybrid proteins made up of distinct domains of different proteins can sometimes be directed from a gene consisting of fragments of two or more genes rather simply spliced together. The structures of proteins can be more subtly modified by introducing simple base changes into the corresponding gene. Genes modified in all of these ways and more can be introduced as foreign genes into transgenic animals so that their functioning, or malfunctioning, and its consequences can be studied in the context of a complete living organism.

Specific factors and signals in gene expression

The pathway by which information flows from the nucleotide sequence of the DNA to the three-dimensional structure of the mature protein involves a series of steps and processes. All of these steps and processes are brought about by interactions between *trans*-acting proteins and *cis*-acting signal elements embedded in the sequence of whichever macromolecule (DNA, RNA or protein) carries the information at that point. All of the *cis*-acting signals, whether present in the DNA, the RNA or the protein, are ultimately encoded in the DNA sequence. Some, such as the signals at the exon–intron boundaries which are involved in RNA processing, are common to most genes; others, such as some of the signals involved in the control of transcription, are not. Some are recognized and quite well characterized, others are not yet well understood, and yet others are unidentified.

Much attention has been focused on transcriptional control signals and the proteins that interact with them, because these interactions underlie the principal all-or-none control mechanism, the primary regulator of tissue-specific expression. The known *cis*-acting signals are quite short DNA sequences which are relatively rather than absolutely invariant. They tend to occur in clusters, each containing a number of different signals, some of which are represented more than once. Three types of cluster are recognized, called promoters, enhancers and silencers. A *promoter*, by definition, is inseparable from the site at which the process of transcription is initiated. Part of its function is to direct the RNA polymerase to that site, and different promoters do this with a greater or lesser degree of tissue-specificity. This is to say that some promoters interact with proteins which are present in different amounts in different cell types. An *enhancer* may be located close to or at some distance from the transcription initiation site, but it must form part of the same DNA molecule. Furthermore, by definition, the position and orientation of an enhancer may be changed experimentally without abolishing its function. Enhancers act to increase transcription from a promoter sequence, and may do so in a tissue-specific manner. The interactions between enhancers and promoters are not direct, but rather they are mediated by interactions between the *trans*-acting proteins which bind to them. *Silencers* are thought to have a dominant role in tissue-specific gene expression, because at least in some instances the action of a silencer shuts down the expression of the gene altogether. Specific *trans*-acting proteins are thought to interact with silencers to suppress transcription, in much the same way as bacterial repressors interact with operators. This area of research has received several recent reviews (Serfling *et al.*, 1985; Maniatis *et al.*, 1987; Ptashne, 1986; Evans and Hollenberg, 1988).

Most of the chromosomal DNA in cells is intimately associated with histones in a highly organized way, and in this form it is relatively resistant to attack by DNase. Interactions between *cis*-acting signals and *trans*-acting proteins cause localized disruptions of the protein-histone complex which can frequently be identified as sites which are sensitive to attack by DNase. In a few cases a correlation of DNase sensitivity with the binding of regulatory proteins has been demonstrated (reviewed by Eissenberg *et al.*, 1985). These so-called *DNase-hypersensitive sites*, putative sites of interaction with regulatory proteins, may be mapped by the Southern blot technique.

Integration of foreign DNA into the chromosomes

The commonest outcome of DNA microinjection is the insertion into one of the chromosomes of a single array of direct tandem repeats (head-to-tail configuration; see Palmiter and Brinster, 1986), but head-to-head and tail-to-tail arrangements, and also rearrangements of the foreign DNA are observed more rarely. An array may be of any length up to several hundred copies, and single-copy insertions are also found. Occasionally arrays are found at two or more chromosomal sites. Furthermore, in about 30% of transgenics not all nuclei carry the foreign genes, suggesting that insertion into a replicated region of the chromosome has occurred. Chromosome replication is proceeding at the time of microinjection and in the following hours before the first postfertilization mitosis. Fragments of chromosomal DNA have been found imbedded in arrays (Covarrubias et al., 1986, 1987), and insertion is sometimes accompanied by rearrangements in the flanking DNA sequences (Wilkie and Palmiter, 1987; Mahon et al., 1988).

The events that accompany DNA integration are not known accurately. However, the observed structures of arrays are consistent with models based on the fate of DNA microinjected or transfected into cells in culture. Molecules become joined together in two ways. The ends of linear molecules are joined so to speak randomly, by a DNA ligation mechanism, and it is probably in this way that foreign DNA fragments are incorporated into the chromosomal DNA. It seems however that tandem arrays are not built up by random joining of ends. This process would generate 25% head-to-head and 25% tail-to-tail arrangements, but in practice these arrangements are found only rarely. Thus it seems that the predominant head-to-tail arrays are mainly built up by some other process, such as repeated cycles of reciprocal exchange. The homogeneous linear DNA that is usually injected cannot form concatemers by reciprocal exchange while it retains its original form. Presumably some of it becomes circularized by the joining together of the ends of the same molecule. Head-to-tail arrays could then be formed by reciprocal exchange between two circular molecules or between one circular and one linear molecule. If the circularized DNA molecules are subsequently broken randomly, they will form a population of circularly permuted linear molecules. Reciprocal exchange between such circularly permuted molecules can also generate direct tandem arrays.

Line-specific gene expression

When several transgenic lines are produced after microinjecting the same DNA construct, the different lines may show different patterns of gene expression. Both qualitative and quantitative differences are found. Usually, even when most lines express the foreign gene, some do not. Similarly, one or a few lines may show an altered pattern of tissue-specific expression. Almost always there are differences between different lines in the level of expression (Table 36.1). In experiments of this sort expression levels have usually been estimated by measuring specific mRNA or protein levels or enzyme activities. Since each comparison is of the same tissue in different lines, it is likely that these reflect differences in rates of transcription.

What is unique about a line is that it carries an array of foreign genes created by a single unique insertion event, and it is to these insertion events that we must look for the likely causes of line-specific phenomena. Insertion events differ from each other in (i) involving a different small set of DNA molecules (typically 1 to 100), (ii) the arrangement of these molecules and (iii) the chromosomal site at which the array is inserted.

Arrays of foreign genes

The arrangement of the molecules in the array (see above) may influence expression through novel interactions between cis-acting signals. Similarly fragments of host DNA incorporated into the array may contain cis-acting signals which perturb the regulation of transcription.

The fact that a small number of DNA molecules goes to make up each array means that the composition of arrays will be sensitive to stochastic events which cause change of any sort, such as mutation and more especially sequence rearrangement by the recombination or end-joining of randomly-broken molecules. Such events may alter or inactivate both coding sequences and cis-acting regulatory signals.

Table 36.1
HSV thymidine kinase activities in male tissues of transgenic mouse lines, from Al-Shawi et al., 1988

Line	Approximate copy number	Thymidine kinase (Units/mg protein)					
		Liver		Preputial gland		Testis	
		n	Mean±SE	n	Mean±SE	n	Mean±SE
40	10	4	0.05±0.04	2	12.0±0.39	3	140±6.7
46	80	8	1.90±0.32	2	83.1±19.9	3	139±17.6
58	3	3	5.82±1.15	3	9.9±2.6	3	31±10.5
64	2	2	1.52±0.71	2	0.04±0.03	2	48±6.0
78	1	3	12.73±1.85	3	32.1±5.3	3	62±8.4

There is a general lack of correlation between the level of expression and the copy number of the foreign gene (see for example Overbeek et al., 1986; Hammer et al., 1987; Davis and MacDonald, 1988). In one study in which the measure of expression was transcription rate the same lack of correlation was observed (Davis and MacDonald, 1988). As a corollary to this, there is a negative relationship between the expression per copy and the copy number (Table 36.2). The effect may come about in either of two ways. First, circumstances may dictate that only a few genes can be active, irrespective of copy number. The remainder would be transcriptionally inactive. Alternatively, the general expression level of the genes may fall as the copy number increases. That is, most genes would be active, but at reduced levels when the copy number is greater. If the response is interactive, the interactions must be restricted to copies at the same site, because when two insertions are brought together by mating transgenic lines their expression is additive. This seems to rule out explanations based on the depletion of trans-acting factors (Davis and MacDonald, 1988). Reduced average gene expression seems then to be a property of the array. It is possible, for example, that many of the gene copies in large arrays have been inactivated mutationally, perhaps during the cycles of homologous exchange which built the array up. Alternatively, genes in large arrays may be susceptible to inactivation by methylation (Palmiter, Chen and Brinster, 1982). As a third alternative, the mere proximity of a large number of transcription units may itself reduce transcription rates, perhaps through illicit interactions between trans-acting proteins bound to the promoters and enhancers of different genes in the array. These alternatives could be at least partially resolved by cloning and characterizing genes from large arrays.

Recently, it has been shown that when the fragment injected carries long flanking regions containing remote DNase-hypersensitive sites (see above) expression becomes proportional to copy number (Grosveld et al., 1987). Whether this is due to the length of the flanking regions or to the DNAase-hypersensitive sites has not been established. One possible explanation is that the long flanking regions abolish a proximity effect. Whatever the basis of the phenomenon, these experiments do carry the promise that it may be possible to regulate expression through control of copy number.

Position effects and methylation

Signals resident in the chromosome in the vicinity of the site of insertion may influence expression.

Table 36.2
Relationship between expression per gene copy and copy number

BS6-tk: Herpes virus thymidine kinase gene controlled by a
mouse major urinary protein gene promoter (Al-Shawi et al.,
1988).

Line	Copies	BS6-tk Expression per gene copy		
		Liver	Preputial gland	Testis
78	1	12.5	30	60
64	2	0.75	0	25
58	3	2	3	10
40	10	0	1.2	14
46	80	0.025	1	2

RSV-CAT: bacterial chloramphenicol acetyl transferase gene controlled by the Rous
sarcoma virus LTR (Overbeek et al., 1986).

Line	Copies	RSV-CAT Expression per gene copy			
		Leg	Abdomen	Foot	Sternum
3	2	60 000	31 000	10 000	12 500
4	6	8000	2500	2000	100
2	11	2400	1200	270	1000
1	13	140	60	130	180
5	180	3	—	0.01	0.4

Genes brought close to heterochromatic DNA are inactivated in drosophila and similar effects are
seen in mammals. But the influence of chromosomal location may be more widespread than this.
Gene expression is believed to depend partly on the organization of chromosomal domains, which
is dictated by sometimes remote DNA signals. A foreign gene inserted into a domain may in some
cases take on the expression properties of the domain, nonexpression perhaps or expression in
inappropriate tissues. One way to test this idea is to introduce genes with different tissue expression
patterns linked together in the same DNA fragment, and ask whether or not they behave
autonomously. In an experiment of this sort one-third of transgenic lines carrying a compound
construction expressed the two genes independently, while two-thirds expressed only one of the two
genes (Einat et al., 1987). Although the arrays in the different lines were partially characterized,
inactivation cannot be ruled out, and several interpretations of the experiments remain open.

In mammals the cytosine residues within the palindromic dinucleotide sequence CpG are liable
to methylation, and when the DNA in expressing and nonexpressing tissues is compared, expression
is found to correlate with a low level of methylation. Where an unmethylated gene, transfected into
cells, is expressed, the same gene after in vitro methylation is not (Busslinger et al., 1983). Also,
transfected methylated DNA quickly takes on other properties of inactive genes (Keshet et al., 1986).
However, in at least one case a methylated gene is expressed, and a trans-acting factor determines

whether or not methylation prevents expression (D.E. Kelley *et al.*, 1988). Thus the effect of methylation can be conditional.

Once a site is methylated, its methylation is propagated through DNA replication by the action of a highly active hemi-methylase, but *de novo* methylation is not well understood. There is evidence that foreign DNA sequences can carry information which determines their methylation status. Within the mammalian genome there are localized regions (methylation-free islands, MFI) that contain clustered CpG nucleotide pairs which show low levels of methylation (Bird *et al.*, 1985). When one of these is introduced into a transgenic mouse it largely preserves the property of under-methylation while flanking DNA sequences within the same foreign DNA fragment become methylated (Kolsto *et al.*, 1986). Thus the MFI seems to resist methylation autonomously. There is also good evidence that the methylation level of a given foreign DNA sequence can sometimes vary both between and within transgenic lines, and the level of expression may correlate negatively with the methylation level (Palmiter, Chen and Brinster, 1982). The experiments of Palmiter *et al.* were carried out with transgenic lines which contained very large tandem arrays of foreign genes (on the order of 100 copies). There is also evidence that the expression of isolated retrovirus proviral DNA sequences correlates negatively with methylation (Feinstein *et al.*, 1982; Jahner and Jaenisch, 1985). Furthermore, by treating mice with 5-azacytidine, a methylated provirus may be demethylated *in vivo*, with concurrent elevation of expression levels (Jaenisch *et al.*, 1985). If methylation is influenced by the site of insertion, this could contribute to position effects.

X-chromosome inactivation is thought to be initiated at a single site on the chromosome and to spread from there into adjacent regions, and seems also to relate to CpG hypermethylation. Some autosomal genes are X-inactivated when translocated to the X, while others are not, depending presumably on whether they possess properties which make them responsive. Similarly, some autosomal foreign genes inserted into the X in transgenic mice are X-inactivated (Krumlauf *et al.*, 1986) while others are not (Goldman *et al.*, 1987). A modified α-fetoprotein gene, inserted into the X, was X-inactivated in liver but not in the extra-embryonic visceral yolk sac (Krumlauf *et al.*, 1986).

Some genes seem to be inactive when inherited from one parent, but not from the other. This phenomenon, known as imprinting, has been studied until recently by genetic techniques and nuclear transplantation into fertilized mouse eggs. Some foreign genes in transgenic animals have now been shown to be methylated when inherited from one parent but not when inherited from the other. In three cases frequent resetting of the methylation status occurred (Swain *et al.*, 1987; Reik *et al.*, 1987; Sapienza *et al.*, 1987). Resetting was not complete in all cases (Sapienza *et al.*, 1987). In most cases inheritance from the mother resulted in methylation, and demethylation in the testis could be demonstrated. In one transgenic line only, inheritance from the father resulted in methylation, and the same foreign gene showed maternal methylation in three other lines (Sapienza *et al.*, 1987). In another study the foreign gene was methylated when it passed through the female germ-line, but it was not demethylated by passing through the male, i.e. methylation was irreversible (Hadchouel *et al.*, 1987). Two studies show a correlation between methylation and lack of expression (Swain *et al.*, 1987; Hadchouel *et al.*, 1987). In one study imprinting was observed in only one of seven lines, suggesting that a position effect may have been involved (Reik *et al.*, 1987). In another, in which the foreign DNA was a quail gene, four out of five lines showed the effect, suggesting that the sequence itself carries signals which lead to imprinting (Sapienza *et al.*, 1987, see above). The same could be true of the first study if the methylation-promoting signal was generated by a chance rearrangement which occurred in only one of the seven lines. The many remaining questions about the method of imprinting will be resolved by transgene-recovery and other techniques, but the observations have obvious immediate implications for any work involving inherited foreign genes in transgenic animals.

Variation within transgenic lines

We and others have observed considerable variation in expression between individuals of a transgenic line (Krumlauf *et al.*, 1985; Hammer *et al.*, 1987; Al-Shawi *et al.*, 1988). Some of this variation is probably due to the segregation of genetic modifiers of gene expression, although it should be noted that this has not been experimentally tested. Most transgenic mouse lines have been established by microinjecting fertilized eggs collected after mating two F_1 offspring of a cross between inbred strains. These embryos, which form the G_0 generation, are F_2 offspring of the inbred lines. The G_1 offspring are thus effectively F_3 offspring of the original mating. If the inbred lines differ in several unlinked modifiers, these will have segregated to generate all possible combinations of genes and alleles in the F_3. Since the foreign gene has usually been brought from a different organism, or has been modified in some way, there will usually be no prior knowledge of the existence of modifiers. Other potential sources of within-line variation are variable methylation-associated inactivation and rearrangement.

Expression of unaltered genes in transgenic animals

When an unaltered cloned gene, along with flanking sequences, is introduced as a foreign gene, it is often expressed in the transgenic animal in the same tissues and at a corresponding time in development as it is when expressed naturally. If the gene is transferred between species, differences in sequence can be used to distinguish between the foreign gene and the endogenous homologue. In most cases human genes have been transferred to mice but genes from other species, such as rat, sheep and chicken, have also been transferred to mice. In a few cases advantage has been taken of variation between inbred mouse strains to transfer mouse genes into the mouse genome (Pinkert *et al.*, 1985; Le Meur *et al.*, 1985; Tronik *et al.*, 1987). Some examples are listed in Table 36.3.

Table 36.3
Some unaltered genes correctly expressed in transgenic mice. For other examples see Palmiter and Brinster (1986), Scangos and Bieberich (1987) and Jaenisch (1988)

Gene	References
Human $\bar{\alpha}$-1-antitrypsin	Kelsey *et al.*, 1987; Ruether *et al.*, 1987; Sifers *et al.*, 1987
Human haptoglobin	Oliviero *et al.*, 1987
Human C-reactive protein	Ciliberto *et al.*, 1987
Human neurofilament protein	Julien *et al.*, 1987
Human CD2	Lang *et al.*, 1988
Rat pancreatic elastase	Swift *et al.*, 1984
Rat α-2u-globulin	da Costa Soares *et al.*, 1987
Sheep ß-lactoglobulin	Simons *et al.*, 1987
Mouse Class II MHC antigen	Pinkert *et al.*, 1985; Le Meur *et al.*, 1985
Mouse renin	Tronik *et al.*, 1987

However, in studies in which a sufficient number of transgenic lines has been examined the foreign DNA fragment is always found to be transcriptionally inert in some of them. At one extreme, some fragments are never expressed or never correctly expressed, due to their lack of crucial transcriptional signals (see below). Others seem to have been inert because they were introduced with vector sequences attached (Townes *et al.*, 1985), although similar vector sequences do not impair the expression of other genes (see the list in Palmiter and Brinster, 1986). At the other extreme some fragments are expressed in most transgenic lines but not in all. There are probably many reasons for sporadic nonexpression, including the inactivation of the incoming genes by mutation or rearrangement, and also position-dependent methylation.

Most of the DNA signals which are known to participate in the control of gene transcription are located in the 5′ flanking region. The distance over which the signals are distributed seems to vary considerably from gene to gene. In some cases signals close to the 5′ end of the coding sequence of the gene are sufficient to direct its expression to the correct tissues (Hanahan, 1985; Ornitz *et al.*, 1985; Overbeek *et al.*, 1985; Pinkert *et al.*, 1987). In some cases signals located a considerable distance away increase the fidelity or the level of expression, or both (Grosveld *et al.*, 1987; Pinkert *et al.*, 1987). Signals that participate in the control of expression have also been located to introns (Banerji *et al.*, 1983; Gillies *et al.*, 1983; Behringer *et al.*, 1987; Brinster *et al.*, 1988; Cohen and Levinson, 1988) and downstream of genes (Krimpenfort *et al.*, 1988; Uematsu *et al.*, 1988). The omission or loss of any such signals from the foreign DNA fragment will affect expression and this may explain most cases of incorrect expression.

Hybrid genes in transgenic animals

One of the major objectives of work with transgenic animals is the targeting of gene expression to novel tissues. This has been accomplished in some cases by coupling the upstream controlling region of one gene to the coding region of another. The expectation is that in whatever tissues it is transcribed the gene will direct the synthesis of the protein specified by the coding region. If it carries the necessary information, and if that information is not confounded in some way (see below), the controlling region should allow transcription only in the tissues in which it is normally active.

Hybrid genes can also be used to direct the synthesis of altered proteins. Many proteins consist of two or more *domains*, each made up of a continuous segment of polypeptide chain, which act autonomously to a degree. The linear polypeptide chain is the manifestation of information stored in the same linear order in the DNA. Thus, segments of DNA molecules code for defined protein domains, and by simple genetic manipulation techniques portions of genes which specify functional protein domains can be brought together. When a compound gene constructed in this way is expressed in a cell, the result is a compound protein containing domains from two or more native proteins. The steroid hormone receptor superfamily of genes and proteins provides an excellent example of the opportunities offered by this approach. The members of the superfamily are homologous, clearly descended from a common ancestor. Each receptor protein contains a receptor domain which recognises a particular hormone (glucocorticoid, oestrogen, androgen, thyroxine or retinoic acid) and a second domain which interacts with a particular set of chromosomal DNA sequences to regulate the expression of genes associated with them. By genetic manipulation of the cloned genes a new gene can be made which directs the synthesis of a protein with a new combination of hormone recognition domain and DNA recognition domain. When such a gene is expressed in a cell, the expression of other genes in the same cell is brought under the control of a hormone to which they were previously indifferent (Giguere *et al.*, 1987; Green and Chambon, 1987).

Some applications of hybrid genes

Pharmaceutical proteins

Milk is an ideal source of animal protein which moreover can be recovered noninvasively. Animal milk is free from human viruses such as HBV and HIV, and it is possible at least in principle to create animals which are completely free of pathogens. The full repertory of post-translational modifications available in the mammary gland is not known, but it is certainly a more likely source of correctly modified proteins than bacteria, fungi or insects. Animal milk is thus a promising source of the many human proteins required for therapeutic use. Consequently, efforts have been made to direct the synthesis of human secretory proteins to the milk of mice (Gordon *et al.*, 1987) and sheep (Simons *et al.*, 1988). To effect this, the promoter of a milk protein gene is coupled to the coding region of a gene which specifies a plasma protein (tissue plasminogen activator, blood clotting Factor IX) and the hybrid gene is introduced into the germ-line. Encouraging results have been obtained in that expression has been correctly directed to the lactating mammary gland, but levels of expression have been low. The recent results showing that the possession of introns (Brinster *et al.*, 1988) and long flanking regions or DNase-hypersensitive sites (Grosveld *et al.*, 1987) increases expression levels and confers copy number dependence promise that very much higher levels of expression will soon be attainable.

Given that the synthesis and secretion of plasma proteins into the milk can be achieved, it may prove to be a relatively short step to achieving the secretion of proteins which are not normally secreted.

Physiology and animal disease models

Attention has been focused to a considerable extent on the fact that foreign genes in transgenic animals are functioning in a natural context. It is equally true, of course, that the consequences of malfunction of the foreign gene ramify through the physiology of an otherwise normal organism. Thus, by manipulating properties of foreign genes which have physiological effects, such fields as developmental biology, endocrinology and neurobiology can now be investigated with a new finesse.

Many alterations to physiology are made with a view to simulating human disease. Animal models of disease may be used to elucidate the origins of disease and to develop and evaluate therapeutic measures. This is a very active research area, and the following discussion is selective. Disease models may be classified as follows. In *addition models* a gene is added to the normal complement. This type of model is technically easiest to attain. In *homologous models* the animal homologue of the gene which is damaged in affected humans is deliberately damaged. In *subtraction models* the gene product which corresponds to the product absent in a human disease is neutralized.

Addition models have been used in attempts to elucidate the origins of cancer (Groner *et al.*, 1987), and the expression of immune potential (Storb, 1987). The success of this approach depends upon the effects of the incoming gene being dominant to the resident genes. Thus deficiencies cannot be modelled in this way.

An illustrative addition model is the modelling of autoimmune diabetes mellitus. Class II major histocompatibility (MHC) molecules have been found on the surface of beta cells of the pancreatic islets in patients suffering from this disease, leading to the hypothesis that this in itself could cause the disease through the presentation of self-antigen causing the destruction of cells by the immune system. In two studies the expression of Class II MHC molecules was directed to the beta cells under the control of the insulin promoter. Two allelic series of MHC Class II genes, *I-A* and *I-E*, occupy different loci. Transgenic mice that expressed either *I-A* or *I-E* allelomorphs in their beta cells displayed the characteristics of insulin-dependent diabetes. In those that expressed *I-A* the islets became atrophied (Sarvetnick *et al.*, 1988) but in those that expressed *I-E*, although insulin production was impaired, an immune attack was not mounted against the islet cells (Lo *et al.*, 1988).

Homologous models of Lesch–Nyhan disease have been made in two instructive ways, both based on the use of ES cells. Spontaneous (Hooper *et al.*, 1987) or retroviral insertion (Kuehn *et al.*, 1987) mutants lacking hypoxanthine-guanosine phosphoribosyl transferase were selected with 6-thioguanine and incorporated into chimaeras, and mutant mice produced by breeding. That these mice did not display the neurological symptoms of Lesch–Nyhan disease does not detract from the importance of the demonstration that models can be made in this way.

Homologous models could be made readily if the incorporation of foreign DNA could be targeted to a particular resident gene. In yeast this can be effected quite readily by homologous recombination between the resident gene and a defective incoming gene (Orr-Weaver and Szostak, 1983). Many opportunities will arise if this sort of approach to transgenic animals can be developed, either after direct microinjection or with the use of ES cells. Mouse homeobox genes have recently been inactivated in this way (Mansour *et al.*, 1988) and mice produced which contain a contribution from the modified ES cells (Zimmer and Gruss, 1989; Joyner *et al.*, 1989).

There are two quite different approaches to *subtraction models*. One of these is cell ablation, the selective destruction of a particular cell type. As in other cases, tissue-specific promoters are used to direct expression to the target cells. The second component of transgenic ablation is a gene product that causes cell death.

One such protein is the diphtheria toxin A subunit (DT-A). Very small amounts of DT-A alone inhibit protein synthesis, but it can be taken up by cells only when complexed with the B subunit. Thus cells that express DT-A will be killed without compromising cells that do not. The elastase promoter has been used to direct DT-A synthesis to the pancreatic exocrine cells (Palmiter *et al.*, 1987) and the growth hormone promoter to direct DT-A synthesis to pituitary somatotroph cells (Behringer *et al.*, 1988a). Both approaches were successful, but in both cases effects additional to the ablation of the target cells were observed; in the first study the death of the pancreatic exocrine cells affected the development of other cell lineages, and in the other a reduction in the number of lactotrophs was observed in addition to the almost complete ablation of somatotrophs.

This approach is promising, and will surely find many applications, but it allows no control of the time of cell death. Expression of the foreign gene at too early a stage in development would sometimes have extensive pleiotropic effects, or indeed cause the death of the animal. A system which promises external control of the timing of cell death is based on the different properties of Herpes simplex virus thymidine kinase (HSV-tk) and the cellular pyrimidine kinases. HSV-tk phosphorylates certain pyrimidine analogues such as acyclovir which, when subsequently incorporated into DNA, inhibit further DNA replication and repair and cause cell death. Transgenic mice expressing HSV-tk in the spleen, thymus and bone marrow are reported to be normal. Upon treatment with acyclovir those cells which express the enzyme are selectively destroyed (Heyman *et al.*, 1988; Borrelli *et al.*, 1988). With methods of this sort, the destruction of cells may in the future be timed precisely.

Unpredictable behaviour of some hybrid genes

Attempts to re-direct protein synthesis to different tissues have sometimes been unsuccessful. Some hybrid genes are not expressed at all, and others are expressed in unexpected cell types, that is, in cell types in which neither of the contributory genes is normally expressed (Table 36.4). Unexpected

Table 36.4
Hybrid genes expressed in unexpected cell types. For other examples see Palmiter and Brinster (1986). Abbreviations: GH, growth hormone; GRF, growth hormone releasing factor; MT-I, metallothionein I; MUP, major urinary protein; tk, thymidine kinase

First gene	Second gene	Expression	Reference
Mouse MT-I	Rat somatostatin	Gonadotrophs	Low *et al.*, 1986
Human GRF	SV40 T antigen	Thymus	Botteri *et al.*, 1987
Mouse protamine	SV40 T antigen	Heart and bone	Behringer *et al.*, 1988b
Mouse MT-I	Rat GH	Hypothalamus	Swanson *et al.*, 1985
Mouse MT-I	Human OTC	Testis	K.A. Kelley *et al.*, 1988
Mouse MT-I	SV40 T antigen	Pancreas	Messing *et al.*, 1985
Mouse MUP	Herpes tk	Preputial gland, testis	Al-Shawi *et al.*, 1988

expression may occur in addition to or instead of normal expression, and in those cases in which expression has been successfully directed to the desired tissues, levels of expression are often unexpectedly low. SV40 T-antigen is prominent among inappropriately expressed proteins. This is undoubtedly due partly to the many attempts that have been made to direct its expression, and partly to the fact that the tumours that it generates, even when the level of expression is low, draw attention to unusual sites.

Several possible explanations of unexpected expression stem from the *cis*-active signals carried by genes and their RNA and protein products. Some of these are recognized and quite well characterized, others are not yet well understood, and yet others are unidentified. Thus ignorance of the nature or even in some cases of the very existence of some *cis*-acting signals goes some way to explaining the behaviour of hybrid genes. Furthermore some signals act cooperatively, through interactions between the *trans*-acting proteins which bind to them. Thus new combinations of fragments from different genes may generate novel forms of cooperation.

Each step in gene expression depends as much on the *trans*-acting proteins as on the *cis*-acting signals. When the expression of a gene is successfully directed to a novel tissue its products may encounter a different set of *trans*-acting proteins with an unexpected outcome. Possible examples of this sort of effect are the comparative failure of fibroblasts to gamma-carboxylate Factor IX (de la Salle *et al.*, 1985) and the aberrant processing of pre-prosomatostatin when it is produced in the kidney or liver (Low *et al.*, 1985).

Hybrid genes have commonly been constructed from the promoter of one gene and a cDNA corresponding to the intronless coding region of another. Recent results show that the substitution of the corresponding gene region with the introns in place for the intronless cDNA sequence may markedly increase expression (Brinster *et al.*, 1988). In these comparisons both the 5′ and 3′ ends of the two sequences were the same, and the only difference between them was in the possession of introns. There are precedents for the presence of tissue-specific enhancer sequences within introns, and recently a point mutation in an intron has been shown to profoundly affect the level of expression of the c-Ha-*ras* gene (Cohen and Levinson, 1988).

Prospects

The study of transgenic animals has been under way for less than a decade, and during this time very satisfactory progress in our ability to direct and control gene expression has been made. It would be pessimistic to suppose that this progress will not continue, or that there will ultimately be any practical limitations to applications of germ-line modification.

Bibliography of Alan Robertson

(Numbers in brackets show chapters in which the paper is cited)

1949

Robertson, A. 1949a. Inbreeding experiments in dairy cattle. *Animal Breeding Abstracts* 17, 1–6. (22)

Robertson, A. 1949b. Crossbreeding experiments with dairy cattle. *Animal Breeding Abstracts* 17, 201–8. (22)

Robertson, A. and Asker, A.A. 1949. The movement of two breeds of pedigree dairy cattle in Great Britain. *Empire Journal of Experimental Agriculture* 17, 125–31. (11, 29)

Robertson, A. and Lerner, I.M. 1949. The heritability of all-or-none traits: Viability of poultry. *Genetics* 34, 395–411. (22, 25, 28)

1950

Rendel, J.M. and Robertson, A. 1950a. Some aspects of longevity in dairy cows. *Empire Journal of Experimental Agriculture* 18, 49–56. (22)

Rendel, J.M. and Robertson, A. 1950b. Estimation of genetic gain in milk yield by selection in a closed herd of dairy cattle. *Journal of Genetics* 50, 1–8. (22, 27, 29)

Rendel, J.M. and Robertson, A. 1950c. Artifical insemination in dairy cattle breeding. *Scottish Agriculture* 30, 79–83. (29)

Robertson, A. 1950. Some observations on experiments with identical twins in dairy cattle. *Journal of Genetics* 50, 32–5.

Robertson, A. and Rendel, J.M. 1950. The use of progeny testing with artificial insemination in dairy cattle. *Journal of Genetics* 50, 21–31.(22, 27, 29, 33)

1951

Rendel, J.M., Robertson, A. and Alim, K.A. 1951. The extent of selection for milk yield in dairy cattle. *Empire Journal of Experimental Agriculture* 19, 295–301.

Rendel, J.M. and Robertson, A. 1951. Genetic variability and artificial insemination. *Nature* 167, 320. (22, 29)

Robertson, A. 1951a. A preliminary report on the herd of Fulani cattle at Shika, Nigeria. *Symposium on Tropical Animal Breeding*, Edinburgh.

Robertson, A. 1951b. The analysis of heterogeneity in the binomial distribution. *Annals of Eugenics* 16, 1–15.

Robertson, A. and Asker, A.A. 1951a. The genetic history and breed-structure of British Friesian cattle. *Empire Journal of Experimental Agriculture* 19, 113–30. (11, 27)

Robertson, A. and Asker, A.A. 1951b. The expansion of a breed of dairy cattle. *Empire Journal of Experimental Agriculture* 19, 191–201.

1952

Carter, T.C. and Robertson, A. 1952. A mathematical treatment of genetical recombination, using a four-strand model. *Proceedings of the Royal Society of London B.* 139, 410–26.

Johansson, I. and Robertson, A. 1952. Progeny testing in the breeding of farm animals. *Proceedings of the British Society of Animal Production* 1952, 79–105. (22)

Robertson, A. 1952. The effect of inbreeding on the variation due to recessive genes. *Genetics* 37, 189–207. (1,12)

1953

Clayton, G., Morris, J.A. and Robertson, A. 1953. Selection for abdominal chaetae in a large population of *Drosophila melanogaster*. Symposium on Genetics of Population Structure. *International Union of*

Biological Sciences, Naples Ser. B, (15) pp.7–15. (12)

Rendel, J. and Robertson, A. 1953. Cattle breeding and artificial insemination: Report to the Agricultural Research Council (London).

Robertson, A. 1953a. A numerical description of breed structure. *Journal of Agricultural Sciences* 43, 334–6. (11, 22, 27)

Robertson, A. 1953b. The use and interpretation of progeny tests in livestock improvement. *Proceedings of the British Society of Animal Production 1953*, 3–16. (22, 29)

1954

Robertson, A. 1954a. Il posto della consanguineita e dell'incrocio nella Zootecnica. *Caryologia* 6, 1273–8.

Robertson, A. 1954b. Artificial insemination and livestock improvement. *Advances in Genetics* 6, 451–72. (29, 35)

Robertson, A. 1954c. Inbreeding and performance in British Friesian cattle. *Proceedings of the British Society of Animal Production 1954*, 87–92.

Robertson, A. and Mason, I.L. 1954. A genetic analysis of the Red Danish breed of cattle. *Acta Agriculturae Scandinavica* 4, 257–65.

Robertson, A. and Rendel, J.M. 1954. The performance of heifers got by artificial insemination. *Journal of Agricultural Sciences* 44, 184–92. (22, 26, 29)

1955

Clayton, G.A. and Robertson, A. 1955. Mutation and quantitative variation. *American Naturalist* 89, 151–8. (9, 12, 13, 15, 16, 17)

Robertson, A. 1955a. Prediction equations in quantitative genetics. *Biometrics* 11, 95–8. (22, 26, 27)

Robertson, A. 1955b. Selection in animals: Synthesis. *Cold Spring Harbour Symposia on Quantitative Biology* 20, 225–9. (1, 13, 15, 18)

Robertson, A. 1955c. Review of *Genetic Homeostasis*. *Journal of Heredity* 46, 176.

Robertson, A. and MacArthur, A.T.G. 1955. Genetic differences between bull-breeding herds. *Proceedings of the British Society of Animal Production 1955*, 94–101. (11, 22)

1956

Falconer, D.S. and Robertson, A. 1956. Selection for environmental variability of body size in mice. *Zeitschrift für induktive Abstammungs- und Vererbungslehre* 87, 385–91. (1, 12)

Knight, G.R., Robertson, A. and Waddington, C.H. 1956. Selection for sexual isolation within a species. *Evolution* 10, 14–22. (1, 12)

Mason, I.L. and Robertson, A. 1956. The progeny testing of dairy bulls at different levels of production. *Journal of Agricultural Sciences* 47, 367–75. (22)

Robertson, A. 1956a. Blood-grouping in dairy dattle improvement. *Proceedings of the VII International Congress on Animal Husbandry*, Madrid; Altamira, pp.1–8.

Robertson, A. 1956b. The effect of selection against extreme deviants based on deviation or on homozygosis. *Journal of Genetics* 54, 236–48. (1, 12, 15)

Robertson, A. 1956c. The genetic effects of radiation – the effects on human populations. *New Biology* 20, 53–9.

Robertson, A. and Mason, I.L. 1956. The progeny testing of dairy bulls: A comparison of special station and field results. *Journal of Agricultural Sciences* 47, 376–81. (22)

Robertson, A., Stewart, A. and Ashton, E.D. 1956. The progeny assessment of dairy sires for milk: The use of contemporary comparison. *Proceedings of the British Society of Animal Production 1956*, 43–50. (22)

Robertson, A., Waite, R. and White, J.C.D. 1956. Variations in the chemical composition of milk with particular reference to the solids-not-fat. II. The effect of heredity. *Journal of Dairy Research* 23, 82–91. (22)

Waite, R., White, J.C.D. and Robertson, A. 1956. Variations in the chemical composition of milk with particular reference to the solids-not-fat. I. The effect of stage of lactation, season of year and age of cow. *Journal of Dairy Research* 23, 65–81.

1957

Clayton, G.A., Morris, J.A. and Robertson, A. 1957. An experimental check on quantitative genetical theory. I. Short-term responses to selection. *Journal of Genetics* 55, 131–51. (12)

Clayton, G.A. and Robertson, A. 1957. An experimental check on quantitative genetical theory. II. The long-term effects of selection. *Journal of Genetics* 55, 152–70. (9, 12, 15, 16, 17)

Clayton, G.A., Knight, G.R., Morris, J.A. and Robertson, A. 1957. An experimental check on quantitative genetical theory. III. Correlated responses. *Journal of Genetics* 55, 171–80. (12)

Knight, G.R. and Robertson, A. 1957. Fitness as a measurable character in Drosophila. *Genetics* 42, 524–30. (1, 12, 13, 14)

Mason, I.L., Robertson, A. and Gjelstad, B., 1957. The genetic connexion between body size, milk production and efficiency in dairy cattle. *Journal of Dairy Research* 24, 135–43. (22)

Rendel, J.M., Robertson, A., Asker, A.A., Khishin, S.S. and Ragab, M.T. 1957. The inheritance of milk production characteristics. *Journal of Agricultural Science* 48, 426–32.

Robertson, A. 1957a. Genetics and the improvement of dairy cattle. *Agricultural Review* 2(8), 10–21.

Robertson, A. 1957b. Optimum group size in progeny testing and family selection. *Biometrics* 13, 442–50. (27)

1958

Robertson, A. 1958a. Modern trends in animal breeding. *Agricultural Progress* 33, 6–16. (35)

Robertson, A. 1958b. Laboratory breeding experiments and animal improvement. *Proceedings of the Xth International Congress on Genetics* 1, 199–205. (12)

Robertson, A. and Khishin, S.S. 1958a. The effect of selection for heifer milk yield on the production level of mature cows. *Journal of Agricultural Science* 50, 12–16.

Robertson, A. and Khishin, S.S. 1958b. An analysis of the variation between dairy-sire progeny-groups for milk yield and fat content. *Proceedings of the British Society of Animal Production 1958*, 19–29.

1959

Gowe, R.S., Robertson, A. and Latter, B.D.H. 1959. Environment and poultry breeding problems. 5. The design of poultry control strains. *Poultry Science* 38, 462–71. (22)

Robertson, A. 1959a. A simple method of pedigree evaluation in dairy cattle. *Animal Production* 1, 167–74.

Robertson, A. 1959b. Experimental design in the evaluation of genetic parameters. *Biometrics* 15, 219–26. (22, 24)

Robertson, A. 1959c. The sampling variance of the genetic correlation coefficient. *Biometrics* 15, 469–85. (22, 24, 33)

Robertson, A. 1959d. Horses for courses? *Journal of the Royal Agricultural Society of England* 120, 88–91.

Robertson, A. 1959e. Populationsgenetik und quantitative Vererbung. In: *Handbuch der Tierzüchtung*. Hamburg, Germany; Verlag Paul Parey, pp.77–107.

1960

Clayton, G.A. and Robertson, A. 1960. Light induction and out-of-season reproduction in the turkey. *British Poultry Science* 1, 17–23.

Latter, B.D.H. and Robertson, A. 1960. Experimental design in the estimation of heritability by regression methods. *Biometrics* 16, 348–53. (24, 33)

Robertson, A. 1960a. The progeny testing of dairy bulls – a comparison of tests on father and son. *Journal of Agricultural Sciences* 54, 100–4. (22)

Robertson, A. 1960b. On optimum family size in selection programmes. *Biometrics* 16, 296–8. (27)

Robertson, A. 1960c. A theory of limits in artificial selection. *Proceedings of the Royal Society of London. B*. 153, 234–49. (12, 17, 18, 22, 27)

Robertson, A. 1960d. Experimental design on the measurement of heritabilities and genetic correlations. In: *Biometrical Genetics* (edited by O. Kempthorne). London, UK; Pergamon Press, pp.101–6. (22)

Robertson, A., O'Connor, L.K. and Edwards, J. 1960. Progeny testing dairy bulls at different management levels. *Animal Production* 2, 141–52.

1961

Neimann-Sørensen, A. and Robertson, A. 1961. The association between blood groups and several production characteristics in three Danish cattle breeds. *Acta Agriculturae Scandinavica* 11, 163–96. (22, 33, 34)

Robertson, A. 1961a. Genetics of dairy cattle. *Journal of Dairy Research* 28, 195–207. (35)

Robertson, A. 1961b. Selection for several characters. *Schriftenreihe des Max-Planck-Instituts für Tierzücht und Tierernährung*, 247–59.

Robertson, A. 1961c. Inbreeding in artificial selection programmes. *Genetical Research* 2, 189–94. (1, 12, 18, 22, 27, 29)

1962

Latter, B.D.H. and Robertson, A. 1962. The effects of inbreeding and artificial selection on reproductive fitness. *Genetical Research* 3, 110–38. (12, 14, 16, 17)

Robertson, A. 1962a. Selection for heterozygotes in small populations. *Genetics* 47, 1291–300. (1, 15)

Robertson, A. 1962b. Weighting in the estimation of variance components in the unbalanced single classification. *Biometrics* 18, 413–7. (22, 23)

Robertson, A. 1962c. The future of animal breeding. *Advancement of Science* No 76 (March), 537–42.

1963

McBride, G. and Robertson, A. 1963. Selection using assortative mating in *Drosophila melanaogaster*. *Genetical Research* 4, 356–69. (12)

Robertson, A. 1963a. The influence of breeding on production efficiency. *Proceedings of the 1st World Conference on Animal Production* 1, 99–110.

Robertson, A. 1963b. Some comments on quantitative genetic theories. In: *Statistical Genetics and Plant Breeding* (edited by W. D. Hanson and H. F. Robinson). Washington DC, USA; Nat. Acad. Sci., Nat. Res. Council Publ. 982, pp.108–115.

Robertson, A. 1963c. Summing up and proposals for future action. In. *British Livestock Breeding – The Way Ahead. Proceedings of the National Livestock Breeding Conference*, 1962, pp. 95–102.

1964

Allan, J.S. and Robertson, A. 1964. The effect of initial reverse selection upon total selection response. *Genetical Research* 5, 68–79. (12)

Clayton, G.A. and Robertson, A. 1964. The effects of X-rays on quantitative characters. *Genetical Research* 5, 410–22. (12, 16, 17)

Robertson, A. 1964a. Genetic aspects of homeostasis. *Symposium of the Society for Experimental Biology* 18, 257–64.

Robertson, A. 1964b. The effect of non-random mating within inbred lines on the rate of inbreeding. *Genetical Research* 5 164–7. (1)

Robertson, A. 1964c. Balanced combinations of polygenes. *Nature* 203, 103.

Sen, B.K. and Robertson, A. 1964. An experimental examination of methods for the simultaneous selection of two characcters using *Drosophila melanogaster*. *Genetics* 50, 199–209. (12)

1965

Robertson, A. 1965a. Variation in scutellar bristle number – an alternative hypothesis. *American Naturalist* 99, 19–23.

Robertson, A. 1965b. The interpretation of genotypic ratios in domestic animal populations. *Animal Production* 7, 319–24. (1, 22)

Robertson, A. 1965c. Synthesis of session:: 'Population genetics – theoretical'. In: *Genetics Today. Proceedings of the XIth International Congress of Genetics* (edited by S.J. Geerts). London, UK; Pergamon Press, Vol. 3, pp. 527–32. (1, 12, 15)

1966

Barker, J.S.F. and Robertson, A. 1966. Genetic and phenotypic parameters for the first three lactations in Friesian cows. *Animal Production* 8, 221–40.

Bohren, B.B., Hill, W.G. and Robertson, A. 1966. Some observations on asymmetrical correlated responses to selection. *Genetical Research* 7, 44–57. (12, 21)

Clayton, G.A. and Robertson, A. 1966. Genetics of changes in economic traits during the laying year. *British Poultry Science* 7 143–51.

Hill, W.G. and Robertson, A. 1966. The effect of linkage on limits to artificial selection. *Genetical Research* 8, 269–94. (1, 6, 7, 12, 27)

Robertson, A. 1966a. Artificial selection in plants and animals. *Proceedings of the Royal Society of London B.* 164, 341–9. (12, 21)

Robertson, A. 1966b. A mathematical model of the culling process in dairy cattle. *Animal Production* 8, 95–108. (13, 22)

Robertson, A. 1966c. Biochemical polymorphisms in animal improvement. In. *Proceedings of the Xth European Congress on Animal Blood Groups and Biochemical Polymorphism of Animals*. Paris, France; Institut National de la Recherche Agronomigue, pp.35–42. (22)

Robertson, A. and Barker, J.S.F. 1966. The correlation between first lactation milk production and longevity in dairy cattle. *Animal Production* 8, 241–52. (22)

Rosado, J.M.C. and Robertson, A. 1966. The genetic control of sex ratio. *Journal of Theoretical Biology* 13, 324–9. (1)

1967

Jamieson, A. and Robertson, A. 1967. Cattle transferrins and milk production. *Animal Production* 9, 491–500. (22, 33).

Ødegård, A.K. and Robertson, A. 1967. A comparison of progeny tests for milk production on fathers and sons. *Acta Agriculturae Scandinavica* 17, 241–6. (22)

Robertson, A. 1967a. The nature of quantitative genetic variation. In: *Heritage from Mendel* (edited by A. Brink). Madison, Wis.; The University of Wisconsin Press, pp.265–80. (12, 15, 17, 18, 21)

Robertson, A. 1967b. Animal breeding. *Annual Review of Genetics* 1, 295–312. (12)

1968

Hill, W.G. and Robertson, A. 1968a. Linkage disequilibrium in finite populations. *Theoretical and Applied Genetics* 38, 226–31. (1, 6, 7, 12)

Hill, W.G. and Robertson, A. 1968b. The effects of inbreeding at loci with heterozygote advantage. *Genetics* 60, 615–28. (1, 6)

Osman, H.E.S. and Robertson, A. 1968. The introduction of genetic material from inferior into superior strains. *Genetical Research* 12, 221–36. (12)

Robertson, A. 1968a. The spectrum of genetic variation. In: *Population Biology and Evolution* (edited by R.C. Lewontin). Syracuse, New York; Syracuse University Press, pp.5–16. (1, 3, 12, 13)

Robertson, A. 1968b. Inbreeding in artificial insemination. In: *The Proceedings of the VIth International Congress on Animal Production and Artificial Insemination* 2, 1317–24.

1969

Narain, P. and Robertson, A. 1969. Limits and duration of response in finite populations. *Indian Journal of Heredity* 1, 1–19. (12)

Robertson, A. 1969a. The theory of animal breeding. *Proceedings of the 12th International Congress of Genetics* 3, 371–7.

Robertson, A. 1969b. The application of genetics to animal breeding. In: *Fifty Years of Genetics* (edited by J.L. Jinks). Edinburgh, UK; Oliver and Boyd, pp.27–36. (22)

1970

McPhee, C.P. and Robertson, A. 1970. The effect of suppressing crossing-over on the response to selection in *Drosophila melanogaster*. *Genetical Research* 16, 1–16. (12, 18)

Robertson, A. 1970a. Molecular biology and animal breeding. *Annales de Génétique et de Séléction animale* 2, 393–402. (21, 33, 34, 35)

Robertson, A. 1970b. A note on disruptive selection experiments in *Drosophila*. *American Naturalist* 104 561–9.

Robertson, A. 1970c. Some optimum problems in individual selection. *Theoretical Population Biology* 1, 120–7. (12, 18, 27)

Robertson, A. 1970d. The reduction in fitness from genetic drift at heterotic loci in small populations. *Genetical Research* 15, 257–259. (1).

Robertson, A. 1970e. A theory of limits in artificial selection with many linked loci. In: *Mathematical Topics in Population Genetics*. (edited by K. Kojima). Berlin, Germany; Springer, pp.246–88. (12, 18, 20)

Robertson, A. 1970f. The state of quantitative genetics in relation to the real world. In: *Proceedings of the 19th National Breeders Roundtable*, Kansas City, pp.2–17.

1971

Robertson, A. 1971. Optimum utilization of genetic material with special reference to cross-breeding in relation to other methods of genetic improvement. In: *Proceedings of the Xth International Congress of Animal Production*, pp.57–67. (11)

Robertson, A. and Narain, P. 1971. The survival of recessive lethals in finite populations. *Theoretical Population Biology* 2, 24–50. (1)

1973

Robertson, A. 1973a. Body size and efficiency. *The Proceedings of the British Society of Animal Production*, 1973 9–14.

Robertson, A. 1973b. Stochastic processes in artificial selection. In: *The Mathematical Theory of the Dynamics of Biological Populations* (edited by M.S. Bartlett and R.W. Hiorns). London; Academic Press, pp. 195–201. (12)

Robertson, A. 1973c. Linkage between marker loci and those affecting quantitative traits. *Behaviour Genetics* 3 389–91. (6, 21, 34)

Robertson, A. 1973d. Growth rate, appetite, body composition and efficiency. *Proceedings of the 15th British Poultry Breeders Roundtable*. (19)

1974

McMillan, I. and Robertson, A. 1974. The power of methods for the detection of major genes affecting quantitative characters. *Heredity* 32, 349–56. (12, 17, 20, 21, 34)

Robertson, A. 1974a. The validity of the optimum model. *Advances in Applied Probability* 6, 17–18. (12, 15)

Robertson, A. 1974b. Symposium on animal genetics. Introduction by the Chairman (*Proceedings of the XIIIth International Congress on Genetics*) *Genetics* 78, 507–10.

1975

Bar-Anan, R. and Robertson, A. 1975. Variation in sex ratio between progeny groups in dairy cattle. *Theoretical and Applied Genetics* 46, 63–5.

Briscoe, D.A., Robertson, A. and Malpica, J.M. 1975. Dominance at Adh locus in response of adult *Drosophila melanogaster* to environmental alcohol. *Nature* 255, 148–9. (1)

Dolan, R. and Robertson, A. 1975. The effect of conditioning the medium in *Drosophila* in relation to frequency-dependent selection. *Heredity* 35, 311–6. (1)

Madalena, F.E. and Robertson, A. 1975. Population structure in artificial selection: Studies with *Drosophila melanogaster*. *Genetical Research* 24, 113–26. (11, 12, 16, 18)

Robertson, A. 1975a. Remarks on the Lewontin–Krakauer test. *Genetics* 80, 396. (1, 3, 11)

Robertson, A. 1975b. Gene frequency distributions as a test of selective neutrality. *Genetics* 81, 775–85. (1, 11)

Robertson, A. 1975c. What has genetics to contribute to horse-breeding? In: *Proceedings of the International Symposium on Genetics and Horse-Breeding*. Dublin, Ireland; Royal Dublin Society, pp.64–70.

Robertson, A. 1975d. Review of: 'The genetic basis of evolutionary change' by R.C. Lewontin. *Nature* 254, 367. (9)

1976

Baptist, R. and Robertson, A. 1976. Asymmetrical responses to automatic selection for body size in *Drosophila melanaogaster.Theoretical and Applied Genetics* 47, 209–13. (12)

Nicholas, F.W. and Robertson, A. 1976. The effect of selection on the standardized variance of gene frequency. *Theoretical and Applied Genetics* 48, 263–8. (11)

Robertson, A. 1976a. Why do we crossbreed? In: *Crossbreeding Experiments and Strategy of Beef Utilization to Increase Beef Production* (edited by I.L. Mason and W. Pabst). EEC Publication EUR 5492e, pp.4–15.

Robertson, A. 1976b. Variation in gene frequency over population as a test for natural selection. *Advances in Applied Probability* 8, 19–20.

1977

Robertson, A. 1977a. Artificial selection with a large number of linked loci. In: *Proceedings of the International Conference on Quantitative Genetics* (edited by E. Pollak, O. Kempthorne and T.B. Bailey). Ames, Iowa; Iowa State University Press, pp.307–22. (6, 12, 18, 21)

Robertson, A. 1977b. The non-linearity of offspring–parent regression. In: *Proceedings of the International Conference on Quantitative Genetics* (edited by E. Pollak, O. Kempthorne and T.B. Bailey). Ames, Iowa; Iowa State University Press, pp.297–304. (12)

Robertson, A. 1977c. The effect of selection on the estimation of genetic parameters. *Zeitschrift für Tierzüchtung* 94, 131–5. (12, 24)

Robertson, A. 1977d. Conrad Hal Waddington. 1905–1975. *Biographical Memoirs of Fellows of the Royal Society* 23, 575–622.

Robertson, A., Briscoe, D.A. and Louw, J.H. 1977. Variations in abdomen pigmentation in *Drosophila melanogaster* females. *Genetica* 47, 73–6. (12)

1978

Robertson, A. 1978. The time of detection of recessive visible genes in small populations. *Genetical Research* 31, 255–64. (1, 12, 16)

1980

Nicholas, F.W. and Robertson, A. 1980. The conflict between natural and artificial selection in finite populations. *Theoretical and Applied Genetics* 56, 57–64. (12, 15)

Robertson, A. (editor) 1980a. Selection experiments in laboratory and domestic animals. Slough, UK; Commonwealth Agricultural Bureaux. (12, 19)

Robertson, A. 1980b. The contribution of computer studies to selection theory. In: *Selection Experiments in Laboratory and Domestic Animals* (edited by A. Robertson). Slough, UK; Commonwealth Agricultural Bureaux, pp.17–20. (12)

Robertson, A. 1980c. Natural selection and continuous variation. In: *Vito Volterra Symposium on Mathematical Models in Biology* (edited by C. Barigozzi). (Lecture Notes in Biomathematical Genetics, No 39) Berlin, Germany; Springer-Verlag, pp.48–55. (12)

1982

Robertson, A. 1982a. Genetic aspects of growth. In: *Proceedings of the World Congress on Sheep and Beef Cattle Breeding*. Palmerston North, New Zealand; Dunsmore Press, pp.427–37. (19)

Robertson, A. 1982b. Genetic engineering in animal improvement. In: *Proceedings of the 2nd World Congress on*

Genetics Applied to Livestock Production. Madrid; Editorial Garsi, Vol. 6, 139–45. (22, 30, 33)
Robertson, A. 1982c. Animal Improvement through genetic engineering. *Span* 25, 53–5. (33)

1983
Robertson, A. and Hill, W.G. 1983a. Identity of different mutations for deleterious genes. *Nature* 301, 176. (1)
Robertson, A. and Hill W.G. 1983b. Population and quantitative genetics of many linked loci in finite populations. *Proceedings of the Royal Society of London B*. 219, 253–64. (1, 3, 6, 12, 18)

1984
Brien, F.D., Sharp, G.L., Hill, W.G. and Robertson, A. 1984. Effects of selection on growth, body composition and food intake in mice. II. Correlated responses in reproduction. *Genetical Research* 44, 73–85.
Robertson, A. 1984. The relevance of molecular biology to animal improvement. *Proceedings of the XVth International Congress of Genetics*, New Delhi, India. Vol. 4, pp.193–200. (22, 30, 33, 34, 35)
Robertson, A. and Hill, W.G. 1984. Deviations from Hardy–Weinberg proportions: sampling variances and use in estimation of inbreeding coefficients. *Genetics* 107, 703–18. (1, 7)
Sharp, G.L., Hill, W.G. and Robertson, A. 1984. Effects of selection on growth, body composition and food intake in mice. I. Responses in selected traits. *Genetical Research* 43, 75–92. (32)

1985
Robertson, A. 1985. Molecular Biology and animal improvement. In: *Biotechnology and Recombinant DNA Technology in the Animal Production Industries*. (edited by R.A. Leng, J.S.F. Barker, D.B. Adams and K.J. Hutchinson). Armidale, New South Wales, Australia; The University of New England Press, pp.3–9.

1986
Robertson, A. 1986a. Molecular biology and animal improvement. In: *Proceedings of the 3rd World Congress on Genetics Applied to Livestock Production*. Lincoln, Nebraska; University of Nebraska, Vol. 12, 32–41. (22, 30, 33, 34, 35)
Robertson, A. 1986b. The analysis of long term selection experiments. In: *Proceedings of the 3rd World Congress on Genetics Applied to Livestock Production*. Lincoln, Nebraska; University of Nebraska, Vol. 12, 141.

1987
Yamaguchi, O., Yamazaki, T., Saigo, K., Mukai, T. and Robertson, A. 1987. Distributions of three transposable elements, *P*, *297*, and copia, in natural populations of *Drosophila melanogaster*. *Japanese Journal of Genetics* 62, 205–16. (8)

1988
Capy, P., David, J.R. and Robertson, A. 1988. Thoracic trident pigmentation in natural populations of *Drosophila simulans*: a comparison with *D.melanogaster*. *Heredity* 61, 263–8.
Shrimpton, A.E. and Robertson, A. 1988a. The isolation of polygenic factors controlling bristle score in *Drosophila melanogaster*. 1. Allocation of third chromosome sternopleural bristle effects to chromosome sections. *Genetics* 118, 437–43. (12, 16, 20, 21, 34)
Shrimpton, A.E. and Robertson, A. 1988b. The isolation of polygenic factors controlling bristle score in *Drosophila melanogaster*. 2. Distribution of third chromosome bristle effects within chromosome sections. *Genetics* 118, 445–459. (12, 16, 17, 20, 21, 31, 34)

(Editors' note: early papers of Alan Robertson on chemistry are not included)

References

(Papers by Alan Robertson are listed in his bibliography. Numbers in brackets show chapters in which the paper is cited)

Abplanalp, H., Lowry, D.C., Lerner, I.M. and Dempster, E.R. 1964. Selection for egg number with X-ray-induced variation. *Genetics* 50, 1083–100. (16)

Aguade, M. 1988. Restriction map variation at the Adh locus of *Drosophila melanogaster* in inverted and non-inverted chromosomes. *Genetics* 119, 135–40. (5)

Ajioka, J.W. and Eanes, W.F. 1989. The accumulation of P-elements on the tipof the X chromosomes of *Drosophila melanogaster*. *Genetical Research* 53, 1–6. (8)

Albers, G.A.A., Burgess, S.E., Adams, D.B., Barker, J.S.F., Le Jambre,L.F. and Piper, L.R. 1984. Breeding *Haemonchus contortus* resistant sheep – problems and prospects. In: *Immunogenetic Approaches to the Control of Endoparasites* (edited by J.K. Dineen and P.M. Outteridge). Melbourne, Australia; CSIRO, Division of Animal Health, pp.41–51.

Allan, J.S. and Robertson, A. 1964. (See AR bibliography)

Allen, P.A. and McCarthy, J.C. 1980. The effects of selection for body weight on the distribution of fat in mice. *Animal Production* 31, 1–11. (19)

Al-Murrani, W.K. 1974. The limits to artificial selection. *Animal Breeding Abstracts* 42, 587–592. (18)

Al-Shawi, R., Burke, J., Jones, C.T., Simons, J.P. and Bishop, J.O.1988. A MUP promoter-thymidine kinase reporter gene shows relaxed tissue-specific expression and confers male sterility upon transgenic mice. *Molecular and Cellular Biology* 8, 482–8. (36)

Anderson, G.B. 1985. Manipulation of the mammalian embryo. *Journal of Animal Science* 61, 1–13. (35)

Anderson, R.L. 1981. Recent developments in designs and estimators for variance components. In: *Statistics and Related Topics* (edited by M. Csorogo, D.A. Dawson, J.N.K. Rao, Md. E. Saleh). Amsterdam; North-Holland, pp.3–24. (24)

Anderson, W.W. 1969. Selection in experimental populations. I. Lethal genes. *Genetics* 62, 653–72. (14)

Antonarakis, S.E., Boehm, C.D., Giardina, P.J. and Kazazian, H.H. 1982. Non-random association of polymorphic restriction sites in the gamma-globin cluster. *Proceedings of the National Academy of Sciences* 79, 137–41. (6)

Aota, S., Gojobori, T., Ishibashi, F., Maruyama, T. and Ikemura, T. 1988. Codon usage tabulated from the GenBank genetic sequence data. *Nucleic Acids Research* 16, r315–r402. (4)

Aota, S., and Ikemura, T. 1986. Diversity in G+C content at the third position of codons in vertebrate genes and its cause. *Nucleic Acids Research* 14, 6345–55. (4)

Aquadro, C.F., Deese, S.F., Bland, M.M., Langley, C.H. and Laurie-Ahlberg, C.C. 1986. Molecular population genetics of the alcohol dehydrogenase gene region of *Drosophila melanogaster*. *Genetics* 114, 1165–90. (3, 5, 6, 8)

Aquadro, C.F. and Greenberg, B.D. 1983. Human mitochondrial DNA variation and evolution: Analysis of nucleotide sequence evolution. *Genetics* 103, 287–312. (3)

Aquadro, C.F., Lado, K.M. and Noon, W.A. 1988. The *rosy* region of *Drosophila melanogaster* and *D. simulans*. I. Contrasting levels of naturally occurring DNA restriction map variation and divergence. *Genetics* 119, 875–88. (3, 5, 6)

Archibald, A.L. and Imlah, P. 1985. The halothane sensitivity locus and its linkage relationships. *Animal Blood Groups and Biochemical Genetics* 16, 253–63. (28)

Arnheim, N. 1983. Concerted evolution of multigene families. In: *Evolution of Genes and Proteins* (edited by M. Nei and R.K. Koehn). Sunderland, Mass., USA; Sinauer, pp.38–61. (6, 9)

Arnold, S.J. and Wade, M.J. 1984a. On the measurement of natural and sexual selection. Theory. *Evolution* 38, 709–19. (13)

Arnold, S.J. and Wade, M.J. 1984b. On the measurement of natural and sexual selection: Applications. *Evolution* 38, 720–34. (13)

Ashton, G.C. and Borecki, I.B. 1987. Further evidence for a gene influencing spatial ability. *Behavior Genetics* 17, 243–56. (20)

Avalos, E. and Smith, C. 1987. Genetic improvement of litter size in pigs. *Animal Production* 44, 153–64. (28)

Avise, J.C. 1986. Mitochondrial DNA and the evolutionary genetics of higher animals. *Philosophical Transactions of the Royal Society of London B* 312, 325–42. (6)

Avise, J.C., Arnold, J., Ball, R.M., Bermingham, E., Lamb, T., Neigel, J.E., Reeb, C.A. and Saunders, N.C. 1987. Intraspecific phylogeography: The mitochondrial bridge between population genetics and systematics. *Annual Review of Ecology and Systematics* 18, 489–522. (6, 11)

Avise, J.C., Gilbin-Davidson, C., Laerm, J., Patton, J.C. and Lansman, R.A. 1979. Mitochondrial DNA clones and matriarchal phylogeny within and among geographic populations of the pocket gopher, *Geomys pinetis*. *Proceedings of the National Academy of Sciences, USA* 76, 6694–8. (3)

Avise, J.C., Lansman, R.A. and Shade, R.O. 1979. The use of restriction endonucleases to measure mitochondrial DNA sequence relatedness in natural populations. I. Population structure and evolution in the genus *Peromyscus*. *Genetics* 92, 279–95. (5)

Baffour-Awuah, O. and Barker, J.S.F. 1988. Selection responses in synthetic populations of *Drosophila melanogaster*. *Journal of Animal Breeding and Genetics* 105, 120–8. (11)

Baker, R.D. and Shea, B.F. 1985. Commercial splitting of bovine embryos. *Theriogenology* 23, 3–12. (35)

Banerji, J., Olson, L. and Schafner, W. 1983. A lymphocyte-specific cellular enhancer is located downstream of the joining region in immmunoglobulin heavy chain genes. *Cell* 33, 729–40. (36)

Banks, R.G. 1986. The effects of population structure on responses to artificial selection: An investigation of the shifting balance theory. PhD Thesis, University of New England, Armidale, Australia. (11)

Baptist, R. and Robertson, A. 1976. (See AR bibliography)

Barker, J.S.F. 1963. The estimation of relative fitness of Drosophila populations. II. Experimental evaluation of factors affecting fitness. *Evolution* 17, 56–71. (14)

Barker, J.S.F. 1988. Quantitative genetics, ecology and evolution. In: *Proceedings of the 2nd International Conference on Quantitative Genetics* (edited by B.S. Weir, E.J. Eisen, M.M. Goodman and G. Namkoong) Sunderland, Mass., USA; Sinauer, pp.596–600. (11)

Barlow, R. 1978. Biological ramifications of selection for pre-weaning growth in cattle: a review. *Animal Breeding Abstracts* 46, 469–94. (19)

Barnes, M.A., Kazmer, G.W., Akers, R.M. and Pearson, R.E. 1985. Influence of selection for milk yield on endogenous hormones and metabolites in Holstein heifers and cows. *Journal of Animal Science* 60, 271–84. (33)

Barrie, P.A., Jeffreys, A.J. and Scott, A.F. 1981. Evolution of the ß-globin gene cluster in Man and the primates. *Journal of Molecular Biology* 149, 319–36. (6)

Barton, N.H. 1986. The maintenance of polygenic variation through a balance between mutation and stabilizing selection. *Genetical Research* 47, 209–16. (1, 15)

Barton, N.H. and Turelli, M. 1987. Adaptive landscapes, genetic distance and the evolution of quantitative characters. *Genetical Research* 49, 157–73. (15)

Bauman, D.E., McCutcheon, S.N., Steinhour, W.D., Eppard, P.J. and Sechen, S.J. 1985. Sources of variation and prospects for improvement of productive efficiency in the dairy cow: a review. *Journal of Animal Science* 60, 583–92. (33)

Bazer, F.W. and Thatcher, W.M. 1986. Foreign travel report. USDA Office of International Co-operation and Development, 21 pp. (28)

Beech, R.N. and Leigh Brown, A.J. 1989. Insertion-deletion variation at the *yellow, achaete-scute* region in two natural populations of *Drosophila melanogaster*. *Genetical Research* 53, 7–15. (6)

Behringer, R.R., Hammer, R.E., Brinster, R.L., Palmiter, R.D. and Townes, T.M. 1987. Two 3' sequences direct adult erythroid-specific expression of human beta-globin genes in transgenic mice. *Proceedings of the National Academy of Sciences, USA* 84, 7056–60. (36)

Behringer, R.R., Mathews, L.S., Palmiter, R.D. and Brinster, R.L. 1988a. Dwarf mice produced by genetic ablation of growth-hormone producing cells. *Genes and Development* 2, 453–61. (36)

Behringer, R.R., Peschon, J.J., Messing, A., Gartside, C.L., Hauschka and S.D., Palmiter, R.D. 1988b. Heart and bone tumours in transgenic mice. *Proceedings of the National Academy of Sciences, USA* 85, 2648–52. (36)

Bell, G.I., Selby, M.J. and Rutter, W.J. 1982. The highly polymorphic region near the human insulin gene is composed of simple tandemly repeating sequences. *Nature* 295, 31–5. (6)

Bener, A., 1982. Further evidence for linkage between blood group Kell and dermatoglyphic sole characters. *Collegium Antropologicum* 6, 155–9. (20)

Bennetzen, J.L. and Hall, B.D. 1982. Codon selection in yeast. *Journal of Biological Chemistry* 257, 3026–31. (4)

Bernardi, G., Olofsson, B., Filipski, J., Zerial, M., Salinas, J., Cuny, G., Meunier-Rotival, M. and Rodier, F. 1985. The mosaic genome of warm-blooded vertebrates. *Science* 228, 953–8. (4)

Betteridge, K.J. (Editor). 1977. *Embryo Transfer in Farm Animals. A Review of Techniques and Applications*. Ottawa, Canada; Agriculture Canada. Monograph 16. (35)

Bichard, M. 1971. Dissemination of genetic improvement through a livestock industry. *Animal Production* 13, 401–11. (27)

Bichard, M., Pease, A.H.R., Swales, P.H. and Özkütük, K. 1973. Selection in a population with overlapping generations. *Animal Production* 17, 215–27. (27)

Bidanel, J.-P. 1988. Bases Zootechniques et génétiques de l'utilisation en élevage itensif des races prolifiques chinoises-cas du porc Meishan. Thesis, Institut National Agronomique Paris-Grignon. (28)

Bidanel, J.-P. and Legault, C. 1986. Experimental and prospective aspects of utilisation of prolific Chinese breeds in Europe. *37th Meeting European Association of Animal Production, Hungary* 24 pp. (28)

Bindon, B.M., Piper, L.R., Cummins, L.J., O'Shea, T., Hillard, M.A., Findlay, J.K. and Robertson, D.M. 1985. Reproductive endocrinology of prolific sheep. In: *Genetics of Reproduction in Sheep* (edited by R.B. Land and D.W. Robinson). London; Butterworths, pp. 217–35. (33)

Bingham, P.M., Kidwell, M.G. and Rubin, G.M. 1982. The molecular basis of *P-M* hybrid dysgenesis: The role of the *P* element, a *P*-strain-specific transposon family. *Cell* 29, 995–1004. (16)

Bird, A., Taggart, M., Frommer, M., Miller, O.J. and Macleod, D. 1985. A fraction of the mouse genome that is derived from islands of nonmethylated, CpG-rich DNA. *Cell* 40, 91–9. (36)

Bishop, D.T. and Cannings, C. 1978. A generalised war of attrition. *Journal of Theoretical Biology* 70, 85–124. (10)

Bishop, G.R., Mayo, O. and Beckman, L. 1987. Estimation of genetical parameters for a quantitative trait subject to major gene influences. *Human Heredity* 36, 182–5. (20)

Blackman, R.K., Grimaila, R., Koehler, M.M. and Gelbart, W.M. 1987 Mobilization of *hobo* elements residing within the decapentaplegic gene complex: Suggestion of a new hybrid dysgenesis system in *Drosophila melanogaster*. *Cell* 49, 497–505. (16)

Boerwinkle, E. and Sing, C.F. 1987. The use of measured genotype information in the analysis of quantitative phenotypes in man. III. Simultaneous estimation of the frequencies and effects of the *apolipoprotein E* polymorphism and residual polygenetic effects on cholesterol, betalipoprotein and triglyceride levels. *Annals of Human Genetics* 51, 211–26. (20)

Bohren, B.B., Hill, W.G. and Robertson, A. 1966. (See AR bibliography)

Borrelli, E., Heyman, M., Hsi, M. and Evans, R.M. 1988. Targeting of an inducible toxic phenotype in animal cells. *Proceedings of the National Academy of Sciences, USA* 85, 7572–6. (36)

Botas, J., Moscoso del Prado, J. and Garcia-Bellido, A. 1982. Gene-dose titration analysis in the search of transregulatory genes in *Drosophila*. *The EMBO Journal* 1, 307–10. (16)

Botstein, D., White, R.L., Skolnick, M. and Davis, B.W. 1980. Construction of a genetic linkage map in man using restriction fragment length polymorphisms. *American Journal of Human Genetics* 32, 314–36. (34)

Botteri, F., van der Putten, H., Wong, D.F., Sauvage, C.A. and Evans, R.M. 1987. Unexpected thymic hyperplasia in transgenic mice harboring a neuronal promoter fused with simian virus 40 large T antigen. *Molecular and Cellular Biology* 7, 3178–84. (36)

Bousquet, D., Goff, A., King, W.A. and Greve, T. 1988. Fertilization *in vitro* of bovine oocytes: analysis of some factors affecting the fertilization rates. *Canadian Journal of Veterinary Research* 52, 277–9. (35)

Bowman, J.C. 1958. Recurrent selection: An experimental study with mice and *Drosophila*. PhD Thesis, University of Edinburgh. (12)

Bowman, J.C. and Falconer, D.S. 1960. Inbreeding depression and heterosis of litter size in mice. *Genetical Research* 1, 262–74. (11)

Box, G.E.P. and Cox, D.R. 1964. An analysis of transformations. *Journal of the Royal Statistical Society* B26, 211–43. (20)

Brackett, B.G. 1983. *In vitro* fertilization in the bovine. *Theriogenology* 19, 1–15. (35)

Bradford, G.E. and Famula, T.R. 1984. Evidence for a major gene for rapid postweaning growth in mice. *Genetical Research* 44, 293–308. (16, 20, 21)

Bradford, G.E. and Kennedy, B.W. 1980. Genetic aspects of embryo transfer. *Theriogenology* 13, 13–26. (29)

Bradford, G.E., Quirke, J.F., Sitorus, P., Inounu, I., Tiesnamurti, B., Bell, F.L., Fletcher, I.C. and Torrell, D.T. 1986. Reproduction in Javanese sheep: evidence for a gene with large effect on ovulation rate and litter size. *Journal of Animal Science* 63, 418–31. (21)

Bradley, A., Evans, M., Kaufman, M.H. and Robertson, E. 1984. Formation of germ- line chimaeras from embryo-derived teratocarcinoma cell lines. *Nature* 309, 255–8. (35, 36)

Brascamp, E.W., Merks, J.W.M. and Wilmink, J.B.M. 1985. Genotype-environment interaction in pig breeding programmes: methods of estimation and relevance of the estimates. *Livestock Production Science* 13, 135–46. (28)

Breese, E.L. and Mather, K. 1957. The organization of polygenic activity within a chromosome in *Drosophila*. 1. Hair characters. *Heredity* 11, 373–95. (20, 21)

Brem, G., Brenig, B., Goodman, H.M., Selden, R.C., Graf, F., Kruff, B., Springman, K., Meyer, J., Winnacker, E-L. and Krasslich, H. 1986. Production of transgenic mice, rabbits and pigs by microinjection. *Theriogenology* 25, 143. (32)

Brindle, K. 1988. ^{31}P NMR magnetisation transfer measurements of flux of genetically modified yeast. *Biochemistry* (in press). (31)

Brinster, R.L. 1974. The effect of cells transferred into the mouse blastocyst on subsequent development. *Journal of Experimental Medicine* 140, 1049–1051. (35)

Brinster, R.L., Allen, J.M., Behringer, R.R., Gelinas, R.E. and Palmiter, R.D. 1988. Introns increase transcriptional efficiency in transgenic mice. *Proceedings of the National Academy of Sciences, USA* 85, 836–40. (36)

Brinster, R., Chen, H.Y., Trumbauer, M., Senear, A.W., Warren, R. and Palmiter, R.D. 1981. Somatic expression of herpes thymidine kinase in mice following injection of a fusion gene into eggs. *Cell* 27, 223–31. (36)

Brinster, R.L., Chen, H.Y., Trumbauer, M.E., Yagle, M.Y. and Palmiter, R.D. 1985. Factors affecting the efficiency of introducing foreign DNA into mice by microinjecting eggs. *Proceedings of the National Academy of Sciences, USA* 82, 4438–42. (36)

Briscoe, D.A., Robertson, A. and Malpica, J.M. 1975. (See AR bibliography)

Brookfield, J.F.Y. 1986. Population biology of transposable elements. *Philosophical Transactions of the Royal Society, London B* 312, 217–26. (8)

Brookfield, J.F.Y., Montgomery, E.A. and Langley, C.H. 1984. Apparent absence of transposable elements related to the P element of *D. melanogaster* in other species of *Drosophila*. *Nature* 310, 330–2. (8)

Brown, W.M., George, M. Jr and Wilson, A.C. 1979. Rapid evolution of animal mitochondrial DNA. *Proceedings of the National Academy of Sciences, USA* 76, 1967–71. (3)

Bryant, E.H., McCommas, S.A. and Combs, L.A. 1986. The effect of an experimental bottleneck upon quantitative genetic variation in the housefly. *Genetics* 114, 1191–211. (12)

Buchanan, D.S. 1987. The crossbred sire: experimental results for swine. *Journal of Animal Science* 65, 117–27. (28)

Bucheton, A., Paro, R., Sang, H.M., Pelisson, A. and Finnegan, D.G. 1984. The molecular basis of *I-R* hybrid dysgenesis in *Drosophila melanogaster*: Identification, cloning, and properties of the *I* factor. *Cell* 38, 153–63. (16)

Bucheton, A. and Picard, G. 1978. Non-Mendelian female sterility in *Drosophila melanogaster*: Hereditary transmission of reactivity levels. *Heredity* 40, 207–23. (16)

Buiting, G.A.J. and Merks, J.W.M. 1986. Lifetime production and reasons for culling of Duroc×Dutch Landrace and Dutch Yorkshire×Dutch Landrace F_1 sows. Zeist Netherlands; *Report B-286*, Institute Schoonoord, 45 pp. (28)Bulfield, G. 1980. Biochemical and genetical determinants of selection for growth in animals. In: *Growth in Animals.* (edited by T.L.J. Lawrence). London; Butterworths, pp.11–24. (32)

Bulfield, G. 1985. The potential improvement of commercial poultry by genetic engineering techniques. In: *Poultry Breeding and Genetics* (edited by W.G. Hill, J. Manson and D. Hewitt). Poultry Science Symposium 18. Harlow, UK; Longman Group. pp.37–46. (32, 34)

Bulfield, G., Isaacson, J.H. and Middleton, R.J. 1988. Biochemical correlates of selection for weight-for-age in chickens: twenty-fold higher muscle ornithine decarboxylaose levels in modern broilers. *Theoretical and Applied Genetics* 75, 432–7. (32)

Bulmer, M.G. 1971. The stability of equilibria under selection. *Heredity* 27, 157–62. (1)

Bulmer, M.G. 1972. The genetic variability of polygenic characters under optimizing selection, mutation and drift. *Genetical Research* 19, 17–25. (15)

Bulmer, M.G. 1973. The maintenance of the genetic variability of polygenic characters by heterozygote advantage. *Genetical Research* 22, 9–12. (15)

Bulmer, M.G. 1976. The effect of selection on genetic variability: a simulation study. *Genetical Research* 28, 101–17. (15)

Bulmer, M.G. 1987. Coevolution of codon usage and transfer RNA abundance. *Nature* 325, 728–730. (4)

Bulmer, M.G. 1988. Are codon usage patterns in unicellular organisms determined by a selection-mutation balance? *Journal of Evolutionary Biology* 1, 15–26. (4)

Bulmer, M.G. 1989. Maintenance of genetic variability by mutation-selection balance: a child's guide through the jungle. *Genome* (in press). (15)

Bundgaard, J. and Christiansen, F.B. 1972. Dynamics of polymorphisms: I. Selection components in an experimental population of *D.melanogaster*. *Genetics* 71, 439–60. (14)

Bürger, R. 1986. On the maintenance of genetic variation: Global analysis of Kimura's continuum of alleles model. *Journal of Mathematical Biology* 24, 341–51. (15)

Bürger, R. 1989. Linkage and the maintenance of heritable variation by mutation-selection balance. *Genetics* 121, 175–84. (15)

Bürger, R., Wagner, G.P. and Stettinger, F. 1989. How much heritable variation can be maintained in finite populations by a mutation-selection balance? *Evolution* (in press). (15)

Burrows, P.M. 1984. Inbreeding under selection from unrelated families. *Biometrics* 40, 357–66. (18)

Burton, R.S. and Place, A.R. 1986. Evolution of selective neutrality – further considerations. *Genetics* 114, 1033–6. (31)

Busslinger, M., Hurst, J. and Flavell, R.A. 1983. DNA methylation and the regulation of globin gene expression. *Cell* 34, 197–206. (36)

Butler, I. von, Willeke, H. and Pirchner, F. 1984. Two-way within-family and mass selection for 8-week body weight in different mouse populations. *Genetical Research* 43, 191–200. (18)

Cameron, J.R., Loh, E.Y. and Davis, R.W. 1979. Evidence for transposition of dispersed repetitive DNA families in yeast. *Cell* 16, 739–51. (8)

Cameron, N.D., Curran, M.K. and Thompson, R. 1988. Estimation of sire with feeding regime interaction in pigs. *Animal Production* 46, 87–95. (28)

Cameron, N.D. and Thompson, R. 1986. Design of multivariate selection experiments to estimate genetic parameters. *Theoretical and Applied Genetics* 72, 466–76. (24)

Campbell, A. 1983. Transposons and their evolutionary significance. In: *Evolution of Genes and Proteins* (edited by M. Nei and R.K. Koehn). Sunderland, Massachusetts. U.S.A.; Sinauer, pp.258–279. (8)

Carbonell, E.A., Frey, J.J. and Bell, A.E.1985. Estimation of maternal, sex-linked and additive x additive epistatic gene effects for body size of *Tribolium*. *Theoretical and Applied Genetics* 70, 133–7. (18)

Castle, W.E. 1922. Genetic studies of rabbits and rats. *Carnegie Institute of Washington* 320, 1–55. (20)

Cepko, C., Roberts, B.E. and Mulligan, R.C. 1984. Construction and applications of a highly transmissible murine retrovirus shuttle vector. *Cell* 37, 1053–62. (36)

Chakraborty, R. and Nei, M. 1982. Genetic differentiation of quantitative characters between populations or species. I. Mutation and random genetic drift. *Genetical Research* 39, 303–14. (15)

Chakravarti, A., Buetow, K.H., Antonarakis, S.E.,Waber, P.G., Boehm, C.D. and Kazazian, H.H. 1984. Nonuniform recombination within the human ß-globin gene cluster. *American Journal of Human Genetics* 36, 1239–58. (6)

Chakravarti, A., Buetow, K.H., Antonarakis, S.E., Waber, P.G., Boehm, C.D. and Kazazian, H.H. 1986. Nonuniform recombination within the human ß-globin gene cluster: a reply to B.S. Weir and W.G. Hill. *American Journal of Human Genetics* 39, 779–81. (6)

Charlesworth, B. 1985. The population genetics of transposable elements. In: *Population Genetics and Molecular Evolution* (edited by T. Ohta and K. Aoki). Berlin, Germany; Springer Verlag, pp.213–32. (8)

Charlesworth, B. 1987. The heritability of fitness. In: *Sexual Selection: Testing the Alternatives* (edited by J.W. Bradbury and M.B. Anderson). New York; John Wiley & Sons, pp.21–40. (13)

Charlesworth, B. 1988. The maintenance of transposable elements in natural populations. In: *International Symposium on Plant Transposable Elements* (edited by O.J. Nelson). New York; Plenum Press, pp.189–212. (8)

Charlesworth, B. and Charlesworth,D. 1983. The population dynamics of transposable elements. *Genetical Research* 42, 1–27. (8)

Charlesworth, B. and Langley, C.H. 1986. The evolution of self-regulated transposition of transposable elements. *Genetics* 112, 359–83. (8)

Charnov, E., Maynard Smith, J. and Bull, J.J. 1976. Why be an hermaphrodite? *Nature* 263, 125–6. (10)

Chauhan, V.P.S. and Thompson, R. 1986. Dairy sire evaluation using a 'rolling months' model. *Journal of Animal Breeding and Genetics* 103, 321–33. (26)

Chevalet, C. and de Rochambeau, H. 1985. Predicting the genetic drift in small populations. *Livestock Production Science* 13, 207–18. (27)

Chia, W., Howes, G., Martin, M., Meng, Y., Moses, Y. and Tsubota, S. 1986. Molecular analysis of the yellow locus in *Drosophila*. *The EMBO Journal* 13, 3597–605. (16)

Christensen, L.G. and Liboriussen, T. 1986. Embryo transfer in the genetic improvement of dairy cattle. In: *Exploiting New Technologies in Animal Breeding* (edited by C.Smith, J.W.B. King and J.C. McKay). Oxford, UK; Oxford Univerity Press, pp.37–46. (29, 33)

Church, R.B. 1987. Embryo manipulation and gene transfer in domestic animals. *Trends in Biotechnology* 5, 13–19. (35)

Church, R.B., Schaefele, F.J. and Meckling, K. 1985. Embryo manipulation and gene transfer in livestock. *Canadian Journal of Animal Science* 65, 527–537. (35)

Church, R.B. and Shea, B.F. 1977. The role of embryo transfer in animal improvement programs. *Canadian Journal of Animal Science* 57, 33–45. (35)

Ciliberto, G., Arcone, R., Wagner, E.F. and Ruether, U. (1987) Inducible and tissue-specific expression of human C-reactive protein in transgenic mice. *The EMBO Journal* 6, 4017–22. (36)

Clark, S.H., Hilliker, A.J. and Chovnick, A. 1988. Recombination can initiate and terminate at a large number of sites within the *rosy* locus of *Drosophila melanogaster*. *Genetics* 118, 261–6. (6)

Clarke, B.C. 1970. Darwinian evolution of proteins. *Science* 168, 1009–11. (4)

Clarke, J.N. 1969. Studies on the genetic control of growth in mice. PhD Thesis, University of Edinburgh. (19)

Clayton, G.A., Knight, G.R., Morris, J.A. and Robertson, A. 1957. (See AR bibliography)

Clayton, G.A., Morris, J.A. and Robertson, A. 1953. (See AR bibliography)

Clayton, G.A., Morris, J.A. and Robertson, A. 1957. (See AR bibliography).

Clayton, G.A. and Robertson, A. 1955. (See AR bibliography)

Clayton, G.A. and Robertson, A. 1957. (See AR bibliography).

Clayton, G.A. and Robertson, A. 1964. (See AR bibliography)

Clutton-Brock, T.H., Guinness, F.E. and Albon, S.D. 1982. *Red Deer: Behavior and Ecology of Two Sexes*. Chicago, Ill., USA; University of Chicago Press. (13)

Cockerham, C.C. 1986. Modifications in estimating the number of genes for a quantitative character. *Genetics* 114, 659–64. (20)

Cockerham, C.C. and Burrows, P.M. 1980. Selection limits and strategies. *Proceedings of the National Academy of Sciences, USA* 77, 546–9. (18)

Cockerham, C.C. and Tachida, H. 1987. Evolution and maintenance of quantitative genetic variation by mutations. *Proceedings of the National Academy of Sciences, USA* 84, 6205–9. (15)

Cohan, F.M. and Graf, J. 1985. Latitudinal cline in *Drosophila melanogaster* for knockdown resistance to ethanol fumes and for rates of response to selection for further resistance. *Evolution* 39, 278–93. (12)

Cohen, J.B. and Levinson, A.D. 1988. A point mutation in the last intron responsible for increased expression and transforming activity of the c-Ha-*ras* oncogene. *Nature* 334, 119–24. (36)

Colleau, J.J. 1985. Efficiency of genetic improvement using embryo transfer in selection nuclei in dairy cattle. *Génétique, Sélection, Évolution* 17, 499–539. (29)

Colleau, J.J. 1986. Genetic improvement by embryo transfer within an open selection nucleus in dairy cattle. *Proceedings of the 3rd World Congress on Genetics Applied to Livestock Production*. Lincoln, Nebraska; University of Nebraska, 12, 127–32. (29)

Collins, R.S. and Weissman, S.M. 1984. Directional cloning of DNA fragments at a large distance from an initial probe. A circularization method. *Proceedings of the National Academy of Sciences* 81, 6812–6. (34)

Cooley, L., Kelley, R. and Spradling, A. 1988. Insertional mutagenesis of the *Drosophila* genome with single P elements. *Science* 239, 1121–8. (16)

Costantini, F. and Lacy, E. 1981. Introduction of a rabbit beta-globin gene into the mouse germ line. *Nature* 294, 92–4. (36)

Covarrubias, L., Nishida, Y. and Mintz, B. 1986. Early postimplantation embryo lethality due to DNA rearrangements in a transgenic mouse strain. *Proceedings of the National Academy of Sciences, USA* 83, 6020–4. (36)

Covarrubias, L., Nishida, Y., Terao, M., D'Eustachio, P. and Mintz, B. 1987. Cellular DNA arrangements and early developmental arrest caused by DNA insertion in transgenic mouse embryos. *Molecular and Cellular Biology* 7, 2243–7. (36)

Cross, S.R.H. and Birley, A.J. 1986. Restriction endonuclease map variation in populations of *Drosophila melanogaster*. *Biochemical Genetics* 24, 415–33. (5, 6)

Crow, J.F. 1954. Breeding structure of populations. II. Effective population number. In: *Statistics and Mathematics in Biology* (edited by O. Kempthorne, T.A. Bancroft, J.W. Gowen and J. Lush). Ames, Iowa, USA; Iowa State College Press, pp. 543–56. (1)

Crow, J.F. 1958. Some possibilities for measuring selection intensities in man. *Human Biology* 30, 1–13. (13)

Crow, J.F. 1972. Some effects of relaxed selection and mutation. *Proceedings of the IVth International Congress of Human Genetics*. Amsterdam; Excerpta Medica, pp. 155–66. (13)

Crow, J.F. and Kimura, M. 1964. The theory of genetic loads. *Proceedings of the XI International Conference of Genetics* 2, 495–505. (15)

Crow, J.F. and Kimura, M. 1970. *An Introduction to Population Genetics Theory*. New York; Harper & Row, (10)

Crow, J.F. and Morton, N.E. 1955. Measurement of gene frequency drift in small populations. *Evolution* 9, 202–214. (1).

Cue, R.I. 1986. Variance-covariance component estimation for uni-variate embedded traits. *Journal of Animal Breeding and Genetics* 103, 334–41. (23)

Curnow, R.N. 1961. The estimation of repeatability and heritability from records subject to culling. *Biometrics* 17: 553–6. (23, 24)

Curnow, R.N. and Smith, C. 1975. Multifactorial models for familial diseases in man. *Journal of the Royal Statistical Society, Series A*, 138, 131–69. (20)

da Costa Soares, V., Gubits, R.M., Feigelson, P. and Constantini, F. 1987. Tissue-specific and hormonally regulated expression of rat alpha2u globulin gene in transgenic mice. *Molecular and Cellular Biology* 7, 3749–58. (36)

Danell, O., Rönningen, K., Ström, H., Andersson, K. and Sundgren, P.E. 1976. An extension of the discounted gene flow method with example in pig breeding. *Acta agriculturae scandinavica* 26, 203–10. (27)

Da Silva, J.M.P. 1961. Limits of response to selection. PhD Thesis, University of Edinburgh. (12, 21)

Davies, R.W. 1971. The genetic relationship of two quantitative characters in *Drosophila melanogaster*. II. Location of the effects. *Genetics* 69, 363–75. (20, 21)

Davis, B.P. and MacDonald, R.J. 1988. Limited transcription of rat elastase I transgene repeats in transgenic mice. *Genes and Development* 2, 13–22. (36)

Davis, G.H. and Hinch, G.H. 1985. Introduction and management of the Booroola gene in sheep flocks in New Zealand. In: *Genetics of Reproduction in Sheep* (edited by R.B. Land and D.W. Robinson). London, UK; Butterworths pp.139–48. (34)

Davis, P.S., Shen, M.W. and Judd, B.H. 1987. Asymmetrical pairings of transposons in and proximal to the white locus of *Drosophila* account for four classes of regularly occurring exchange products. *Proceedings of the National Academy of Sciences, USA* 84, 174–8. (8)

Dean, A.M., Dykhuizen, D.E. and Hartl, D.L. 1986. Fitness as a function of ß-galactosidase activity in *Escherichia coli*. *Genetical Research* 48, 1–8. (31)

Dean, A.M., Dykhuizen, D.E. and Hartl, D.L. 1988. Theories of metabolic control in quantitative genetics. *Proceedings of the 2nd International Conference on Quantitative Genetics* (edited by B.S. Weir, E.J. Eisen, M.M. Goodman and J. Namkoong), Sunderland, Mass., USA; Sinauer, pp. 536–48. (31)

de Fehilly, C.B., Willadsen, S.M. and Tucker, E.M. 1984. Interspecific chimaerism between sheep and goat. *Nature* 307, 634–6. (35)

de la Salle, H., Altenburger, W., Elkaim, R., Dott, K., Dieterle, A., Drillien, R., Cazenave, J.-P., Tolstoshev, P. and Lecoq, J.-P. 1985. Active gamma-carboxylated factor IX expressed using recombinant DNA techniques. *Nature* 316, 268–70. (36)

Del-Bosque-Gonzales, A.S. and Kinghorn, B.P. 1987. Consequences of genotype X environment interaction in group breeding schemes. In: *Merino Improvement Programs in Australia* (edited by B.J. McGuirk). Melbourne; Australian Wool Corporation, pp.437–42. (27)

Demenais, F., Lathrop, M. and Lalouel, J.M. 1986. Robustness and power of the unified model in the analysis of quantitative measurements. *American Journal of Human Genetics* 38, 228–34. (20)

Dempfle, L. 1974. A note on increasing the limit of selection through selection within families. *Genetical Research* 24, 127–35. (18, 27)

Dempfle, L. 1977. Relation entre BLUP (Best Linear Unbiased Prediction) et estimateurs bayesiens. *Annales de Génétique et de Sélection Animale* 9, 27–32. (26)

Dempster, A.P., Laird, N.M. and Rubin, D.B. 1977. Maximum likelihood from incomplete data via the EM algorithm. *Journal of the Royal Statistical Society B* 39, 1–38. (23)

Dempster, A.P., Selwyn, M.R., Patel, C.M. and Roth, A.J. 1984. Statistical and computational aspects of mixed model analysis. *Applied Statistics* 33, 203–14. (23)

Dempster, E.R. and Lerner, I.M. 1950. Heritability of threshold characters. *Genetics* 35, 121–236. (25)

Dente, L., Ruether, U., Tripodi, M., Wagner, E.F. and Cortese, R. 1988. Expression of human alpha1-acid glycoprotein genes in cultured cells and in transgenic mice. *Genes and Development* 1, 259–66. (36)

Dickerson, G.E. 1969. Experimental approaches in utilising breed resources. *Animal Breeding Abstracts* 37, 191–202. (11)

Dickerson, G.E. and Hazel L.N. 1944. Effectiveness of selection on performance as a supplement to earlier culling of livestock. *Journal of Agricultural Research* 69, 459–476. (22, 27, 29)

Dickerson, R.E. 1971. The structure of cytochrome c and the rate of molecular evolution. *Journal of Molecular Evolution* 1, 26–45. (3)

Dobzhansky, Th. 1948. Genetics of natural populations. XVI. Altitudinal and seasonal changes produced by natural selection in certain populations of *Drosophila pseudoobscura* and *Drosophila persimilis. Genetics* 33, 158–76. (14)

Dolan, R. and Robertson, A. 1975. (See AR bibliography)

Donis-Keller, H. *et al.* (32 authors). 1987. A genetic linkage map of the human genome. *Cell* 51, 319–37. (34)

Doolittle, W.F. and Sapienza, C. 1980. Selfish genes, the phenotype paradigm, and genome evolution. *Nature* 284, 601–7. (8)

Dover, G.A. (1982). Molecular drive: a cohesive mode of species evolution. *Nature* 299, 111–7. (9)

Dudley, J.W. 1977. 76 generations of selection for oil and protein percentage in maize. In: *Proceedings of the 2nd International Conference on Quantitative Genetics* (edited by E. Pollak, O. Kempthorne, and T.B. Bailey). Ames, Iowa; Iowa State University, pp.459–73. (16, 17)

Duncan, R. and McConkey, E.H. (1982). How many proteins are there in a typical mammalian cell? *Clinical Chemistry* 28, 749–55. (32)

Durrant, A. and Mather, K. 1954. Heritable variation in a long inbred line of *Drosophila. Genetica* 27, 97–119. (16)

Dykhuizen, D.E., Dean, A.M. and Hartl, D.L. 1987. Metabolic flux and fitness. *Genetics* 115, 25–31. (31)

Eggleston, W.B., Johnson-Schlitz, D. and Engels, W.R. 1988. P-M hybrid dysgenesis does not mobilize other transposable elements in *D. melanogaster. Nature* 331, 368–70. (8)

Ehrenberg, M., Kurland, C.G. and Ruusala, T. 1986. Counting cycles of EF-Tu to measure proofreading in translation. *Biochimie* 68, 261–73. (4)

Einat, P., Berman, Y., Yaffe, D. and Shani, M. 1987. Expression in transgenic mice of two genes of different tissue specificity integrated into a single chromosomal site. *Genes and Development* 1, 1075–84. (36)

Eisen, E.J. 1967. Mating designs for estimating direct and maternal genetic variances and direct maternal covariances. *Canadian Journal of Genetics and Cytology* 9, 13–22. (24)

Eisen, E.J. 1975. Population size and selection intensity effects on long-term selection response in mice. *Genetics* 79, 305–23. (12)

Eisen, E.J. 1980. Conclusions from long-term selection experiments with mice. *Journal of Animal Breeding and Genetics* 97, 305–19. (17, 18)

Eissenberg, J.C., Cartwright, I.L., Thomas, G.H. and Elgin, S.C.R. 1985. Selected topics in chromatin structure. *Annual Review of Genetics* 19, 485–536. (36)

Eklund, J. and Bradford, G.E. 1977. Genetic analysis of a strain of mice plateaued for litter size. *Genetics* 85, 529–42. (17)

Elsen, J.M. 1980. Diffusion du progrès génétique dans les populations avec générations imbriquées: quelques propriétés d'un modèle de prévision. *Annales de Génétique et de Sélection Animale* 12, 49–80. (27)

Elsen, J.M. and Mocquot, J.C. 1974. Recherches pour une rationalisation technique et économique des schémas de sélection des bovins et ovins. *Bulletin Technique de Département de Génétique Animale* No 17, INRA, France. (27)

Elston, R.C. 1988. A general linkage method for the detection of major genes. In: *Advances in Statistical Methods for Genetic Improvement of Livestock* (edited by D. Gianola and K. Hammond). Heidelberg, Germany; Springer Verlag, (in press) (21)

Elston, R.C. and Stewart, J. 1971. A general model for the genetic analysis of pedigree data. *Human Heredity* 21, 523–42. (20)

Endler, J.A. 1986. *Natural Selection in the Wild.* Princeton, New Jersey, USA; University Press, (14)

Enfield, F.D. 1970. Effect of population structure on progress from selection in Tribolium. *Journal of Animal Science* 31, 163. (11)

Enfield, F.D. 1977. Selection experiments in tribolium designed to look at gene action issues. In: *Proceedings of* the International Conference in Quantitative Genetics (edited by E. Pollack, O. Kempthorne and T.B. Bailey Jr). Ames, Iowa, USA; Iowa State University Press, pp.77–190. (16)

Enfield, F.D. 1980. Long-term effects of selection; the limits to response. In: *Selection Experiments in Laboratory and Domestic Animals* (edited by A. Robertson). Slough, UK; Commonwealth Agricultural Bureaux, pp.69–86. (16, 17)

Enfield, F.D. and Anklesaria, F. 1986. An evaluation of multiple peak epistasis and population structure in directional selection program. *Proceedings of the 3rd World Congress on Genetics Applied to Livestock Production.* Lincoln, Nebraska; University of Nebraska, Vol. 12, 283–94. (11, 18)

Engels, W.R. 1981. Estimating genetic divergence and genetic variability with restriction endonucleases. *Proceedings of the National Academy of Sciences* 78, 6329–33. (6)

Engels, W.R. 1983. The *P* family of transposable elements in *Drosophila. Annual Review of Genetics* 17, 315–44. (14)

Engels, W.R. 1986. On the evolution and population genetics of hybrid-dysgenesis causing transposable elements in *Drosophila. Philosophical Transactions of the Royal Society, London B* 312, 205–15. (8)

Engels, W.R. 1988. *P* elements in *Drosophila.* In: *Mobile DNA* (edited by D.E. Berg and M.M. Howe). Washington, D.C.; American Society for Microscopy (in press). (16)

Engels, W.R. and Preston, C.R. 1984. Formation of chromosome rearrangements by *P* factors. *Genetics* 107, 657–78. (8)

Evans, M.J. and Kaufman, M.H. 1981. Establishment in culture of pluripotential cells from mouse embryos. *Nature* 292, 154–5. (35)

Evans, R.M. and Hollenberg, S.M. 1988. Zinc fingers: gilt by association. *Cell* 52, 1–3. (36)

Ewens, W.J. 1972. The sampling theory of selectively neutral alleles. *Theoretical Population Biology* 3, 87–112. (6, 8)

Ewens, W.J., Spielman, R.S. and Harris, H. 1981. Estimation of genetic variation at the DNA level from restriction endonuclease data. *Proceedings of the National Academy of Sciences* 78, 3748–50. (6)

Falconer, D.S. 1954. Asymmetrical response in selection experiments. In: *Symposium on Genetics of Population Structure.* Naples: International Union of Biological Sciences, Naples, Series B, No 15, pp.16–41. (17)

Falconer, D.S. 1960a. *Introduction to Quantitative Genetics.* Edinburgh; Oliver and Boyd. (12, 24)

Falconer, D.S. 1960b. Selection of mice for growth on high and low planes of nutrition. *Genetical Research* 1, 422–30. (19)

Falconer, D.S. 1965. The inheritance of liability to certain diseases, estimated from the incidence among relatives. *Annals of Human Genetics* 29, 51–76. (25)

Falconer, D.S. 1971. Improvement of litter size in a strain of mice at a selection limit. *Genetical Research* 17, 215–35. (16, 17)

Falconer, D.S. 1973. Replicated selection for body weight in mice. *Genetical Research* 22, 291–321. (18, 32)

Falconer, D.S. 1981. *Introduction to Quantitative Genetics,* 2nd edition, London; Longman. (13, 16, 17, 23)

Falconer, D.S. and Latyszewski. 1952. The environment in relation to selection for size in mice. *Journal of Genetics* 51, 67–80. (19)

Falconer, D.S. and Robertson, A. 1956. (See AR bibliography)

Famula, T.R. 1986. Identifying single genes of large effect in quantitative traits using best linear unbiased prediction. *Journal of Animal Science* 63, 68–76. (20, 34)

Famula, T.R., Calvert, C.C. and Bradford, G.E. 1986. The major gene for rapid postweaning growth in mice: a review. *Proceedings of the 3rd World Congress on Genetics Applied to Livestock Production.* Lincoln, Nebraska; University of Nebraska, Vol. 11, 383–5. (20)

Fassler, J.S. and Winston,F. 1988. Isolation and analysis of a novel class of suppressor of *Ty* insertion mutations in *Saccharomyces cerevisiae. Genetics* 118, 203–12. (16)

Feinstein, S.C., Ross, S.R. and Yamamoto, K.R. 1982.Chromosomal position effects determine transcriptional potential of integrated MMTV DNA. *Journal of Molecular Biology* 156, 549–65. (36)

Fell, D.A. and Sauro, H.M. 1985. Metabolic control and its analysis. Additional relationships between elasticities and control coefficients. *European Journal of Biochemistry* 148, 555–61. (31)

Felsenstein, J. 1965. The effect of linkage on directional selection. *Genetics* 52, 349–63. (1)

Felsenstein, J. 1974. The evolutionary advantage of recombination. *Genetics* 78, 737–56. (1)

Felsenstein, J. 1976. The theoretical population genetics of variable selection and migration. *Annual Review of Genetics* 10, 253–80. (11)

Felsenstein, J. 1986. Population differences in quantitative characters and gene frequencies: A comment on papers by Lewontin and Rogers. *American Naturalist* 127, 731–2. (11)

Fernando, R.L. and Gianola, D. 1986. Optimal properties of the conditional mean as a selection criterion. *Theoretical and Applied Genetics* 72, 822–5. (27)

Finnegan, D.J. 1985. Transposable elements in eukaryotes. *International Review of Cytology* 93, 281–326. (16)

Finnegan, D.J. and Fawcett, D.H. 1986. Transposable elements in *Drosophila melanogaster*. *Oxford Surveys on Eukaryotic Genes* 3, 1–62. (2, 8)

Fisher, A.W.G., Meckling, K. and Church, R.B. 1984. Caudal dysplasis in two Hereford calves. *Proceedings of the Anatomical Society* 49, 20–1. (35)

Fisher, R.A. 1918. The correlation between relatives on the supposition of Mendelian inheritance. *Transactions of the Royal Society Edinburgh* 52, 399–433. (12, 20, 21)

Fisher, R.A. 1930. *The Genetical Theory of Natural Selection*. Oxford, UK; Clarendon Press. (1, 10, 13)

Fitzpatrick, B.J. and Sved, J.A. 1986. High level of fitness modifiers induced by hybrid dysgenesis in *Drosophila melanogaster*. *Genetical Research* 48, 89–94. (8, 16)

Flint, H.J., Tateson, R.W., Barthelmess, I.B., Porteous, D.J., Donachie, W.D. and Kacser, H. 1981. Control of the flux in the arginine pathway of *Neurospora crassa*: modulations of enzyme activity and concentration. *Biochemical Journal* 200, 231–46. (21, 31)

Forrest, P.A. and Bichard, M. 1974. Analysis of production records from a lowland sheep flock. 2. Phenotypic and genetic parameters for reproductive performance. *Animal Production* 19, 33–45. (25)

Foulley, J.L., Im, S., Gianola, D. and Hoschele, Ina 1987. Empirical Bayes estimation of parameters for n polygenic binary traits. *Génétique, Sélection, Evolution* 19, 197–224. (25)

Fowler, R.E. and Edwards, R.G. 1961. 'Midget', a new dwarfing gene in the house mouse dependent on a genetic background of small body size for its expression. *Genetical Research* 2, 272–82. (17)

Fowler, V.R., Bichard, M. and Pease, A. 1976. Objectives in pig breeding. *Animal Production* 23, 365–387. (28)

Fox, G.E., Magrum, L.J., Balch, W.E., Wolfe, R.S. and Woese, C.R. 1977. Classification of methogenic bacteria by 16S ribosomal RNA characterization. *Proceedings of the National Academy of Sciences, USA* 74, 4537–41. (3)

Frankham, R. 1977. Optimum selection intensities in artificial selection programmes: an experimental evaluation. *Genetical Research* 30, 115–9. (12, 18)

Frankham, R. 1980a. Origin of genetic variation in selection lines. In: *Selection Experiments in Laboratory and Domestic Animals* (edited by A. Robertson). Slough, UK; Commonwealth Agricultural Bureaux, pp.56–68. (16)

Frankham, R. 1980b. The founder effect and response to artificial selection in *Drosophila*. In: *Selection Experiments in Laboratory and Domestic Animals* (edited by A. Robertson). Slough, UK; Commonwealth Agricultural Bureaux, pp.87–90. (12)

Frankham, R. 1982. Contributions of *Drosophila* research to quantitative genetics and animal breeding. *Proceedings of the 2nd World Congress on Genetics Applied to Livestock Production* 5 43–56. (12)

Frankham, R. 1988. Exchanges in the rRNA multigene family as a source of genetic variation. In: *Proceedings of the 2nd International Conference on Quantitative Genetics* (edited by B.S. Weir, E.J. Eisen, M.M. Goodman and G. Namkoong). Sunderland, Mass., USA; Sinauer, pp. 236–42. (12)

Frankham, R., Briscoe, D.A. and Nurthen, R.K. 1978. Unequal crossing over at the rRNA locus as a source of quantitative genetic variation. *Nature* 272, 80–1. (9, 12, 15, 16)

Frankham, R., Jones, L.P. and Barker, J.S.F. 1968. The effects of population size and selection intensity in selection for a quantitative character in *Drosophila*. III. Analysis of the lines. *Genetical Research* 12, 267–83.

(16, 17, 18)

Frankham, R., Yoo, B.H. and Sheldon, B.L. 1988. Reproductive fitness and artificial selection in animal breeding: culling on fitness prevents a decline in reproductive fitness in lines of *Drosophila melanogaster* selected for increased inebriation time. *Theoretical and Applied Genetics* 76, 909–14. (12)

Fraser, A.S. 1957. Simulation of genetic systems. I. Introduction. *Australian Journal of Biological Sciences* 10,484–91. (12)

Fröhlander, N. 1987. Haptoglobin groups and serum cholesterol levels. *Human Heredity* 37, 323–5. (20)

Gadgil, M.D. and Solbrig, O.T. 1972. The concept of r- and K-selection: evidence from wild flowers and some theoretical considerations. *American Naturalist* 106, 14–31. (10)

Gahne, B. and Juneja, R.K. 1985. Predictions of the halothane (Hal) genotypes of pigs by deducing Hal, Phi, PO2, Pgd haplotypes of parents and offspring: results from a large-site practice in Swedish breeds. *Animal Blood Groups Biochemical Genetics* 16, 265–84. (34)

Gallego, A. and García-Dorado, A. 1986. Evolution de l'effectif génétique de lignées de *Drosophila melanogaster* soumises à une séléction artificielle. *Génétique, Sélection, Évolution* 18, 249–60. (18)

Gallego, A. and López-Fanjul, C. 1983. The number of loci affecting a quantitative trait in *Drosophila melanogaster* revealed by artificial selection. *Genetical Research* 42, 137–49. (18)

García-Dorado, A. and López-Fanjul, C. 1985. Optimum selection strategies: studies with *Drosophila melanogaster*. *Genetical Research* 46, 101–5. (18)

García-Dorado, A. and López-Fanjul, C. 1987. Balanced polymorphism at the selection limit in *Drosophila melanogaster*. *Journal of Heredity* 78, 110–1. (18)

Garel, J.P. 1974. Functional adaptation of tRNA population. *Journal of Theoretical Biology* 43, 211–25. (4)

Garnett, I. 1976. The genetic relationship between the *Hbb* locus and body size in a population of mice divergently selected for six-week body weight. *Canadian Journal of Genetics and Cytology* 18, 519–23. (32)

Garnett, I. and Falconer, D.S. 1975. Protein variation in strains of mice differing in body weight. *Genetical Research* 25, 45–7.(32)

Gavora, J.S. and Spencer, J.L. 1983. Breeding for immuneresponsiveness and disease resistance. *Animal Blood Groups Biochemical Genetics* 14, 159–80. (34)

Geldermann, H. 1975. Investigations on inheritance of quantitative characters in animals by gene markers. I. Methods. *Theoretical and Applied Genetics* 46, 319–30. (34)

Geldermann, H., Pieper, U. and Roth, B. 1985. Effects of marked chromosome sections on milk performance in cattle. *Theoretical and Applied Genetics* 70, 138–46. (34)

George, V.T. and Elston, R.C. 1987. Testing the association between polymorphic markers and quantitative traits. *Genetic Epidemiology* 4, 193–201. (20)

Georges, M., Lequarre, A.S., Castelli, M., Hanset, R. and Vassart, G. 1988. DNA fingerprinting in domestic animals using four different mini-satellite probes. *Cytogenetics and Cell Genetics* 47, 127–31. (34)

Gerhard, D., Kidd, K.K., Kidd, J.R., Egeland, J.A. and Housman, D.E. 1984. Identification of a recent recombination event within the human gamma-globin gene cluster. *Proceedings of the National Academy of Sciences* 81, 7875–9. (6)

Gianola, D. and Fernando, R.L. 1986. Bayesian methods in animal breeding. *Journal of Animal Science* 63: 217–44. (23)

Gianola, D. and Foulley, J.L. 1983. Sire evaluation for ordered categorical data with a threshold model. *Genetique, Selection, Evolution* 15, 201–24. (25)

Giguere, V., Ong, E.S., Segui, P. and Evans, R.M. 1987. Identification of a receptor for the morphogen retinoic acid. *Nature* 330, 624–9. (36)

Gilbert, W. 1978. Why genes in pieces? *Nature* 271, 501. (9)

Gill, J.L. and Jensen, E.L. 1968. Probability of obtaining negative estimates of heritability. *Biometrics* 24, 517–26. (25)

Gill, P.E., Murray, W. and Wright, M.H. 1981. *Practical Optimization*. New York; Academic Press. (23)

Gillespie, J.H. 1978. A general model to account for enzyme variation in natural populations. V. SAS-CFF model.

Theoretical Population Biology 14, 1–45. (3)

Gillespie, J.H. 1984a. The molecular clock may be an episodic clock. *Proceedings of the National Academy of Sciences, USA* 81, 8009–13. (3)

Gillespie, J.H. 1984b. Molecular evolution over the mutational landscape. *Evolution* 38, 1116–29. (3)

Gillespie, J.H. 1984c. Pleiotropic overdominance and the maintenance of genetic variation in polygenic characters. *Genetics* 107, 321–30. (15)

Gillespie, J.H. 1986a. Variability of evolutionary rates of DNA. *Genetics* 113, 1077–99. (3)

Gillespie, J.H. 1986b. Natural selection and the molecular clock. *Molecular Biology and Evolution* 3, 138–55. (3)

Gillespie, J.H. and Langley, C.H. 1979. Are evolutionary rates really variable? *Journal of Molecular Evolution* 13, 27–34. (3)

Gillies, S.D., Morrison, S.L., Oi, V.T. and Tonegawa, S. 1983. A tissue-specific enhancer element is located in the major intron of a rearranged immunoglobulin heavy chain gene. *Cell* 33, 717–28. (36)

Gilmour, A.R., Anderson, R.D. and Rae, A.L. 1985. The analysis of binomial data in a generalised linear mixed model. *Biometrika* 72, 593–9. (25)

Goddard, M.E. 1987. Policy of selecting bulls to breed bulls. *Animal Production* 44, 29–38. (27)

Goddard, M.E. and Ahmed, A.M. 1982. The use of the genetic distance between cattle breeds to predict the heterosis in crosses. *Proceedings of 2nd World Congress of Genetics Applied to Livestock Production.* Madrid; Editorial Garsi, Vol. 8, 377–82. (11)

Goddard, M.E. and Nicholas, F.W. 1987. Optimal breeding structures for the Australian dairy industry. *Proceedings of the Australian Association of Animal Breeding and Genetics* 6, 131–8. (29)

Gojobori, T. and Yokoyama, S. 1985. Rates of evolution of the retroviral oncogene of Moloney murine sarcoma virus and of its cellular homologues. *Proceedings of the National Academy of Sciences, USA* 82, 4198–201. (2)

Goldberg, M.L., Shen, J.-Y, Gehring, W.J. and Green, M.M. 1983. Unequal crossing-over associated with asymmetrical synapsis between nomadic elements of the *Drosophila* genome. *Proceedings of the National Academy of Sciences, USA* 80, 5017–21. (8)

Golding, G.B., Aquadro, C.F. and Langley, C.H. 1986. Sequence evolution within populations under multiple types of mutation. *Proceedings of the National Academy of Sciences, USA* 83, 427–31. (6)

Goldman, M.A., Stokes, K.R., Idzerda, R.J., McKnight, G.S., Hammer, R.E., Brinster, R.L. and Gartler, S.M. 1987. A chicken transferrin gene in transgenic mice escapes X-chromosome inactivation. *Science* 236, 593–5. (36)

Goldstein, J.L. and Brown, M.S. 1973. Familial hypercholesterolaemia. Identification of a defect in the regulation of 3-hydroxy-3-methylglutaryl coenzyme A reductase activity associated with overproduction of cholesterol. *Proceedings of the National Academy of Sciences, USA* 70, 2804–8. (20)

Goodman, M. 1976. Protein sequences in phylogeny. In: *Molecular Evolution* (edited by F.J. Ayala). Sunderland, Mass., USA; Sinauer, pp.141–59. (9)

Goodwill, R. 1974. Comparison of three selection programs using *Tribolium castaneum*. *Journal of Heredity* 65, 8–14. (11, 18)

Gordon, J.W. and Ruddle, F.H. 1981. Integration and stable germ-line transmission of genes injected into mouse pronuclei. *Science* 214, 1244–6. (36)

Gordon, J.W. and Ruddle, F.H. 1985. DNA-mediated genetic transformation of mouse embryos and bone marrow – a review. *Gene* 33, 121–36. (36)

Gordon, J.W., Scangos, G.A., Plotkin, D.J., Barbosa, J.A. and Ruddle, F.H. 1980. Genetic transformation of mouse embryos by microinjection of purified DNA. *Proceedings of the National Academy of Sciences, USA* 77, 7380–4. (36)

Gordon, K., Lee, E., Vitale, J.A., Smith, A.E., Westphal, H. and Hennighausen, L. 1987 Production of human tissue plasminogen activator in transgenic mouse milk. *Biotechnology* 5, 1183–7. (35, 36)

Gossler, A., Doetschman, T., Korn, R., Serfling, E. and Kemler, R. 1986. Trans-genesis by means of blastocyst-derived embryonic stem cell lines. *Proceedings of the National Academy of Sciences, USA* 83, 9065–9069. (36)

Goto, K., Kajihara, S., Kosaka, M., Koba, Y., Nakanishi, Y. and Ogawa, K. 1988. Pregnancies after co-culture of cumulos cells with bovine embryos derived from *in vitro* fertilization of *in vitro* matured follicular oocytes. *Journal of Reproduction and Fertility* 83, 753–8. (35)

Gould, S.J. and Lewontin, R.C. 1979. The spandrels of San Marco and the Panglossian paradigm: a critique of the adaptationist programmes. *Proceedings of the Royal Society, B* 205, 581–98. (10)

Gouy, M. and Gautier, C. 1982. Codon usage in bacteria: correlation with gene expressivity. *Nucleic Acids Research* 10, 7055–73. (4)

Gowe, R.S. 1983. Lessons from selection studies in poultry for animal breeders. *Proceedings of the 33rd Annual Breeders' Roundtable*, pp.22–50. (12)

Gowe, R.S., Robertson, A. and Latter, B.D.H. 1959. (See AR bibliography)

Graf, F. 1984. Of what value are blood tests (physiological characterisation) to estimate the production capacity and reliability of cattle? *Animal Research and Development* 19, 118–25. (33)

Grafen, A. 1979. A hawk–dove game played between relatives. *Animal Behaviour* 27, 905–7. (10)

Graml, R. and Pirchner, F. 1984. Relation of genetic distance between cattle breeds and heterosis of resulting crosses. *Animal Blood Groups Biochemical Genetics* 15, 173–80. (11)

Grantham, R., Gautier, C., Gouy, M., Mercier, R. and Pave, A. 1980. Codon catalog usage and the genome hypothesis. *Nucleic Acids Research* 8, r49–r62. (4)

Graser, H.-U., Smith, S.P. and Tier, B. 1987. A derivative-free approach for estimating variance components in animal models by Restricted Maximum Likelihood. *Journal of Animal Science* 64, 1362–70. (23)

Green, M.C. 1981. *Catalog of mutant genes and polymorphic loci*. In: *Genetic Variants and Strains of the Laboratory Mouse* (edited by M.C. Green). Stuttgart, Germany; Fisher. (21)

Green, S. and Chambon, P. 1987. Oestradiol induction of a glucocorticoid-responsive gene by a chimaeric receptor. *Nature* 325, 75–8. (36)

Gridley, T., Soriano, P. and Jaenisch, R. 1987. Insertional mutagenesis in mice. *Trends in Genetics* 3, 162–6. (36)

Groen, A.K., Wanders, R.J.A., Westerhoff, H.V., van der Meer, R. and Tager, J.M. 1982. Quantification of the contribution of various steps to the control of mitochondrial respiration. *Journal of Biological Chemistry* 257, 2754–7. (31)

Groner, B., Schoenenberger, C.A. and Andres, A.C. 1987. Targeted expression of the ras abd myc genes in transgenic mice. *Trends in Genetics* 3, 306–8. (36)

Grosveld, F., van Assendelft, G.B., Greaves, D.R. and Kollias, G. 1987. Position-independent, high-level expression of the human beta-globin gene in transgenic mice. *Cell* 51, 975–85. (36)

Guy, D.R. and Smith, C. 1981. Derivation of improvement lags in a livestock industry. *Animal Production* 32, 333–6. (27)

Hadchouel, M., Farza, H., Simon, D., Tiollais, P. and Pourcel, C. 1987. Maternal inhibition of hepatitis B surface antigen gene expression in transgenic mice correlates with de novo methylation. *Nature* 329, 454–6. (36)

Hagenbuch, P. and Hill, W.G. 1978. Effectiveness of sequential selection in pig improvement. *Animal Production* 27, 21–7. (27)

Haldane, J.B.S. 1954. The measurement of natural selection. *Proceedings of the 9th International Congress of Genetics* 1, 480–7. (15)

Hale, L.R. and Singh, R.S. 1987.Mitochondrial DNA variation and genetic structure in populations of *Drosophila melanogaster*. *Molecular Biology and Evolution* 4, 622–37. (5)

Haley, C.S., Avalos, E. and Smith, C. 1988. Selection for litter size in the pig. *Animal Breeding Abstracts* 56, 317–32. (28)

Hamilton, W.D. 1967. Extraordinary sex ratios. *Science* 156, 477–488. (10)

Hammer, R.E., Krumlauf, R., Camper,S.A., Brinster, R.L. and Tilghman, S.M. 1987. Diversity of alpha-fetoprotein gene expression· in mice is generated by a combination of separate enhancer elements. *Science* 235, 53–8. (36)

Hammer, R.E., Pursel, V.G., Rexroad, C.E.Jr., Wall, R.J., Bolt, D.J., Ebert, K.M., Palmiter, R.D. and Brinster,

R.L. 1985. Production of transgenic rabbits, sheep and pigs by microinjection. *Nature* 315, 680–3. (35, 36)

Hammond, K. 1974. *Drosophila* as an experimental model in animal breeding research. *Proceedings of the 1st World Congress on Genetics Applied to Livestock Production* 1, 425–37. (12)

Hammond, K. and James, J.W. 1970. Genes of large effect and the shape of the distribution of a quantitative character. *Australian Journal of Biological Sciences* 23, 867–76. (20)

Hammond, K. and James, J.W. 1972. The use of higher degree statistics to estimate the number of loci which contribute to a quantitative character. *Heredity* 28, 146–7. (20)

Hanahan, D. 1985. Heritable formation of pancreatic B-cell tumours in transgenic mice expressing recombinant insulin/simian virus 40 oncogenes. *Nature* 315, 115–22. (36)

Hanrahan, J.P. and Owen, J.B. 1985. Variation and repeatability of ovulation rate in Cambridge ewes. *Animal Production* 40, 529. (Abstr.). (21)

Hanset, R. 1982. Major genes in animal production, examples and perspectives: Cattle and pigs. *Proceedings of the 2nd World Congress on Genetics Applied to Livestock Production* 6, 439–53. (16)

Hanset, R. and Michaux, C. 1985. On the genetic determinism of muscular hypertrophy in the Belgian White and Blue cattle breed. *Génétique, Sélection, Évolution* 17, 359–86. (34)

Harris, H. 1966. Enzyme polymorphism in man. *Proceedings of the Royal Society, London, B* 164, 298–310. (3)

Harris, H. 1975. *Principles of Human Biochemical Genetics* 2nd edition. London; North Holland. (21)

Hartl, D.L. 1980. *Principles of Population Genetics*. Sunderland, Mass., USA; Sinauer. (11)

Hartl, D.L., Dean, A.M. and Dykhuizen, D.E. 1986. The molecular biology of natural selection: Reply to Burton and Place. *Genetics* 114, 1037–9. (31)

Hartl, D.L., Dykhuizen, D.E. and Dean, A.M. 1985. Limits of adaptation: the evolution of selective neutrality. *Genetics* 111, 655–74. (31)

Hartl, D.L. and Jungen, H. 1979. Estimation of average fitness of populations of *Drosophila melanogaster* and the evolution of fitness in experimental populations. *Evolution* 33, 371–80. (14)

Hartl, D.L., Medhora, M., Green, L. and Dykhuizen, D.E. 1986. The evolution of DNA sequences in *Escherichia coli*. *Philosophical Transactions of the Royal Society, London B* 312, 191–204. (8)

Hartmann, W. 1985. The effect of selection and genetic factors on resistance to disease in fowls. *World's Poultry Science Journal* 41, 20–35. (33)

Harvey, W.R. 1960. Least squares analysis of data with unequal subclass numbers. *United States Department of Agriculture, ARS.* 20–8. (23)

Harvey, W.R. 1977. Users guide for LSML76. Mixed model least squares and maximum likelihood computer program. *United States Department of Agriculture, ARS.* (23)

Harville, D.A. 1977. Maximum Likelihood approaches to variance component estimation and to related problems. *Journal of the American Statistical Association* 72, 320–38. (23)

Harville, D.A. and Callanan, T.P. 1988. Computational aspects of likelihood-based inference for variance components. In: *Advances in Statistical Methods for Genetic Improvement of Livestock* (edited by D. Gianola and K. Hammond). Heidelberg, Germany; Springer Verlag (in press). (23)

Hayashida, H., Toh, H., Kikuno, R. and Miyata, T. 1985. Evolution of influenza virus genes. *Molecular Biology and Evolution* 2, 289–303.(2)

Hayes, J.F. and Hill, W.G. 1980. A reparameterisation of a genetic selection index to locate its sampling properties. *Biometrics* 36, 237–48. (23)

Hayes, J.F. and McCarthy, J.C. 1976. The effects of selection at different ages for high and low body weight on the pattern of fat deposition in mice. *Genetical Research* 27, 389–433. (19)

Haymer, D.S. and Hartl, D.L. 1982. The experimental assessment of fitness in *Drosophila*. I. Comparative measures of competitive reproductive success. *Genetics* 102, 455–66. (12)

Haymer, D.S. and Hartl, D.L. 1983. The experimental assessment of fitness in *Drosophila*. II. A comparison of competitive and noncompetitive measures. *Genetics* 104, 343–52. (12, 14)

Hazel, L.N. and Lush, J.L. 1943. The efficiency of three methods of selection. *Journal of Heredity* 33, 393-9. (12)

Hedrick, P.W. 1983. *Genetics of Populations*. Boston, Mass. USA; Science Books. (11)

Hedrick, P.W., Jain, S. and Holden, L. 1978. Multilocus systems in evolution. *Evolutionary Biology* 11, 101–82. (6)

Hedrick, P.W. and Thomson, G. 1983. Evidence for balancing selection at HLA. *Genetics* 104, 449–56. (6)

Hegele, R.A. and Breslow, J.L. 1987. Apolipoprotein genetic variation in the assessment of atherosclerosis susceptibility. *Genetic Epidemiology* 4, 163–84. (20)

Heldin, C.H. and Westermark, B. 1984. Growth factors: mechanisms of action and relation to proto-oncogenes. *Cell* 37, 9–20. (32)

Henderson, C.R. 1953. Estimation of variance and covariance components. *Biometrics* 9: 226–52. (23)

Henderson, C.R. 1973. Sire evaluation and genetic trends. *Proceedings of the Animal Breeding and Genetics Symposium in honour of J.L. Lush Blackburg*, Champaign, Illinois; American Society for Animal Science. pp.10–41. (23, 26)

Henderson, C.R. 1975. Best linear unbiased estimation and prediction under a selection model. *Biometrics* 31, 423–47. (27)

Henderson, C.R. 1976. A simple method for computing the inverse of a numerator relationship matrix used in prediction of breeding values. *Biometrics* 32, 69–83. (23, 26)

Henderson, C.R. 1980. A simple method for unbiased estimation of variance components in the mixed model. *Journal of Animal Science* 58 Suppl. 1, 119 (Abstr.). (23)

Henderson, C.R., Carter, H.W. and Godfrey, J.T. 1954. Use of contemporary herd average in appraising progeny tests of dairy bulls. *Journal of Animal Science* 13, 949 (Abstr.). (26)

Hernández, J.L. and Weir, B.S. 1989a. A disequilibrium coefficient approach to testing for Hardy-Weinberg equilibrium. *Biometrics* (in press). (7)

Hernández, J.L. and Weir, B.S. 1989b. A disequilibrium coefficient approach to testing for linkage equilibrium. (in preparation). (7)

Hershberger, W.K. and Iwamoto, R.N. 1985. *Genetics manual and guidelines for the Pacific salmon hatcheries of Washington*, University of Washington, College of Fisheries WH-10, Seattle, Washington. (34)

Hetzel, D.J.S. and Nicholas, F.W. 1986. Growth, efficiency and body composition of mice selected for post-weaning weight gain on *ad libitum* or restricted feeding. *Genetical Research* 48, 101–4. (19)

Heyman, R., Borrelli, E., Anderson, D., Hsi, M., Lesley, J., Hyman, R. and Evans, R. 1988. Inducible cell-specific destruction in transgenic mice by expression of herpes thymidine kinase. *Journal of Cellular Biochemistry*, Suppl. 12B, 190. (36)

Hill, A.P. 1975. Quantitative linkage: a statistical procedure for its detection and estimation. *Annals of Human Genetics* 38, 439–50. (20)

Hill, W.G. 1963. Cyclical Inbreeding with Selection in *Drosophila melanogaster*. MS Thesis, University of California, Davis. (11)

Hill, W.G. 1970. Design of experiments to estimate heritability by regression of offspring on selected parents. *Biometrics* 26, 566–71. (24)

Hill, W.G. 1971a. Investment appraisal for national breeding programmes. *Animal Production* 13, 37–50. (27)

Hill, W.G. 1971b. Design and efficiency of selection experiments for estimating genetic parameters. *Biometrics* 27, 293–311. (24)

Hill, W.G. 1972. Effective size of populations with overlapping generations. *Theoretical Population Biology* 3, 278–89. (27)

Hill, W.G. 1974a. Prediction and evaluation of response to selection with overlapping generations. *Animal Production* 18, 117–40. (27)

Hill, W. G. 1974b. Estimation of linkage disequilibrium in random mating populations. *Heredity* 33, 229–39. (7)

Hill, W.G. 1975. Tests for association of gene frequencies at several loci in random mating diploid populations. *Biometrics* 31, 881–8. (7)

Hill, W.G. 1976. Order statistics of correlated variables and implications in genetic selection programmes. *Biometrics* 32, 889–902. (26)

Hill, W.G. 1979. A note on effective population size with overlapping generations. *Genetics* 92, 317–22. (27)

Hill, W.G. 1980. Design of quantitative genetic selection experiments. In: *Selection Experiments in Laboratory and*

Domestic Animals (edited by A. Robertson). Slough, UK; Commonwealth Agricultural Bureaux, pp.1–13. (29)

Hill, W.G. 1982a. Rates of change in quantitative traits from fixation of new mutations. *Proceedings of the National Academy of Sciences USA* 79, 142–5. (15, 16)

Hill, W.G. 1982b. Predictions of response to artificial selection from new mutations. *Genetical Research* 40, 255–78. (12, 15, 16, 17, 18)

Hill, W.G. 1985. Detection and genetic assessment of physiological criteria of merit within breeds. In: *Genetics of Reproduction in Sheep* (edited by R.B. Land and D.W. Robinson). London, UK; Butterworths. pp. 319–31 (24, 33)

Hill, W.G. 1986. Population size and design of breeding programmes. *Proceedings of the 3rd World Congress on Genetics Applied to Livestock Production*. Lincoln, Nebraska; University of Nebraska, Vol. 12, 245–56. (18)

Hill, W.G. 1988. Considerations in the design of animal breeding experiments. In: *Advances in Statistical Methods for Genetic Improvement of Livestock* (edited by K. Hammond and D. Gianola). Heidelberg, Germany; Springer Verlag (in press). (24)

Hill, W.G. and Avery, P.J. 1978. On estimating the number of genes by genotype assay. *Heredity* 40, 397–403. (20)

Hill, W.G. and Bishop, S.C. 1986. Genetic control of growth, carcass composition and food utilisation in laboratory animals. *Proceedings of the Third World Congress on Genetics Applied to Livestock Production* 11, 355–66. (28)

Hill, W.G. and Keightley, P.D. 1988. Interrelations of mutation, population size, artificial and natural selection. In: *Proceedings of the 2nd International Conference on Quantitative Genetics* (edited by B.S. Weir, E.J. Eisen, M.M. Goodman and G. Namkoong). Sunderland, Mass., USA; Sinauer, pp.57–70. (15)

Hill, W.G. and Knott, S. 1988. Identification of genes with large effects. In: *Advances in Statistical Methods for Genetical Improvement of Livestock* (edited by D. Gianola and K. Hammond). Heidelberg, Germany; Springer Verlag (in press). (21, 34)

Hill, W.G. and Nicholas, F.W. 1974. Estimation of heritability by both regression of offspring on parent and intra-class correlation of sibs in one experiment. *Biometrics* 30, 447–68. (23)

Hill, W.G. and Rasbash, J. 1986a. Models of long-term artificial selection in finite populations. *Genetical Research* 48, 41–50. (17, 18)

Hill, W.G. and Rasbash, J. 1986b. Models of long-term artificial selection in finite populations with recurrent mutation. *Genetical Research* 48, 125–31. (16)

Hill, W.G. and Robertson, A. 1966. (See AR bibliography)

Hill, W.G. and Robertson, A. 1968a. (See AR bibliography)

Hill, W.G. and Robertson, A. 1968b. (See AR bibliography)

Hill, W.G. and Thompson, R. 1977. Design of experiments to estimate offspring-parent regression using selected parents. *Animal Production* 24, 163–8. (24)

Hill, W.G. and Weir, B.S. 1988. Variances and covariances of squared linkage disequilibria in finite populations. *Theoretical Population Biology* 33, 54–78. (7)

Hines, W.G.S. 1980. Strategy stability in complex populations. *Journal of Applied Probability* 17, 600–10. (10)

Hines, W.G.S. and Maynard Smith, J. 1979. Games between relatives. *Journal of Theoretical Biology* 79, 19–30. (10)

Hinks, C.J.M. 1972. The effects of continuous sire selection on the structure and age composition of dairy cattle populations. *Animal Production* 15, 103–10. (27)

Hiraizumi, Y. 1960. Negative correlation between rate of development and female fertility in *Drosophila melanogaster*. *Genetics* 46, 615–24. (13)

Hiraizumi, Y. and Crow, J.F. 1960. Heterozygous effects on viability, fertility, rate of development and longevity of Drosophila chromosomes that are lethal when homozygous. *Genetics* 45, 1071–83. (13)

Hoekema, A., Kastelein, R.A., Vasser, M. and de Boer, H.A. 1987. Codon replacement in the PGK1 gene of *Saccharomyces cerevisiae*: experimental approach to study the role of biased codon usage in gene expression. *Molecular and Cellular Biology* 7, 2914–24. (4)

Hofmeyr, J.S., Kacser, H. and van der Meer, K.J. 1986. Metabolic control analysis of moiety conserved cycles. *European Journal of Biochemistry* 155, 631–41. (31)

Hollingdale, B. 1971. Analyses of some genes from abdominal bristle number selection lines in *Drosophila melanogaster*. *Theoretical and Applied Genetics* 41, 292–301. (16)

Hollingdale, B. and Barker, J.S.F. 1971. Selection for increased abdominal bristle number in *Drosophila melanogaster* with concurrent irradiation. I. Populations derived from an inbred lines. *Theoretical and Applied Genetics* 41, 208–15. (16)

Holmquist, G., Gray, M., Porter, T. and Jordan, J. 1982. Characterization of giemsa dark- and light-band DNA. *Cell* 31, 121–9. (4)

Hood, L., Campbell, J.H. and Elgin, S.C.R. 1975. The organization, expression and evolution of antibody genes and other multigene families. *Annual Review of Genetics* 9, 305–53. (9)

Hooper, M., Hardy, K., Handyside, A., Hunter, S. and Monk, M. 1987. HPRT-deficient (Lesch-Nyhan) mouse embryos derived from germline colonization by cultured cells. *Nature* 326, 293–5. (36)

Hopkins, I.R. and James, J.W. 1977. Some optimum selection strategies and age structures with overlapping generations. *Animal Production* 25, 111–32. (27)

Hopkins, I.R. and James, J.W. 1978. Theory of nucleus breeding schemes with overlapping generations. *Theoretical and Applied Genetics* 53, 17–24. (27)

Hopkins, I.R. and James, J.W. 1979. Genetic responses in the early years of selection programmes using genetic differences between generations. *Animal Production* 28, 65–77. (27)

Hori, H. and Osawa, S. 1979. Evolutionary chages in 5S RNA secondary structure and a phylogenetic tree of 54 5S RNA species. *Proceedings of the National Academy of Sciences, USA* 76, 381–5. (3)

Hori, H. and Osawa, S. 1987. Origin and evolution of organisms as deduced from 5S ribosomal RNA sequences. *Molecular Biology and Evolution* 4, 445–72. (3)

Howard, R.D. 1979. Estimating reproductive success in natural populations. *American Naturalist* 114, 221–31. (13)

Howe, R.R. and James, J.W. 1973. Response to selection in synthetic lines of *Drosophila melanogaster*. *Australian Journal of Biological Science* 26, 613–23. (11)

Hudson, R.R. 1982. Estimating genetic variability with restriction endo-nucleases. *Genetics* 100, 711–9. (6)

Hudson, R.R. 1983. Testing constant-rate neutral allele model with protein sequence data. *Evolution* 37, 203–17. (3)

Hudson, R.R. and Kaplan, N. 1988. The coalescent process in models with selection and recombination. *Genetics* 120, 831–40. (3, 5)

Hudson, R.R., Kreitman, M. and Aguade, M. 1987. A test of neutral molecular evolution based on nucleotide data. *Genetics* 116, 153–9. (3)

Hull, P. 1960. Genetic relations between carcass fat and body weight in mice. *Journal of Agricultural Science* 55, 317–21. (19)

Hunter, T. (1987). A thousand and one protein kinases. *Cell* 50, 823–9. (9)

Ikemura, T. 1985a. Codon usage and tRNA content in unicellular and multicellular organisms. *Molecular Biology and Evolution* 2, 13–34. (4)

Ikemura, T. 1985b. Codon usage, tRNA content, and rate of synonymous substitution. In: *Population Genetics and Molecular Evolution* (edited by K. Aoki and T. Ohta). Berlin, Germany; Springer-Verlag, pp.385–406. (4)

Im, S. and Gianola, D. 1988. Mixed models for binomial data with an application to lamb mortality. *Applied Statistics* 37, 196–204. (23)

Imakawa, K., Anthony, R.V., Kazemi, M.,Marotti, K.R., Polites, H.G. and Roberts, R.M. 1987. Interferon-like sequence of ovine trophoblast protein secreted by embryonic trophectoderm. *Nature* 330, 377–379. (34)

Ingham, P.W., Pinchin, S.M., Howard, K.R. and Ish-Horowicz, D. 1985. Genetic analysis of the hairy locus in *Drosophila melanogaster*. *Genetics* 111, 463–486. (16)

Jackson, N., Lax, J. and Wilson, R.L. 1986. Sex and selection for fleece weight in Merino sheep. *Zeitschrift für Tierzüchtung und Zühtungsbiologie* 103, 97–115. (27)

Jackson, N. and Turner, H.N. 1972. Optimal structure for a cooperative breeding system. *Proceedings of the Australian Society of Animal Production* 9, 55–69. (27)

Jaenisch, R. 1976. Germ line integration and Mendelian transmission of the exogenous Moloney leukemia virus. *Proceedings of the National Academy of Sciences, USA* 73, 1260–4. (36)

Jaenisch, R. 1977. Germ line integration of Moloney leukaemia virus: effect of homozygosity at the M-MuLV locus. *Cell* 12, 691–6. (36)

Jaenisch, R. 1988. Transgenic animals. *Science* 240, 1468–74. (35, 36)

Jaenisch, R., Schnieke, A. and Harbers, K. 1985. Treatment of mice with 5-azacytidine efficiently activates silent retroviral genomes in different tissues. *Proceedings of the National Academy of Sciences, USA* 82, 1451–5. (36)

Jahner, D. and Jaenisch, R. 1985. Chromosomal position and specific demethylation in enhancer sequences of germ-line transmitted retroviral genomes during mouse development. *Molecular and Cellular Biology* 5, 2212–20. (36)

James, J.W. 1962. Conflict between directional and centripetal selection. *Heredity* 17, 487–99. (12, 15)

James, J.W. 1971. The founder effect and response to artificial selection. *Genetical Research* 16, 241–50. (12)

James, J.W. 1972. Optimum selection intensity in breeding programmes. *Animal Production* 14, 1–9. (27)

James, J.W. 1977. Open nucleus breeding systems. *Animal Production* 24, 287–303. (27)

James, J.W. 1978. Effective population size in open nucleus breeding schemes. *Acta agriculturae scandinavica* 28, 387–92. (27)

James, J.W. 1979. Optimum family size in progeny testing when prior information is available. *Zeitschrift für Tierzüchtung und Züchtungsbiologie* 95, 194–203. (27)

James, J.W. 1986. Cumulative selection differentials and realized heritabilities with overlapping generations. *Animal Production* 42, 411–5. (24)

James, J.W. 1987. Determination of optimal selection policies. *Zeitschrift für Tierzüchtung und Züchtungsbiologie* 104, 23–7. (27)

James, J.W. and McGuirk, B.J. 1982. Regression of offspring on parent for all-or-none traits. *Zeitschrift für Tierzüchtung und Züchtungsbiologie* 99, 308–14. (25)

Jamieson, A. and Robertson, A. 1967. (See AR bibliography)

Jeffreys, A.J. 1979. DNA sequence variants in the $^{G}\gamma$-, $^{A}\gamma$-, δ- and ß-globin genes of man. *Cell* 18, 1–10. (3, 5, 6)

Jeffreys, A.J., Wilson, V., Kelly, R., Taylor, B.A. and Bulfield, G. (1987). Mouse DNA 'fingerprints': analysis of chromosome localization and germ–line stability in recombinant inbred strains. *Nucleic Acids Research* 15, 2823–36. (32)

Jeffreys, A.J., Wilson, V. and Thein, S.L. 1985. Hypervariable 'mini satellite' regions in human DNA. *Nature* 314, 67–73. (6, 34)

Jenkins, N.A. and Copeland, N.G. 1985. High frequency germline acquisition of ecotropic MuLV proviruses in SWR/J-RF/J hybrid mice. *Cell* 43, 811–9. (16)

Jenkins, N.A., Copeland, N.G., Taylor, B.A. and Lee, B.K., 1981. Dilute (*d*) coat colour mutation of DBA/2J mice is associated with the site of integration of an ecotropic MuLV genome. *Nature* 293, 370–4. (16)

Jiang, C., Gibson, J.B., Wilks, A.U. and Freeth, A.L. 1987. Restriction endonuclease variation in the region of the alcohol dehydrogenase gene: a comparison of null and normal alleles from natural populations of *Drosophila melanogaster*. *Heredity* 60, 101–7. (5)

Jinks, J.L. and Towey, P. 1976. Estimating the number of genes in a polygenic system by genotype assay. *Heredity* 37, 69–81. (20)

Jódar, B. and López-Fanjul, C. 1977. Optimum proportions selected with unequal sex numbers. *Theoretical and Applied Genetics* 50, 57–61. (18)

Johansson, I. and Robertson, A. 1952. (See AR bibliography)

Johnson, D.L. 1977a. Inbreeding in populations with overlapping generations. *Genetics* 87, 581–91. (27)

Johnson, D.L. 1977b. Variance-covariance structure of group means with overlapping generations. In: *Proceedings of the International Conference on Quantitative Genetics* (edited by E. Pollack, O. Kempthorne and T.B. Bailey, Jr). Ames, Iowa, USA; Iowa State University Press, pp.851–8. (27)

Johnson, T.K. and Judd, B.H. 1979. Analysis of the cut locus of *Drosophila melanogaster*. *Genetics* 92, 485–502. (16)

Jones, L.P. 1969. Effects of artificial selection on rates of inbreeding in populations of *Drosophila melanogaster*. I. Effect in early generations. *Australian Journal of Biological Sciences* 22, 143–55. (18, 27)

Jones, L.P., Frankham, R. and Barker, J.S.F. 1968. The effects of population size and selection intensity in selection for a quantitative character in *Drosophila*. II. Long-term response to selection. *Genetical Research* 12, 249–66. (12)

Jones, S.J., Aberle, E.D. and Judge, M.D. 1986. Estimation of fractional breakdown rates of myofibrillar proteins from quantitation of 3MH excretion. *Poultry Science* 65, 2142–7. (33)

Jonmundsson, J.V. and Adalsteinsson, S. 1985. *Single genes for fecundity in Icelandic sheep*. In: *Genetics of Reproduction in Sheep* (edited by R.B. Land and D.W. Robinson). London; Butterworths, pp.159–68. (21)

Jorde, L.B. and Durbize, P. 1986. Opportunity for natural selection in the Utah Mormons. *Human Biology* 58, 97–114. (13)

Joyner, A.L., Skarnes, W.C. and Rossant, J. 1989. Production of a mutation in mouse *En-2* gene by homologous recombination in embryonic stem cells. *Nature* 338, 153–5. (36)

Jukes, T.H. 1985. A change in the genetic code in *Mycoplasma capricolum*. *Journal of Molecular Evolution* 22, 361–2. (2)

Jukes, T.H. and Kimura, M. 1984. Evolutionary constraints and the neutral theory. *Journal of Molecular Evolution* 21, 90–2. (4)

Julien, J.-P., Tretjakoff, I., Beaudet, L. and Peterson, A. 1987. Expression and assembly of a human neurofilament protein in transgenic mice provide a novel neuronal marking system. *Genes and Development* 1, 1085–95. (36)

Kacser, H. 1983a. The control of enzyme systems *in vivo*: Elasticity analysis of the steady state. *Biochemical Society Transactions* a11 35–40. (31)

Kacser, H. 1983b. Biochemical control and the improvement of farm animals. In: *Molecular genetic manipulation of farm animals*. Edinburgh, UK; AFRC Animal Breeding Research Organisation; pp.6–9. (34)

Kacser, H. 1987. Control of metabolism. In: *The Biochemistry of Plants* (edited by D.D. Davies). New York, USA; Academic Press, Vol.11, pp.39–67. (31)

Kacser, H. and Burns, J.A. 1973. *The control of flux* In: *Rate Control of Biological Processes* (edited by D.D. Davies). Cambridge, UK; Cambridge University Press, pp.65–104. (21, 31)

Kacser, H. and Burns, J.A. 1979. Molecular Democracy: Who shares the controls? *Biochemical Society Transactions* 7, 1149–60. (21, 30, 31, 34)

Kacser, H. and Burns, J.A. 1981. The molecular basis of dominance. *Genetics* 97, 639–66. (21, 31)

Kacser, H. and Porteous, J.W. 1987. Control of metabolism: what do we have to measure? *Trends in Biochemical Science* 12, 5–14. (31)

Kahler, A.L. and Wehrhahn, C.F. 1986. Associations between quantitative traits and enzyme loci in the F_2 population of a maize hybrid. *Theoretical and Applied Genetics* 72, 15–26. (34)

Kalmus, H. and Seedburgh, D. 1975. Correlated odour threshold bimodality of two out of three synthetic musks. *Annals of Human Genetics* 38, 495–500. (20)

Kaplan, N.L. and Brookfield, J.F.Y. 1983a. The effect on homozygosity of selective differences between sites of transposable elements. *Theoretical Population Biology* 23, 273–80. (8)

Kaplan, N.L. and Brookfield, J.F.Y. 1983b. Transposable elements in Mendelian populations. III. Statistical results. *Genetics* 104, 485–95. (8)

Kaplan, N., Darden, T. and Hudson, R.R. 1988. The coalescent process in models with selection. *Genetics* 120, 819–29. (3)

Kaplan, N.L., Darden, T. and Langley, C.H. 1985. Evolution and extinction of transposable elements in Mendelian populations. *Genetics* 109, 459–80. (8)

Karlin, S., Williams, P.T. and Carmelli, D. 1981. Structured exploratory data analysis (SEDA) for determining mode of inheritance of quantitative traits. I. Simulation studies on the effect of background distributions. *American Journal of Human Genetics* 33, 262–81. (20)

Karp, M.L. 1936. On the cooperation of bristles' genes in D. melanogaster. *Comptes Rendus (Doklady) de l'Academie des Sciences de l'URSS Vol. 1(x)*, No 1(78) 43–7. (21)

Kashi, Y., Soller, M., Hallerman, E. and Beckmann, J.S. 1986. Restriction fragment length polymorphisms in dairy cattle improvement. *3rd World Congress on Genetics Applied to Livestock Production*. Lincoln, Nebraska; University of Nebraska, Vol. 12, 57–64. (34)

Katz, A.J. and Enfield, F.D. 1977. Response to selection for increased pupa weight in *Tribolium castaneum* as related to population structure. *Genetical Research* 30, 237–46. (11, 18)

Katz, A.J. and Young, S.S.Y. 1975. Selection for high adult body weight in Drosophila populations with different structures. *Genetics* 81, 163–75. (11)

Kaufman, P.K., Enfield, F.D. and Comstock, R.E. 1977. Stabilizing selection for pupal weight in *Tribolium castaneum*. *Genetics* 87, 327–41. (15, 18)

Kazazian, H.H., Chakravarti, A., Orkin, S.H. and Antonarakis, S.E. 1983. DNA polymorphisms in the human ß-globin gene cluster. In: *Evolution of genes and proteins* (edited by M. Nei and R.K. Koehn). Sunderland, Mass., USA; Sinauer, pp.137–46. (3, 6)

Keightley, P.D. and Hill, W.G. 1988. Quantitative genetic variability maintained by mutation-stabilizing selection balance in finite populations. *Genetical Research* 52, 33–43. (15)

Keightley, P.D. and Kacser, H. 1987 Dominance, pleiotropy and metabolic structure. *Genetics* 117, 319–29. (31)

Kelley, K.A., Chamberlain, J.W., Nolan, J.A., Horwich, A.L., Kalousek, F., Eisenstadt, J., Herrup, K. and Rosenberg, L.E. 1988. Meiotic expression of human ornithine transcarbamylase in the testes of transgenic mice. *Molecular and Cellular Biology* 8, 1821–1825. (36)

Kelley, D.E., Pollock, B.A., Atchison, M.L. and Perry, R.P. 1988. The coupling between enhancer activity and hypomethylation of kappa immunoglobulin genes is developmentally regulated. *Molecular and Cellular Biology* 8, 930–7. (36)

Kelsey, G.D., Povey, S., Bygrave, A.E. and Lovell-Badge, R.H. 1987. Species- and tissue-specific expression of human alpha-1-antitrypsin in transgenic mice. *Genes and Development* 1, 161–71. (36)

Kennedy, B. W. 1988. Use of mixed model methodology in analysis of designed experiments. In: *Advances in Statistical Methods for Genetic Improvement of Livestock* (edited by D. Gianola and K. Hammond). Heidelberg, Germany; Springer Verlag (in press). (23, 34)

Kennedy, B.W., Schaeffer, L.R. and Sorensen, D.A. 1988. Genetic properties of animal models. *Journal of Dairy Science* 71 Suppl. 2, 17–26. (23)

Kennedy, B.W. and Sorensen, D.A. 1988. Properties of mixed model methods for prediction of genetic merit under different genetic models in selected and unselected populations. *Proceedings of the 2nd International Conference in Quantitative Genetics* (edited by B.S. Weir, E.J. Eisen, M.M. Goodman and G. Namkoong). Sunderland, Mass., USA; Sinauer, pp. 91–103. (34)

Keshet, I., Lieman-Hurwitz, J. and Cedar, H. 1986. DNA methylation affects the formation of active chromatin. *Cell* 44, 535–43. (36)

Kidwell, M.G., Kidwell, J.F. and Sved, J.A. 1977. Hybrid dysgenesis in *Drosophila melanogaster*: A syndrome of aberrant traits including mutation, sterility, and male recombination. *Genetics* 86, 813–33. (16)

Kimura, M. 1964. Diffusion models in population genetics. *Journal of Applied Probability* 1, 177–232. (2)

Kimura, M. 1965. A stochastic model concerning the maintenance of genetic variability in quantitative characters. *Proceedings of the National Academy of Sciences, USA* 54, 731–6. (15)

Kimura, M. 1968. Evolutionary rate at the molecular level. *Nature* 217, 624–6. (2, 3)

Kimura, M. 1969. The number of heterozygous nucleotide sites maintained in a finite population due to steady flux of mutations. *Genetics* 61, 893–903. (15)

Kimura, M. 1977. Preponderance of synonymous changes as evidence for the neutral theory of molecular evolution.

Nature 267, 275–6. (2)

Kimura, M. 1983. *The Neutral Theory of Molecular Evolution*. Cambridge, UK; Cambridge University Press. (2, 3, 4, 9, 15)

Kimura, M. 1987. Molecular evolutionary clock and the neutral theory. *Journal of Molecular Evolution* 26, 24–33. (2)

Kimura, M. and Crow, J.F. 1963. On the maximum avoidance of inbreeding. *Genetical Research* 4, 399–415. (1)

Kimura, M. and Crow, J.F. 1964. The number of alleles that can be maintained in a finite population. *Genetics* 49, 725–38. (2)

Kimura, M. and Ohta, T. 1969. The average number of generations until fixation of a mutant gene in a finite population. *Genetics* 61, 763–71. (2)

Kimura, M. and Ohta, T. 1973. Eukaryote-prokaryote divergence estimated by 5S ribosomal RNA sequences. *Nature* 243, 199–200. (3)

Kimura, M. and Ohta, T. 1974. On some principles governing molecular evolution. *Proceedings of the National Academy of Sciences, USA* 71, 2848–52. (2)

King, J.L. and Jukes, T.H. 1969. Non-Darwinian evolution. *Science* 164, 788–98. (4)

King, J.W.B., Curran, M.K., Standal, N., Power, P. Heaney, I.H., Kallweit, E., Schroder, J., Maijala, K., Kangasniemi, R. and Walstra, P. 1975. An international comparison of pig breeds using a common control stock. *Livestock Production Science* 2, 367–79. (28)

Kinghorn, B.P. 1986. Mating plans for selection across breeds. *Proceedings of 3rd World Congress on Genetics Applied to Livestock Production*. Lincoln, Nebraska; University of Nebraska, Vol. 12, 233–44. (11, 27)

Kingman, J.F.C. 1978. A simple model for the balance between selection and mutation. *Journal of Applied Probability* 15, 1–12. (15)

Kitagawa, O. 1967. The effects of X-ray irradiation on selection response in *Drosophila melanogaster*. *Japanese Journal of Genetics* 42, 121–37. (16)

Kleckner, N. 1981. Transposable elements in prokaryotes. *Annual Review of Genetics* 15, 341–404. (8)

Klein, J. 1986. *Natural History of the Major Histocompatibility Complex*. Chichester. Wiley. (6)

Klitz, W., Thomson, G. and Baur, M.P. 1986. Contrasting evolutionary histories among tightly linked HLA loci. *American Journal of Human Genetics* 39, 340–9. (6)

Klose, J. 1982. Genetic variability of soluble proteins studied by two-dimensional gel electrophoresis on different inbred strains and on different mouse organs. *Journal of Molecular Evolution* 18, 315–28. (32)

Knight, G.R. and Robertson, A. 1957. (See AR bibliography)

Knight, G.R., Robertson, A. and Waddington, C.H. 1956. (See AR bibliography).

Koenig, M., Hoffman, E.P., Bastelon, C.J., Monaco, A.P., Feener, C. and Kunkel, L.M. 1987. Complete cloning of the Duchenne Muscular Dystrophy (DMD) cDNA and preliminary genomic organisation of the DMD gene in normal and affected individuals. *Cell* 50, 509–17. (34)

Kolsto, A.-B., Kollias, G., Giguere, V., Isobe, K.-I., Prydz, H. and Grosveld, F. 1986. The maintenance of methylation-free islands in transgenic mice. *Nucleic Acids Research* 14, 9667–78. (36)

Kreitman, M. 1983. Nucleotide polymorphism at the alcohol dehydrogenase locus of *Drosophila melanogaster*. *Nature* 304, 412–7. (1, 3, 5, 6)

Kreitman, M. and Aguade, M. 1986a. Genetic uniformity in two populations of *Drosophila melanogaster* as revealed by filter hybridization of 4-nucleotide-recognizing restriction enzyme digests. *Proceedings of the National Academy of Sciences, USA* 83, 3562–6. (5, 6)

Kreitman, M. and Aguade, M. 1986b. Excess polymorphism at the Adh locus in *Drosophila melanogaster*. *Genetics* 114, 93–110. (3, 5)

Krimpenfort, P., de Jong, R., Uematsu, Y., Dembic, Z., Ryser, S., von Bohmer, H. Steinmetz, M. and Berns, A. 1988. Transcription of T cell receptor beta-chain is controlled by a downstream element. *EMBO Journal* 7, 745–50. (36)

Krumlauf, R., Chapman, V.M., Hammer, R.E., Brinster, R. and Tilghman, S.M. 1986. Differential expression of alpha-fetoprotein genes on the inactive X-chromosome in extra embryonic and somatic tissues of a transgenic mouse line. *Nature* 319, 224–6. (36)

Krumlauf, R., Hammer, R.E., Tilghman, S.M. and Brinster, R.L. 1985. Developmental regulation of alpha-fetoprotein genes in transgenic mice. *Molecular and Cellular Biology* 5, 1639–48. (36)

Kuehn, M., Bradley, A., Robertson, E.J. and Evans, M.J. 1987. A potential animal model for Lesch–Nyhan syndrome through introduction of HPRT mutations into mice. *Nature* 326, 295-8. (36)

Lado, K.M., Noon, W.A. and Aquadro,C.F. 1988. Naturally occurring restriction map variation in the rosy region of *Drosophila simulans*: contrasts with *Drosophila melanogaster*. *Genetics* (in press). (5)

Lalouel, J.M. and Morton, N.E. 1981. Complex segregation analysis with pointers. *Human Heredity* 31, 312–21. (20)

Lalouel, J.M., Rao, D.C., Morton, N.E. and Elston, R.C. 1983. A unified model for complex segregation analysis. *American Journal of Human Genetics* 35, 816–26. (20)

Land, R.B. and Hill, W.G. 1975. The possible use of superovulation and embryo transfer in cattle to increase response to selection. *Animal Production* 21, 1–12. (29)

Lande, R. 1975. The maintenance of genetic variability by mutation in a polygenic character with linked loci. *Genetical Research* 26, 221–35. (15)

Lande, R. 1980a. The genetic covariance between characters maintained by pleiotropic mutation. *Genetics* 94, 203–15. (15)

Lande, R. 1980b. Genetic variation and phenotypic evolution during allopatric speciation. *American Naturalist* 116, 463–77. (1)

Lande, R. 1981. The minimum number of genes contributing to quantitative variation between and within populations. *Genetics* 99, 541–53. (17, 20)

Lande, R. 1988. Quantitative genetics and evolutionary theory. In: *Proceedings of the 2nd International Conference on Quantitative Genetics* (edited by B.S.Weir, E.J. Eisen, M.M. Goodman, and G. Namkoong). Sunderland, Mass., USA; Sinauer, pp. 71–84. (17)

Lande, R. and Arnold, S.J. 1983. The measurement of selection on correlated characters. *Evolution* 37, 1210–26. (15)

Lang, G., Wotton, D., Owen, M.J., Sewell, W.A., Brown, M.H., Mason, D.Y., Crumpton, M.J. and Kioussis, D. 1988. The structure of the human CD2 gene and its expression in transgenic mice. *The EMBO Journal* 7, 1675–82. (36)

Langley, C.H. 1977. Non-random associations between allozymes in natural populations of *Drosophila melanogaster*. In: *Measuring Selection in Natural Populations* (edited by F.B. Christiansen and T.M. Fenchel). Berlin, Germany; Springer-Verlag, pp.265–73. (6)

Langley, C.H. and Aquadro, C.F. 1987. Restriction map variation in natural populations of *Drosophila melanogaster*: *white*-locus region. *Molecular Biology and Evolution* 4, 651–63. (5, 6, 8)

Langley, C.H., Brookfield, J.F.Y. and Kaplan, N.L. 1983. Transposable elements in Mendelian populations. I. A theory. *Genetics* 104, 457–72. (8)

Langley, C.H., Montgomery, E.A., Hudson, R.H., Kaplan, N.L. and Charlesworth, B. 1988. On the role of unequal exchange in the containment of transposable element copy number. *Genetical Research* 52, 223–35. (3, 8)

Langley, C.H., Montgomery, E.A. and Quattlebaum, W.F. 1982. Restriction map variation in the *Adh* region of *Drosophila*. *Proceedings of the National Academy of Sciences, USA* 79, 5631–5. (5, 6, 8)

Lathe, R., Clark, A.J., Archibald, A.L., Bishop, J.O., Simons, P. and Wilmut, I. 1986. Novel products from livestock. In: *Exploiting New Technologies in Animal Breeding: Genetic Developments* (edited by C. Smith, J.W.B. King and J.C. McKay). Oxford, UK; Oxford University Press. (34)

Latter, B.D.H. 1960. Natural selection for an intermediate optimum. *Australian Journal of Biological Sciences* 13, 30–5. (15)

Latter, B.D.H. 1965. The response to artificial selection due to autosomal genes of large effect. II. The effects of linkage on limits to selection in finite populations. *Australian Journal of Biological Sciences* 18, 1009–23. (12)

Latter, B.D.H. 1970. Selection in finite populations with multiple alleles. II. Centripetal selection, mutation and isoallelic variation. *Genetics* 66, 165–86. (15)

Latter, B.D.H. and Robertson, A. 1960. (See AR bibliography)

Latter, B.D.H. and Robertson, A. 1962. (See AR bibliography)

Laurie-Ahlberg, C.C., Wilton, A.N., Curtsinger, J.W. and Emigh, T.H. 1982. Naturally occurring enzyme activity variation in *Drosophila melanogaster*. I. Sources of variation for 23 enzymes. *Genetics* 102, 191–206. (16)

Leclerc, B., Simon, J. and Ricard, F.H. 1987. Effects of selection for high and low plasma glucose concentration in chickens. *British Poultry Science* 28, 557–67. (33)

Lee, G.J. and Land, R.B. 1985. Testis size and LH response to LH-RH as male criteria of female reproductive performance. In *Genetics of Reproduction in Sheep* (edited by R.B. Land and D.W. Robinson). London, UK; Butterworths, pp. 333–41. (33)

Leigh Brown, A.J. 1983. Variation at the 87A heat shock locus in *Drosophila melanogaster*. *Proceedings of the National Academy of Sciences USA* 80, 5350–4. (6, 8)

Leigh Brown, A.J. and Ish-Horowicz, D. 1981. Evolution of the 87A and 87C heat shock loci in *Drosophila*. *Nature* 290, 677–82. (6)

Leigh Brown, A.J. and Moss, J.E. 1987. Transposition of the *I* element and *copia* in a natural population of *Drosophila melanogaster*. *Genetical Research* 49, 121–8. (8, 16)

Le Meur, M., Gerlinger, P., Benoist, C. and Mathis, D. 1985. Correcting an immune-response deficiency by creating E_{alpha} gene transgenic mice. *Nature* 316, 38–42. (36)

Lerner, I.M. 1954. *Genetic Homeostasis*. Edinburgh; Oliver and Boyd: New York; Dover, 1970. (1)

Levene, H. 1949. On a matching problem in genetics. *Annals of Mathematical Statistics* 20, 91–4. (1)

Levins, R. 1962. Theory of fitness in a heterogeneous environment. I. The fitness set and adaptive function. *American Naturalist* 96, 361–73. (10)

Lewin, R. 1987. The surprising genetics of bottlenecked flies. *Science* 235, 1325–7. (12)

Lewontin, R.C. 1964. The interaction of selection and linkage. II. Optimum models. *Genetics* 50, 757–82. (1)

Lewontin, R.C. 1974. *The Genetic Basis of Evolutionary Change*. New York; Columbia University Press. (3, 5, 6, 9, 15)

Lewontin, R.C. 1984. Detecting population differences in quantitative characters as opposed to gene frequencies. *American Naturalist* 123, 115–24. (11)

Lewontin, R.C. 1986. A comment on the comments of Rogers and Felsenstein. *American Naturalist* 127, 733–4. (11)

Lewontin, R.C. 1989. Inferring evolutionary events from nucleotide sequence differences. *Molecular Biology and Evolution* 6, 15–32. (5)

Lewontin, R.C. and Hubby, J.L. 1966. A molecular approach to the study of genic heterozygosity in natural populations. II. Amount of variation and degree of heterozygosity in natural populations of *Drosophila pseudoobscura*. *Genetics* 54, 595–609. (3)

Lewontin, R.C. and Krakauer, J. 1973. Distribution of gene frequency as a test of the theory of the selective neutrality of polymorphisms. *Genetics* 74, 175–95. (1, 3, 11)

Lewontin, R.C. and Krakauer, J. 1975. Testing the heterogeneity of F values. *Genetics* 80, 397–8. (1)

Li, W.-H. 1985. Accelerated evolution following gene duplication and its implication for the neutralist-selectionist controversy. In: *Population Genetics and Molecular Evolution* (edited by T. Ohta, and K. Aoki). Tokyo; Scientific Soc. Press; Berlin, Springer, pp.333–52. (9)

Li, W.-H. 1987. Models of nearly neutral mutations with particular implications for nonrandom usage of synonymous codons. *Journal of Molecular Evolution* 24, 337–45. (4)

Li, W.-H., Gojobori, T. and Nei, M. 1981. Pseudogenes as a paradigm of neutral evolution. *Nature* 292, 237–9. (2)

Li, W.-H., Tanimura, M. and Sharp, P.M. 1987. An evaluation of the molecular clock hypothesis using mammalian DNA sequences. *Journal of Molecular Evolution* 25, 330–42. (4)

Li, W.-H., Wu, C-I. and Luo, C-C. 1985. A new method for estimating synonymous and nonsynonymous rates of nucleotide substitution considering the relative likelihood of nucleotide and codon changes. *Molecular Biology and Evolution* 2, 150–74. (4)

Lin, C.Y. 1987. Application of singular value decomposition to Restricted Maximum Likelihood estimation of variance components. *Journal of Dairy Science* 70, 2680–4. (23)

Lindsley, D.L. and Sandler, L. 1977. The genetic analysis of meiosis in female *Drosophila*. *Philosophical Transactions of the Royal Society, London B* 277, 295–312. (8)

Linney, R., Barnes, B.W. and Kearsey, M.J. 1971. Variation for metrical characters in *Drosophila* populations. III. The nature of selection. *Heredity* 27, 163–74. (15)

Lo, D., Burkly, L.C., Widera, G., Cowing, C., Flavell, R.A., Palmiter, R.D. and Brinster, R.L. 1988. Diabetes and tolerance in transgenic mice expressing class II MHC moleculaes in pancreatic beta cells. *Cell* 53, 159–68. (36)

Louw, J.H. 1966. An analysis of gene effects in a quantitative character in *Drosophila*. PhD Thesis, University of Edinburgh. (12, 21)

Low, M.J., Hammer, R.E., Goodman, R.H., Habener, J.F., Palmiter, R.D. and Brinster, R.L. 1985. Tissue-specific post-translational processing of pre-prosomatostatin encoded by a metallothionein-somatostatin fusion gene in transgenic mice. *Cell* 41, 211–9. (36)

Low, M.J., Lechan, R.M., Hammer, R.E., Brinster, R.L., Habener, J.F., Mandel, G. and Goodman, R.H. 1986. Gonadotroph-specific expression of metallothionein fusion genes in pituitaries of transgenic mice. *Science* 231, 1002–4. (36

Lush, J.L., Lamoreux, W.F. and Hazel, L.N. 1948. The heritability of resistance to death in poultry. *Poultry Science* 27, 375–88. (25)

Lynch, M. 1988. The rate of polygenic mutation. *Genetical Research* 51, 137–48. (13, 15, 16)

Lynch, M. and Hill, W.G. 1986. Phenotypic evolution by neutral mutation. *Evolution* 40, 915–35. (15, 16)

MacArthur, A.T.G. 1954. The assessment of progeny tests of dairy bulls made under farm conditions. *Proceedings of the British Society of Animal Production* 1954, 75–82. (29)

MacArthur, J.W. 1949. Selection for small and large body size in the house mouse. *Genetics* 34, 194–209. (17)

McBride, G. and Robertson, A. 1963. Selection using assortative mating in *Drosophila melanogaster*. *Genetical Research* 4, 356–69. (12)

McCarthy, J.C. 1982. The laboratory mouse as a model for animal breeding: a review of selection for increased body weight and litter size. *Proceedings of the 2nd World Congress on Genetics Applied to Livestock Production*. Madrid; Editorial Garsi. Vol. 5, 66–83. (19)

McCarthy, J.C. 1983. The nature of genetic improvement in the efficiency of meat and milk production. *Proceedings of the 6th World Congress of Food Science and Technology* 5, (edited by J.V. McLoughlin and B.M. McKenna). Dublin; Boole Press, pp. 23–32. (19)

McDaniel, B.T. and Cassell, B.G. 1981. Effects of embryo transfer on genetic change in dairy cattle. *Journal of Dairy Science* 64, 2484–92. (29)

McDonald, J.F. and Ayala, F.J. 1978. Genetic and biochemical basis of enzyme activity variation in natural populations. I. Alcohol dehydrogenase in *Drosophila melanogaster*. *Genetics* 89, 371–88. (16)

McKay, J. 1981. Variation in activity and thermostability of alcohol dehydrogenase in *Drosophila melanogaster*. *Genetical Research* 37, 227–37. (1)

Mackay, T.F.C. 1984. Jumping genes meet abdominal bristles: Hybrid dysgenesis-induced quantitative variation in *Drosophila melanogaster*. *Genetical Research* 44, 231–7. (16)

Mackay, T.F.C. 1985a. A quantitative genetic analysis of fitness and its components in *Drosophila melanogaster*. *Genetical Research* 47, 59–70. (17)

Mackay, T.F.C. 1985b. Transposable element-induced response to artificial selection in *Drosophila melanogaster*. *Genetics* 111, 351–74. (8, 16, 21)

Mackay, T.F.C. 1986. Transposable element-induced fitness mutations in *Drosophila melanogaster*. *Genetical Research* 48, 77–87. (8, 16, 21)

Mackay, T.F.C. 1987. Transposable element-induced polygenic mutations in *Drosophila melanogaster*. *Genetical

Research 49, 225–33. (16, 17)

Mackay, T.F.C. 1988. Transposable element-induced quantitative genetic variation in Drosophila. In: *Proceedings of the 2nd International Conference in Quantitative Genetics* (edited by B.S. Weir, E.J. Eisen, M.M. Goodman and G. Namkoong). Sunderland, Mass., USA; Sinaeur, pp.219–35. (16)

MacLean, C.J., Morton, N.E., Elston, R.C. and Yee, S. 1976. Skewness in commingled distributions. *Biometrics* 32, 695–9. (20)

MacLean, C.J., Morton, N.E. and Yee, S. 1984. Combined analysis of genetic segregation and linkage under an oligogenic model. *Computers in Biomedical Research* 17, 471–80. (20)

McMillan, I. and Robertson, A. 1974. (See AR bibliography)

McPhee, C.P. 1981. Selection for efficient lean growth in a pig herd. *Australian Journal of Agricultural Research* 32, 681–90. (28)

McPhee, C.P. 1985. Experiments on selection for efficiency of lean growth in pigs. *Proceedings of 5th Conference of Australian Association of Animal Breeding and Genetics* pp.310–2. (28)

McPhee, C.P. and Neill, A.R. 1976. Changes in the body composition of mice selected for high and low eight-week weight. *Theoretical and Applied Genetics* 47, 21–6. (19)

McPhee, C.P. and Robertson, A. 1970. (See AR bibliography)

McPhee, C.P. and Trappett, P.C. 1987. Growth and body composition changes in mice selected for high post-weaning weight gain on two levels of feeding. *Theoretical and Applied Genetics* 73, 926–31. (19)

McPhee, C.P., Trappett, P.C., Neill, A.R. and Duncalfe, F. 1980. Changes in growth, appetite, food conversion efficiency and body composition in mice selected for high post-weaning weight gain on restricted feeding. *Theoretical and Applied Genetics* 57, 49–56. (19)

McPhee, H.C. and Wright, S. 1925. Mendelian analysis of the pure breeds of livestock. III. The Shorthorns. *Journal of Heredity* 16, 205–15. (11)

McPhee, H.C. and Wright, S. 1926. Mendelian analysis of the pure breeds of livestock. IV. The British Dairy Shorthorns. *Journal of Heredity* 17, 397–401. (11)

McWhir, J., Evans, M. and Church, R.B. 1988. Polyoma early region transforms a porcine multipotential embryonic cell line. *Genome* 30, 444. (35)

McWhir, J. and Wilton, J. 1987. Heritabilities of predicted performance measures in beef bull tests. *Journal of Animal Science* 64, 1323–31. (35)

Madalena, F.E. and Hill, W.G. 1972. Population structure in artificial selection programmes: simulation studies. *Genetical Research* 20, 75–99. (11, 12, 18)

Madalena, F.E. and Robertson, A. 1975. (See AR bibliography)

Mahon, K.A., Overbeek, P.A. and Westphal, H. 1988. Prenatal lethality in a transgenic mouse line is the result of a chromosomal translocation. *Proceedings of the National Academy of Sciences, USA* 85, 1165–9. (36)

Maki-Tanila, A. 1982. The validity of the heritability concept in quantitative genetics. PhD Thesis, University of Edinburgh. (12)

Mandryk, J.A. 1978. Effects of chromosome homozygosity in *Drosophila melanogaster*: viability, fertility and overall fitness. PhD Thesis, University of Sydney. (14)

Maniatis, T., Goodbourn, S. and Fischer, J.A. 1987. Regulation of inducible and tissue-specific gene expression. *Science* 236, 1237–45. (36)

Mansour, S.L., Thomas, K.R. and Capecchi, M.R. 1988. Disruption of the proto-oncogene *int-2* in mouse embryo-derived stem cells: a general strategy for targeting mutations to non-selectable genes. *Nature* 336, 348–52. (36)

Markow, T.A. 1975. A genetic analysis of phototactic behavior in *Drosophila melanogaster*. I. Selection in the presence of inversions. *Genetics* 79, 527–34. (12)

Martin, A., Gross, W.B. Dunnington, E.A., Briles, R.W., Briles, W.E. and Siegel, P.B. 1986. Resistance to natural and controlled exposures to *Eimeria tenella*. *Poultry Science* 65, 1847–52. (33)

Martin, F.G. and Cockerham, C.C. 1960. High speed selection studies. In: *Biometrical Genetics* (edited by O. Kempthorne). New York; Pergamon Press, pp. 35–45. (12)

Martin, N.G. Clark, P., Ofulue, A.F., Eaves, L.J., Corey, L.A. and Nance, W.E. 1987. Does the PI polymorphism alone control alpha-1-antitrypsin expression? *American Journal of Human Genetics* 40, 267–77. (20)

Maruyama, T. 1970. On the fixation probability of mutant genes in a subdivided population. *Genetical Research* 15, 221–7. (18)

Mason, I.L. and Robertson, A. 1956. (See AR bibliography)

Mason, J.L., Robertson, A. and Gjelstade, B. 1957. (See AR bibliography)

Mather, K. and Wigan, L.G. 1942. The selection of invisible mutations. *Proceedings of the Royal Society, London* B 131, 50–64. (16)

Matthysse, S., Lange, K. and Wagener, D.K. 1979. Continuous variation caused by genes with graduated effects. *Proceedings of the National Academy of Sciences* 76, 2862–5. (20)

Maynard Smith, J. 1958. The effects of temperature and of egg-laying on the longevity of Drosophila subobscura. *Journal of Experimental Biology* 35, 832–42. (10)

Maynard Smith, J. 1978. Optimisation theory in evolution. *Annual Review of Ecological Systems* 9, 31–56. (10)

Maynard Smith, J. 1981. Will a sexual population evolve to an ESS? *American Naturalist* 117, 1015–8. (10)

Maynard Smith, J. 1982. *Evolution and the Theory of Games*. Cambridge, UK; Cambridge University Press. (10)

Maynard Smith, J. 1988. Can a mixed strategy be stable in a finite population? *Journal of Theoretical Biology* 130, 247–51. (10)

Maynard Smith, J. and Brown, R.L.W. 1986. Competition and body size. *Theoretical Population Biology* 30, 166–79. (10)

Maynard Smith, J., Clarke, J.M. and Hollingsworth, M.J. 1955. The expression of hybrid vigour in *Drosophila subobscura*. *Proceedings of the Royal Society, London* B144, 159–171. (10)

Maynard Smith, J. and Haigh, J. 1974. The hitch-hiking effect of a favourable gene. *Genetical Research* 23, 23–35. (6)

Maynard Smith, J. and Price, G.R. 1973. The logic of animal conflict. *Nature* 246, 15–18. (10)

Maynard Smith, J. and Savage, R.J.G. 1956. Some locomotory adaptations in mammals. *Journal of the Linnean Society, Zoology* 42, 603–22. (10)

Mayo, O., Eckert, S.R. and Waego Hado Nugroho, 1983. Properties of the major gene index and related functions. *Human Heredity* 33, 205–12. (20)

Mayo, O., Eckert, S.R. and Waego Hado Nugroho, 1985. Models of gene effects for a quantitative trait in man. In: *Proceedings Indian Statistical Institute Golden Jubilee Conferences* (edited by K.C. Malhotra and Amitabha Basu), pp. 479–89. (20)

Mayo, O., Fraser, G.R. and Stamatoyannopoulos, G. 1969. Genetic influences on serum cholesterol in two Greek villages. *Human Heredity* 19, 86–99. (20)

Mayo, O. and Hancock, T.W. 1985. Fluctuating rates of evolution. *Experientia* 41, 1080–2. (20)

Mayo, O., Hancock, T.W. and Baghurst, P. 1980. Influence of major genes on variance within sibships for a quantitative trait. *Annals of Human Genetics* 43, 419–21. (20)

Mayo, O. and Hopkins, A.M. 1985. Problems in estimating the minimum number of genes contributing to quantitative variation. *Biometrical Journal* 27, 181–7. (20)

Mayo, O., Wiesenfeld, S., Stamatoyannopoulos, G. and Fraser, G.R. 1971. Genetical influences on serum-cholesterol level. *The Lancet* ii, 554–5. (20)

Meijering, A. and Gianola, D. 1985. Linear versus non-linear methods of sire evaluation for categorical traits: a simulation study. *Génétique, Séléction, Évolution* 17, 115–32. (25)

Mellor, A.L., Weiss, E.H., Ramachandran, K. and Flavell, R.A. 1983. A potential donor gene for the *bml* gene conversion event in the c57BL mouse. *Nature* 306, 792–5. (6)

Mendell, M. and Elston, R.C. 1971. Use of tetrachoric correlation coefficient in the estimation of heritability of quasi-continuous traits. *Biometrics* 27, 483–4. (25)

Mérat, P. 1968. Distributions de fréquences, interprétation du determinisme génétique des caractères quantitatifs et recherche de 'gènes majeurs'. *Biometrics* 24, 277–93. (20)

Mérat, P. and Ricard, F.H. 1974. Étude d'un gène de nanisme lié au sexe chez la poule: importance de l'état d'engraissement et gain de poids chez l'adulte. *Annals de Génétique et de Séléction animale* 6, 211–7. (21)

Mercer, J.T. and Hill W.G. 1984. Estimation of genetic parameters for skeletal defects in broiler chickens. *Heredity* 53, 193–203. (25)

Merks, J.W.M. 1988. Genotype x environment interaction in pig breeding programmes. Thesis, University of Wageningen. (28)

Messing, A., Chen, H.Y., Palmiter, R.D. and Brinster, R.L. 1985. Peripheral neuropathies, hepatocellular carcinomas and islet cell adenomas in transgenic mice. *Nature* 316, 461–3. (36)

Meyer, H.H. and Enfield, F.D. 1975. Experimental evidence on limitations of the heritability parameter. *Theoretical and Applied Genetics* 45, 268–73. (17)

Meyer, K. 1983. Maximum likelihood procedures for estimating genetic parameters for later lactations of dairy cattle. *Journal of Dairy Science* 66, 1988–97. (23)

Meyer, K. 1985. Maximum Likelihood estimation of variance components for a multivariate mixed model with equal design matrices. *Biometrics* 41, 153–65. (23)

Meyer, K. 1986. Between algorithms : A 'shortcut' Restricted Maximum Likelihood procedure to estimate variance components. *Journal of Dairy Science* 69, 1904–16. (23, 24)

Meyer, K. 1987a. Restricted Maximum Likelihood (REML) programs for the analysis of animal breeding data. Program notes prepared for a workshop in Orange, NSW, 2–6 March, 1987. Mimeo. 55pp. (23)

Meyer, K. 1987b. Restricted Maximum Likelihood to estimate variance components for mixed models with two random factors. *Génétique, Séléction, Évolution* 19, 49–68. (23)

Meyer, K. 1988a. DFREML – a set of programs to estimate variance components under an Individual Animal Model. *Journal of Dairy Science* 71, Suppl. 2, 33–4 (Abstr.). (23)

Meyer, K. 1988b. Restricted Maximum Likelihood to estimate variance components for animal models with several random effects using a derivative-free algorithm. *Génétique, Séléction, Évolution* (in press). (23)

Meyer, K. 1988c. Estimating variances and covariances for multivariate animal models by Restricted Maximum Likelihood. *Génétique, Séléction, Évolution* (submitted). (23)

Meyer, K. and Thompson, R. 1984. Bias in variance and covariance component estimators due to selection on a correlated trait. *Journal of Animal Breeding and Genetics* 101, 33–50. (23)

Middleton, R.J. and Kacser, H. 1983. Enzyme variation, metabolic flux and fitness: alcohol dehydrogenase in *Drosophila melanogaster*. *Genetics* 105, 633–50. (31)

Miklos, G.L.G. 1985. Localised highly repetitive DNA sequences in vertebrate and invertebrate genomes. In: *Molecular Evolutionary Genetics* (edited by R.J. MacIntyre). New York; Plenum Press, pp.240–321. (6)

Miklos, G.L.G., Healy, M.J., Pain, P., Howells, A.J. and Russell, R.J. 1984. Molecular genetic studies on the euchromatin–heterochromatin junction in the X chromosome of *Drosophila melanogaster*. I. A cloned entry point near to the uncoordinated (*unc*) locus. *Chromosoma* 89, 218–27. (8)

Miklos, G.L.G., Yamamoto, M., Davies, J. and Pirrotta, V. 1988. Microcloning reveals a high frequency of repetitive sequences characteristic of chromosome 4 and the ß-heterochromatin of *Drosophila melanogaster*. *Proceedings of the National Academy of Sciences, USA* 85, 2051–5. (8)

Mikus, M.D. and Petes, T.D. 1982. Recombination between genes located on nonhomologous chromosomes in *Saccharomyces cerevisiae*. *Genetics* 101, 369–404. (8)

Miller, G.F. 1962. The evaluation of eigenvalues of a differential equation arising in genetics. *Proceedings of the Cambridge Philosophical Society* 58, 588–93. (1)

Mills, L.E.P., Batterham, P., Alegre, J., Starmer, W.T. and Sullivan, D.T. 1986. Molecular genetic characterisation of a locus that contains duplicate Adh genes in *Drosophila mojavensis* and related species. *Genetics* 112, 295–310. (6)

Mirmirani, M. and Oster, G. 1978. Competition, kin selection and evolutionary stable strategies. *Theoretical Population Biology* 13, 304–39. (10)

Mitchell, G., Smith, C., Makower, M. and Bird, P.J.W.N. 1982. An economic appraisal of pig improvement in Great Britain. 1. Genetic and production aspects. *Animal Production* 35, 215–24. (28)

Mitchell, J.A. 1977. Fitness effects of EMS-induced mutations on the X chromosome of *Drosophila melanogaster*. I. Viability effects and heterozygous fitness effects. *Genetics* 87, 763–74. (13, 16)

Mitchell, J.A. and Simmons, M.J. 1977. Fitness effects of EMS-induced mutations on the X chromosome of *Drosophila melanogaster*. II Hemizygous fitness effects. *Genetics* 87, 775–83. (16)

Miyashita, N. and Langley, C.H. 1988. Molecular and phenotypic variation of the *white* locus region in *Drosophila melanogaster*. *Genetics* 120, 199–212. (3)

Miyata, T., Miyazawa, S. and Yasunaga, T. 1979. Two types of amino acid substitution in protein evolution. *Journal of Molecular Evolution* 12, 219–36. (3)

Miyata, T. and Yasunaga, T. 1981. Rapidly evolving mouse alpha-globin-related pseudogene and its evolutionary history. *Proceedings of the National Academy of Sciences, USA* 78, 450–3. (2)

Miyata, T., Yasunaga, T. and Nishida, T. 1980. Nucleotide sequence divergence and functional constraint in mRNA evolution. *Proceedings of the National Academy of Sciences, USA* 77, 7328–32. (4)

Modolell, J., Bender, W. and Meselson, M. 1983. *Drosophila melanogaster* mutations suppressible by the *suppressor-of-Hairy wing* are insertions of a 7.3 kilobase mobile element. *Proceedings of the National Academy of Sciences, USA* 80, 1678–1682. (16)

Montgomery, E.A., Charlesworth, B. and Langley, C.H. 1987. A test for the role of natural selection in the stabilisation of transposable element copy number in a population of *Drosophila melanogaster*. *Genetical Research* 49, 31–41. (6, 8, 16)

Montgomery, E.A. and Langley, C.H. 1983. Transposable elements in Mendelian populations. II. Distribution of three copia-like elements in a natural population of *Drosophila melanogaster*. *Genetics* 104, 473–83. (8, 16)

Moran, C. 1984. Sex-linked effective population size in control populations, with particular reference to honeybees (*Apis mellifera* L.) *Theoretical and Applied Genetics* 67, 317–22. (29)

Moritz, C., Dowling, T.E. and Brown, W.M. 1987. Evolution of animal mitochondrial DNA: Relevance for population biology and systematics. *Annual Review of Ecology and Systematics* 18, 269–92. (6, 11)

Morley, F.H.W. 1952. Selection for economic characters in Australian Merino sheep. 2. Relative efficiency of certain aids to selection. *Australian Journal of Agricultural Research* 3, 409–18. (27)

Morley, F.H.W. 1954. Selection for economic characters in Australian Merino sheep. 4. The effect of inbreeding. *Australian Journal of Agricultural Research* 5, 305–16. (27)

Morton, N.E. and MacLean, C.J. 1974. Analysis of family resemblance. III Complex segregation analysis of quantitative traits. *American Journal of Human Genetics* 26, 489–503. (20)

Morton, N.E., Williams, W.R. and Lew, R. 1982. Trials of structured exploratory data analysis. *American Journal of Human Genetics* 34, 489–500. (20)

Mostafa, M.A. 1963. The distribution of gene effects in quantitative variation. PhD Thesis, Edinburgh University. (21)

Mouches, C., Pasteur, N., Berge, J.B., Hyrien, O., Raymond, M., de Saint Vincent, B.R., de Silvestri, M. and Georghiou, G.P. 1986. Amplification of an esterase gene is responsible for insecticide resistance in a California *Culex* mosquito. *Science* 233, 778–80. (16)

Mount, S.M., Green, M.M. and Rubin, G.M. 1988. Partial revertants of the transposable element-associated suppressible allele *white apricot* in *Drosophila melanogaster*: structures and responsiveness to genetic modifiers. *Genetics* 118, 221–34. (16)

Mourao, C.A. and Ayala, F.J. 1971. Competitive fitness in experimental populations of *Drosophila willistoni*. *Genetica* 42, 65–78. (14)

Mueller, J.P. 1984. Single and two-stage selection on different indices in open nucleus breeding systems. *Génétique, Séléction, Évolution* 16, 103–19. (27)

Mueller, J.P. and James, J.W. 1984. Development in open nucleus breeding systems. *Proceedings of 2nd World Congress of Sheep and Beef Cattle Breeding*. Bloemfontein, South Africa; South African Stud Book and Livestock Improvement Association, pp.204–13. (11, 27)

Mueller, J.P. and James, J.W. 1985. Phenotypic response to selection for traits with direct and maternal components when generations overlap. *Theoretical and Applied Genetics* 70, 123–7. (24)

Mukai, T. 1964. The genetic structure of natural populations of *Drosophila melanogaster*. I. Spontaneous mutation rate of polygenes controlling viability. *Genetics* 50, 1–19. (16)

Mukai, T. 1979. Polygenic mutation. In: *Quantitative Genetic Variation* (edited by J.N. Thompson, Jr and J.M. Thoday). New York; Academic Press, pp. 177–96. (16)

Mukai, T. 1985. Experimental verification of the neutral theory. In: *Population Genetics and Molecular Evolution* (edited by T. Ohta and K. Aoki). Berlin; Springer-Verlag, pp. 125–45. (13)

Mukai, T., Cardellino, A.K., Watanabe, T.K. and Crow, J.F. 1974. The genetic variance for viability and its components in a population of *Drosophila melanogaster*. *Genetics* 78, 1195–208. (13)

Mukai, T., Chigusa, S.I., Mettler, L.E. and Crow, J.F. 1972. Mutation rate and dominance of genes affecting viability in *Drosophila melanogaster*. *Genetics* 72, 335–55. (13, 14, 16)

Mukai, T. and Cockerham, C.C. 1977. Spontaneous mutation rates at enzyme loci in *Drosophila melanogaster*. *Proceedings of the National Academy of Sciences, USA* 74, 2514–7. (16)

Mukai, T. and Yamaguchi, O. 1974. The genetic structure of natural populations of *Drosophila*. *Genetics* 82, 63–82. (8)

Müller, E. 1986. Physiological and biochemical indicators of growth and composition. In: *Exploiting New Technologies in Animal Breeding* (edited by C.Smith, J.W.B. King and J.C. McKay). Oxford, UK; Oxford University Press, pp.132–9. (33)

Muto, A. and Osawa, S. 1987. The guanine and cytosine content of genomic DNA and bacterial evolution. *Proceedings of the National Academy of Sciences, USA* 84, 166–9. (4)

Muto, A., Yamao, F., Kawauchi, Y. and Osawa, S. 1985. Codon usage in *Mycoplasma capricolum*. *Proceedings of the Japanese Academy* 61, Ser.B, 12–15. (2)

Technologies in Animal Breeding (edited by C.Smith, J.W.B. King and J.C. McKay). Oxford, UK; Oxford University Press, pp.132–9. (33)

Muto, A. and Osawa, S. 1987. The guanine and cytosine content of genomic DNA and bacterial evolution. *Proceedings of the National Academy of Sciences, USA* 84, 166–9. (4)

Muto, A., Yamao, F., Kawauchi, Y. and Osawa, S. 1985. Codon usage in *Mycoplasma capricolum*. *Proceedings of the Japanese Academy* 61, Ser.B, 12–15. (2)

Nadeau, J.H., Collins, R.L. and Klein, J. 1982. Organisation and evolution of the mammalian genome. I. Polymorphism of H-2 linked loci. *Genetics* 102, 583–98. (6)

Nagylaki, T. (1984). The evolution of multigene families under intra- chromosomal gene conversion. *Genetics* 106, 529–48. (9)

Nakamura, Y., Leppert, M., O'Connell,P., Wolff, R., Holm, T., Culver, M., Martin, C., Fujimoto, E., Hoff, M., Kumlin, E. and White, R. 1987. Variable number of tandem repeat (VNTR) markers for human gene mapping. *Science* 235, 1616–22. (34)

Narain, P. and Robertson, A. 1969. (See AR bibliography)

Neel, J.V. 1970. Lessons from a 'primitive' people. *Science* 170, 805–22. (13)

Nei, M. 1971. Fertility excess necessary for gene substitution in regulated populations. *Genetics* 68, 169–84. (4)

Nei, M. and Graur, D. 1985. Extent of protein polymorphism and the neutral mutation theory. *Evolutionary Biology* 17, 73–118. (4)

Nei, M. and Li, W.-H. 1979. Mathematical model for studying genetic variation in terms of restriction endonucleases. *Proceedings of the National Academy of Sciences* 76, 5269–73. (6)

Nei, M. and Maruyama, T. 1975. Lewontin–Krakauer test for neutral genes. *Genetics* 80, 395 (1).

Nei, M. and Murata, M. 1966. Effective population size when fertility is inherited. *Genetical Research* 8, 257–60. (1)

Neimann-Sorenson, A. and Robertson, A. 1961. (See AR bibliography)

Newgard, C.B., Nakano, K., Hwang, P.K. and Fletterick, R.J. 1986. Sequence analysis of the cDNA encoding human liver glycogen phosphorylase reveals tissue-specific codon usage. *Proceedings of the National Academy of Sciences, USA* 83, 8132–6. (4)

Nicholas, F.W. 1977. The potential effect of multiple ovulationand embryo transfer on response to selection in dairy cattle. *Proceedings of the 3rd International Congress of the Society for the Advancement of Breeding Researches in Asia and Oceania SABRAO, Animal Breeding Papers*. pp.9-1 to 9-5. (29)

Nicholas, F.W. 1979. The genetical implications of multiple ovulation and embryo transfer in small dairy herds. *Proceedings of the 30th Annual Meeting of the European Association of Animal Production, Harrogate, England*. paper CG1.11. (29)

Nicholas, F.W. 1980. Size of population required for artificial selection. *Genetical Research* 35, 85–105. (29)

Nicholas, F.W. and Robertson, A. 1976. (See AR bibliography)

Nicholas, F.W. and Robertson, A. 1980. (See AR bibliography)

Nicholas, F.W. and Smith, C. 1983. Increased rates of genetic change in dairy cattle by embryo transfer and splitting. *Animal Production* 36, 341–53. (27, 29)

Nieto, B., Salgado, C. and Toro, M.A. 1986. Optimization of artificial selection response. *Journal of Animal Breeding and Genetics* 103, 199–204. (18)

Nordskog, A.W. 1959. Note on optimum group size for progeny tests. *Biometrics* 15, 513–517. (27)

Oakeshott, J.G., Gibson, J.B., Anderson, W.R., Knibb, W.R., Anderson, D.G. and Chambers, G.K. 1982. Alcohol dehydrogenase and glycerol-3-phosphate dehydrogenase clines in *Drosophila melanogaster* on different continents. *Evolution* 36, 86–96. (14)

Ochman, H. and Wilson, A.C. 1987. Evolution in bacteria: evidence for a universal substitution rate in cellular genomes. *Journal of Molecular Evolution* 26, 74–86. (4)

Ødegård, A.K. and Robertson, A. 1967. (See AR bibliography)

O'Donald, P. 1971. The distribution of genotypes produced by alleles segregating at a number of loci. *Heredity* 26, 233–41. (20)

Ohta, T. 1980. Evolution and variation of multigene families. In: *Lecture Notes in Biomathematics* Vol. 37. Berlin, Germany; Springer-Verlag. (9)

Ohta, T. 1981. Population genetics of selfish DNA. *Nature* 292, 648–9. (8)

Ohta, T. 1983a. Theoretical study on the accumulation of selfish DNA. *Genetical Research* 41, 1–16. (8)

Ohta, T. 1983b. On the evolution of multigene families. *Theoretical Population Biology* 23, 216–40. (9)

Ohta, T. 1985. A model of duplicative transposition and gene conversion for repetitive DNA families. *Genetics* 110, 513–24. (9)

Ohta, T. 1986. Population genetics theory of multigene families with emphasis on genetic variation contained in the family. In: *Evolutionary Processes and Theory* (edited by S. Karlin and E. Nevo). London, UK; Academic Press, pp.239–53 (9)

Ohta, T. 1987. Simulating evolution by gene duplication. *Genetics* 115, 207–13. (9)

Ohta, T. 1988a. Multigene and supergene families. In: *Oxford Surveys of Evolutionary Biology* 5 (edited by P. Harvey and L. Partridge). Oxford, UK; Oxford University Press. pp.41–65. (6, 9)

Ohta, T. 1988b. Further simulation studies on evolution by gene duplication. *Evolution* 42, 375–86. (9)

Ohta, T. and Kimura, M. 1969a. Linkage disequilibrium due to random genetic drift. *Genetical Research* 13, 47–55. (1)

Ohta, T. and Kimura, M. 1969b. Linkage disequilibrium at steady state determined by random genetic drift and recurrent mutation. *Genetics* 63, 229–38. (6)

Ohta, T. and Kimura, M. 1970a. Genetic loads at a polymorphic locus which is maintained by frequency-dependent selection. *Genetical Research* 16, 145–50. (1)

Ohta, T. and Kimura, M. 1970b. Development of associative overdominance through linkage disequilibrium in finite populations. *Genetical Research* 16, 165–77. (6)

Ohta, T. and Kimura, M. 1971a. Functional organization of genetic material as a product of molecular evolution. *Nature* 233, 118–9. (3)

Ohta, T. and Kimura, M. 1971b. On the constancy of the evolutionary rate of cistrons. *Journal of Molecular Evolution* 1, 18–25. (3)

Ohta, T. and Kimura, M. 1981. Some calculations on the amount of selfish DNA. *Proceedings of the National Academy of Sciences, USA* 78, 1129–32. (8)

Oliviero, S., Morrone, G. and Cortese, R. 1987. The human haptoglobin gene: transcriptional regulation during development and acute phase induction. *The EMBO Journal* 6, 1905–12. (36)

Ollausson, A. and Ronningen, K. 1975. Estimation of genetic parameters for threshold traits. *Acta agriculturae scandinavica* 25, 201–8. (25)

Ollivier, L. 1974. Optimum replacement rates in animal breeding. *Animal Production* 19, 257–71. (27)

Orgel, L.E. and Crick, F.H.C. 1980. Selfish DNA: the ultimate parasite. *Nature* 284, 604–7. (8)

Ornitz, D.M., Palmiter, R.D., Hammer, R.E., Brinster, R.L., Swift, G.H. and MacDonald, R.J. 1985. Specific expression of an elastase-human growth hormone fusion gene in pancreatic acinar cells of transgenic mice. *Nature* 313, 600–3. (36)

Orr-Weaver, T.L. and Szostak, J.W. 1983. Yeast recombination: the association between double-strand gap repair and crossing-over. *Proceedings of the National Academy of Sciences, USA* 80, 4417–21. (36)

Osman, H.E.S. and Robertson, A. 1968. (See AR bibliography)

Outteridge, P.M., Windon, R.G. and Dineen, J.K. 1985. An association between a lymphocyte antigen in sheep and the response to vaccination against the parasite *Trichostrongylus colubriformis*. *International Journal for Parasitology* 15, 121–7. (21)

Ouweltjes, W., Schaeffer, L.R. and Kennedy, B.W. 1988. Sensitivity of methods of variance component estimation to culling type of selection. *Journal of Dairy Science* 71, 773–779. (23)

Overbeek, P.A., Chepelinsky, A.B., Khillan, J.S., Piatigorsky, J. and Westphal, H. 1985. Lens-specific expression and developmental regulation of the bacterial chloramphenicol acetyltransferase gene driven by the murine alphaA-crystallin promoter in transgenic mice. *Proceedings of the National Academy of Sciences, USA* 82, 7815–7819. (36)

Overbeek, P.A., Lai, S.-P., Van Quill, K.R. and Westphal, H. 1986. Tissue-specific expression in transgenic mice of a fused gene containing RSV terminal sequences. *Science* 231, 1574–6. (36)

Ozil, J.P. 1983. Production of identical twins by bisection of blastocysts in the cow. *Journal of Reproduction andFertility* 69, 463–8. (35)

Palmer, J.D. 1985. Evolution of chloroplast and mitochondrial DNA in plants and algae. In: *Molecular Evolutionary Genetics* (edited by R.J. MacIntyre). New York; Plenum Press, pp.131–240 (6)

Palmiter, R.D., Behringer, R.R., Quaife, C.J., Maxwell, F., Maxwell, I.H. and Brinster, R.L. 1987. Cell lineage ablation in transgenic mice by cell-specific expression of a toxin gene. *Cell* 50, 435–43. (36)

Palmiter, R.D. and Brinster, R.L. 1986. Germline transformation of mice. *Annual Review of Genetics* 20, 465–99. (32, 34, 35, 36)

Palmiter, R.D., Brinster, RL., Hammer, R.E. Trumbauer, M.E., Rosenfeld, M.G. Birnberg, N.C. and Evans, R.M. 1982. Dramatic growth of mice that develop from eggs microinjected with metallothionein-growth hormone fusion genes. *Nature* 300, 611–5. (30, 32, 35)

Palmiter, R.D., Chen, H.Y. and Brinster,R.L. 1982. Differential regulation of metallothionein-thymidine kinase fusion genes in transgenic mice and their offspring. *Cell* 29, 701–10. (36)

Patterson, H.D. and Thompson, R. 1971. Recovery of inter-block information when block sizes are unequal. *Biometrika* 58, 545–54. (23)

Paxman, G.J. 1957. A study of spontaneous mutation in *Drosophila melanogster*. *Genetics* 29, 39–57. (16)

Pearson, K. 1902. On the mathematical theory of errors of judgement with special reference to the personal equation. *Philosophical Magazine* 2, 559–72. (20)

Pedersen, S. 1984. *Escherichia coli* ribosomes translate *in vivo* with variable rate. *European Molecular Biology Organization Journal* 3, 2895–8. (4)

Perry, M.M. (1988). A complete culture system for the chicken embryo. *Nature* 331, 70–2. (32)

Perutz, M.F. and Lehman, H. 1968. Molecular pathway of human haemoglobin. *Nature* 219, 902–9. (3)

Pinkert, C.A., Ornitz, D.M., Brinster, R.L. and Palmiter, R.D. 1987. An albumin enhancer located 10 kb upstream functions along with its promoter to direct efficient, liver-specific expression in transgenic mice. *Genes and Development* 1, 268–76. (36)

Pinkert, C.A., Widera, G., Cowing, C., Heber-Katz, E., Palmiter, R.D., Flavell, R.A. and Brinster, R.L. 1985. Tissue-specific, inducible and functional expression of the E_{alpha}^{d} MHC class II gene in transgenic mice. *The EMBO Journal* 4, 2225–30. (36)

Piper, L.R. 1972. The isolation of genes underlying continuous variation. PhD Thesis, University of Edinburgh. (12, 21, 34)

Piper, L.R. and Bindon, B.M. 1982. Genetic segregation for fecundity in Booroola Merino sheep. In: *Proceedings of the World Congress on Sheep and Beef Cattle Breeding* Vol. 1. (edited by R.A. Barton and W.C. Smith). Palmerston North, New Zealand; Dunmore Press, pp.315–31 (12, 20, 21).

Piper, L.R. and Bindon, B.M. 1985. The single gene inheritance of the high litter size of the Booroola Merino. In: *Genetics of Reproduction in Sheep* (edited by R.B. Land and D.W. Robinson). London, UK; Butterworths, p.115–25. (34)

Piper, L.R., Bindon, B.M., Davis, G.H. and Elsen, J.M. 1988. Control of litter size: Major genes and industry utilisation of the Boorool Merino. In: *Proceedings of the 3rd World Congress on Sheep and Beef Cattle Breeding*. Paris, France; Institute Nationale de Researche Agronomique, Vol. 2. pp.589–609. (21)

Polge, C. 1985. How does embryo manipulation fit into present and future pig production? *Journal of Reproduction and Fertility* 33, 93–100. (35)

Popescu, C.P., Cotinot, C., Boscher, J. and Kirszenbaum, M. 1988. Chromosomal localization of a bovine male specific probe. *Annales de Génétique* 31, 39–42. (35)

Poutous, M. and Vissac, B. 1962. Recherche theorique des conditions de rentabilite maximum de l'epreuve de descendance des taureaux d'insemination artificielle. *Annales de Zootechnie* 11, 233–56. (27)

Preston, C.R. and Engels, W.R. 1984. Movement of P elements within a P strain. *Drosophila Information Service* 60, 169–170. (16)

Priestley, G.C. and Robertson, M.S.M. 1973. Protein and nucleic acid metabolism in organs from mice selected for larger and smaller body size. *Genetical Research* 22, 255–78. (32)

Prout, T. 1971. The relation between fitness components and population prediction in *Drosophila*. I. The estimation of fitness components. *Genetics* 68, 127–49. (14)

Ptashne, M. 1986. Gene regulation by proteins acting nearby and at a distance. *Nature* 322, 697–701. (36)

Pumo, D.E., Goldin, E.Z., Elliot, B., Phillips, C.J. and Genoways, H. 1988. Mitochondrial DNA polymorphism in three Antillean island populations of the fruit bat, *Artibeus jamaicensis*. *Molecular Biology and Evolution* 5, 79–89. (5)

Quaas, R.L. 1976. Computing the diagonal elements and inverse of a large numerator relationship matrix. *Biometrics* 32, 949–53. (23, 26)

Quaas, R.L. and Pollak, E.J. 1980. Mixed model methodology for farm and ranch beef cattle testing programs. *Journal of Animal Science* 51, 1277–87. (26)

Rabesa, S.L. 1950. Genetical reduction of a reproductive unit in relation to the male-female ratio. *Nature* 116, 821–2. (22)

Radcliffe, J.D. and Webster, A.J.F. 1976. Regulation of food intake during growth in fatty and lean Zucker rats given diets of different protein content. *British Journal of Nutrition* 36, 457–69. (19)

Rao, C.R. 1972. Estimation of variance and covariance components in linear models. *Journal of the American Statistical Association* 67, 112–5. (23)

Rao, D.C., Laskarzewski, P.M., Morrison, J.A., Khoury, P., Kelly, K., Wette, R., Russell, J. and Glueck, C.J. 1982. The Cincinnati Lipid Research Clinic Family Study: cultural and biological determinants of lipids and lipoprotein concentrations. *American Journal of Human Genetics* 34, 888–903. (20)

Rathie, K.A. and Nicholas, F.W. 1980. Artificial selection with differing population structures. *Genetical Research* 36, 117–31. (11, 17, 18)

Rawlings, J.O. 1976. Order statistics for a special class of unequally correlated multinormal variates. *Biometrics* 32, 875–87. (26)

Reik, W., Collick, A., Norris, M.L., Barton, S.C. and Surani, M.A. 1987. Genomic imprinting determines methylation of parental alleles in transgenic mice. *Nature* 328, 248–51. (36)

Rendel, J.M. 1959. Optimum group size in half-sib family selection. *Biometrics* 15, 376–81. (27)

Rendel, J.M. and Robertson, A. 1950a. (See AR bibliography)

Rendel, J.M. and Robertson, A. 1950b. (See AR bibliography)

Rendel, J.M. and Robertson, A. 1950c. (See AR bibliography)

Rendel, J.M. and Robertson, A. 1951. (See AR bibligraphy)

Rich, S.S., Green, A., Morton, N.E. and Barbosa, J. 1987. A combined segregation and linkage analysis of insulin-dependent diabetes mellitus. *American Journal of Human Genetics* 40, 237–49. (20)

Riley, M. 1988. Nucleotide sequence of the Xdh region in *D. pseudoobscura* and an analysis of the evolution of synonymous codons. *Molecular Biology and Evolution* (in press). (5)

Ritossa, F. 1976. The *bobbed* locus. In: *The genetics and biology of Drosophila* Volume 1b. (edited by M. Ashburner and E. Novitski). New York; Academic Press, pp.801–46. (16)

Roberts, R.C. 1966. The limits to artificial selection for body weight in the mouse. II. The genetic nature of the limits. *Genetical Research* 8, 361–75. (16, 21)

Roberts, R.C. 1967. The limits to artificial selection for body weight in the mouse. IV. Sources of new genetic variance-irradiation and outcrossing. *Genetical Research* 9, 87–98. (16)

Roberts, R.C. 1979. Side-effects of selection for growth in laboratory animals. *Livestock Production Science* 6, 93–104. (19)

Roberts, R.C. 1981. The growth of mice selected for large and small size in relation to food intake and the efficiency of conversion. *Genetical Research* 38, 9–24. (19)

Roberts, R.C. and Smith, C. 1982. Genes with large effects. Theoretical aspects in livestock breeding. *2nd World Congress on Genetics Applied to Livestock Production*. Madrid; Editorial Garsi, Vol. 6, 420–38. (16, 34)

Robertson, D.E. 1969. Selection in small populations. PhD Thesis, University of New South Wales. (12)

Robertson, F.W. 1954. Studies in quantitative inheritance. V. Chromosome analysis of crosses between selected and unselected lines of different body size in *Drosophila melanogaster*. *Journal of Genetics* 52, 494–520. (11)

Robertson, F.W. 1955. Selection response and the properties of genetic variation. *Cold Spring Harbor Symposia on Quantitative Biology* 20, 166–177. (15, 17)

Robertson, F.W. and Reeve, E. 1952. Studies in quantitative inheritance. I. The effects of selection of wing and thorax length in *Drosophila melanogaster*. *Journal of Genetics* 50, 414–48. (17)

Robinson, J.A.B. 1988. Multiple lactation variance component estimation using Restricted Maximum Likelihood. PhD thesis, Cornell University, Ithaca, NY. (23)

Roeder, G.S. 1983. Unequal crossing over between yeast transposable elements. *Molecular and General Genetics* 190, 117–21. (8)

Rogers, A.R. 1986. Population differences in quantitative characters as opposed to gene frequencies. *American Naturalist* 127, 729–30. (11)

Rogers, J. 1985. Origins of repeated DNA. *Nature* 317, 765–6. (8)

Rollins, W.C., Tanaka, M., Nott, C.F.G. and Theissen, R.B. 1972. On the mode of inheritance of double-muscled conformation in bovines. *Hilgardia* 41, 433–56. (21)

Rosado, J.M.C. and Robertson, A. 1966. (See AR bibliography)

Rose, M.R. 1982. Antagonistic pleiotropy, dominance and genetic variation *Heredity* 48, 63–78. (13)

Rose, M. and Charlesworth, B. 1980. A test of evolutionary theories of senescence. *Nature* 287, 141–2. (10)

Rothschild, M.F. 1985. Selection for disease resistance in the pig. *Pig News Information* 6, 277–80. (28)

Ruane, J. 1988. Review of the use of embryo transfer in the genetic improvement of dairy cattle. *Animal Breeding Abstracts* 56, 437–46. (29)

Ruano, R.G., Orozco, F. and López-Fanjul, C. 1975. The effect of different selection intensities on selection response in egg-laying of *Tribolium castaneum*. *Genetical Research* 25, 17–27. (18)

Rubin, G.M. 1983. Dispersed repetitive DNA's in Drosophila. In: *Mobile Genetic Elements* (edited by J.A. Shapiro). New York; Academic Press, pp.329–61. (16)Ruether, U., Tripodi, M., Cortese, R. and Wagner, E.F. 1987. The human alpha-1-antitrypsin gene is efficiently expressed from two tissue-specific promoters in transgenic mice. *Nucleic Acids Research* 15, 7519–29. (36)

Russell, W.A., Sprague, G.F. and Penny, L.H. 1963. Mutations affecting quantitative characters in long-time inbred lines of maize. *Crop Science* 3, 175–8. (15)

Rutledge, A.R. 1970. The survival of epistatic gene complexes in subdivided populations. PhD Thesis, Columbia University, New York. (11)

Saghai-Maroof, M.A., Soliman, K.M., Jorgensen, R.A. and Allard, R.W. 1984. Ribosomal DNA spacer-length polymorphism in barley: Mendelian inheritance, chromosomal location and population dynamics. *Proceedings of the National Academy of Sciences, USA* 81, 8014–8. (5)

Saitou, N. and Nei, M. 1986. Polymorphism and evolution of influenza A virus genes. *Molecular Biology and Evolution* 3, 57–74. (2)

Sales, J. and Hill, W.G. 1976a. Effect of sampling errors on efficiency of selection indices. 1. Use of information from relatives for single trait improvement. *Animal Production* 22, 1–17. (24)

Sales, J. and Hill, W.G. 1976b. Effect of sampling errors on efficiency of selection indices. 2. Use of information on associated traits for improvement of a single important trait. *Animal Production* 23, 1–14. (24, 33)

Salter, M., Knowles, R.G. and Pogson, C.I. 1986. Quantitation of the importance of individual steps in the control of aromatic amino acid metabolism. *Biochemical Journal* 234, 635–47. (31)

Sapienza, C., Peterson, A.C., Rossant, J. and Balling, R. 1987. Degree of methylation of transgenes is dependent on gamete of origin. *Nature* 328, 251–4. (36)

Sarich, V.M. and Wilson, A.C. 1966. Immunological time scale for homonid evolution. *Science* 158, 1200–3. (3)

Sarvetnick, N., Liggitt, D., Pitts, S.L., Hansen, S.E. and Stewart, T.A. 1988. Insulin-dependent diabetes induced in transgenic mice by ectopic expression of class II MHC and interferon gamma. *Cell* 52, 773–82. (36)

Sauro, H.M., Small, R. and Fell, D.A. 1987. Metabolic control and its analysis: Extensions to the theory and matrix method. *European Journal of Biochemistry* (in press). (31)

Scangos, G. and Bieberich, C. 1987. Gene transfer into mice. *Advances in Genetics* 24, 285–322. (36)

Schaeffer, L.R. 1979. Estimation of variance and covariance components for average daily gain and backfat thickness in swine. In: *Proceedings of the Conference in Honor of C.R. Henderson on Variance Components in Animal Breeding* pp.123–37. (23)

Schaeffer, L.R. 1986a. Estimation of variances and covariances within the allowable parameter space. *Journal of Dairy Science* 69, 187–94. (23)

Schaeffer, L.R. 1986b. Pseudo expectation approach to variance component estimation. *Journal of Dairy Science* 69, 2884–9. (23)

Schaeffer, L.R., Wilton, J.W. and Thompson, R. 1978. Simultaneous estimation of variance and covariance components from multitrait mixed model equations. *Biometrics* 34, 199–208. (23)

Schaeffer, S.W. and Aquadro, C.F. 1987. Nucleotide sequences of the alcohol dehydrogenase region of *D. pseudoobscura*: evolutionary changes and evidence for an ancient duplication. *Genetics* 117, 61–73. (5)

Schaeffer, S.W., Aquadro, C.F. and Anderson, W.W. 1987. Restriction map variation in the alcohol dehydrogenase region of *D. pseudoobscura*. *Molecular Biology and Evolution* 4, 254–65. (5)

Schaeffer, S.W., Aquadro, C.F. and Langley, C.H. 1988. Restriction-map variation in the *notch* region of *Drosophila melanogaster*. *Molecular Biology and Evolution* 5, 30–40. (5, 6)

Schmid, C.W. and Shen, C.-K.J. 1985. The evolution of interspersed repetitive DNA sequences in mammals and other vertebrates. In: *Molecular Evolutionary Genetics* (edited by R.J. MacIntyre). New York; Plenum Press, pp.323–58. (8)

Schnee, F.B. and Thompson, J.N.Jr 1984. Conditional polygenic effects in the sternopleural bristle system of *Drosophila melanogaster*. *Genetics* 108, 409–24. (21)

Scossiroli, R.E. and Scossiroli, S. 1959. On the relative role of mutation and recombination in responses to selection for polygenic traits in irradiated populations of *D.melanogaster*. *International Journal of Radiation Biology* 1, 61–9. (16)

Searle, S.R. 1971. *Linear Models*. New York; Wiley. (23)

Seidel, G. 1984. Applications of embryo transfer and related technologies to cattle. *Journal of Dairy Science* 67, 2786–96. (35)

Sejrsen, K., Larsen, F. and Anderson, B.B. 1984. Use of plasma hormone and metabolite levels to predict breeding value of young bulls for butter fat production. *Animal Production* 39, 335–44. (33)

Sejrsen, K. and Lovendahl, P. 1986. Criteria identifying genetic merit for milk production. In: *Exploiting New Technologies in Animal Breeding* (edited by C. Smith, J.W.B. King and J.C. McKay). Oxford, UK; Oxford University Press, pp.142–52. (33)

Selander, R.K. and Whittam, T.S. 1983. Protein polymorphism and the genetic structure of populations. In: *Evolution of Genes and Proteins* (edited by M. Nei and R.K. Koehn). Sunderland, Mass.; Sinauer, pp.89-114. (11)

Sellier, P. and Legault, C. 1986. The Chinese prolific breeds of pigs: examples of extreme genetic stocks. In *Exploiting New Technologies in Animal Breeding* (edited by C. Smith, J.W.B. King and J.C.McKay). Oxford, UK; Oxford University Press, pp.153–6. (33)

Sellwood, R., Gibbons, R.A., Jones, G.W. and Rutter, J.M. 1975. Adhesions of enteropathogenic *Escherichia coli* to pig intestinal brush borders; the existence of two phenotypes. *Journal of Medical Microbiology* 8, 405–11. (34)

Sen, B.K. and Robertson, A. 1964. (See AR bibliography)

Serfling, E., Jasin, M. and Schaffner, W. 1985. Enhancers and eukaryotic gene transcription. *Trends in Genetics* 1, 224–30. (36)

Shah, D.M. and Langley, C.H. 1979. Inter- and intraspecific variation in restriction maps of *Drosophila* mitochondrial DNAs. *Nature* 281, 696–9. (3, 5)

Shapiro, J.A. 1983. *Mobile Genetic Elements*. New York; Academic Press. (8)

Shapiro, J.A. and Cordell, B. 1982. Eukaryotic mobile and repeated genetic elements. *Biology of the Cell* 43, 31–54. (16)

Sharp, G.L., Hill, W.G. and Robertson, A. 1984. (See AR bibliography)

Sharp, P.M., Cowe, E., Higgins, D.G., Shields, D.C., Wolfe, K.H. and Wright, F. 1988. Codon usage patterns in *Escherichia coli*, *Bacillus subtilis*, *Saccharomyces cerevisiae*, *Schizosaccharomyces pombe*, *Drosophila melanogaster* and *Homo sapiens*; a review of the considerable within-species diversity. *Nucleic Acids Research* 16, 8207–11. (4)

Sharp, P.M. and Li, W-H. 1986a. Codon usage in regulatory genes in *Escherichia coli* does not reflect selection for 'rare' codons. *Nucleic Acids Research* 14, 7737–49. (4)

Sharp, P.M. and Li, W-H. 1986b. An evolutionary perspective on synonymous codon usage in unicellular organisms. *Journal of Molecular Evolution* 24, 28–38. (4)

Sharp, P.M. and Li, W-H. 1987a. The codon adaptation index – a measure of directional synonymous codon usage bias, and its potential applications. *Nucleic Acids Research* 15, 1281–95. (4)

Sharp, P.M. and Li, W-H. 1987b. The rate of synonymous substitution in enterobacterial genes is inversely related to codon usage bias. *Molecular Biology and Evolution* 4, 222–30. (4)

Sharp, P.M., Tuohy, T.M.F. and Mosurski, K.R. 1986. Codon usage in yeast: cluster analysis clearly differentiates highly and lowly expressed genes. *Nucleic Acids Research* 14, 5125–43. (4)

Sharp, P.M. and Wright, F. 1989. Codon selection in *Schizosaccharomyces pombe* (in preparation). (4)

Shaw, R.G. 1987. Maximum-likelihood approaches to quantitative genetics of natural populations. *Evolution* 41, 812–26. (23)

Sheldon, B.L. and Evans, M.K. 1981. Studies on the scutellar bristles of *Drosophila melanogaster*. III. Long-term selection for high bristle number in three further lines derived from the Oregon-RC strain, correlated responses in abdominal bristles, and changes in regulation of the scute locus. *Australian Journal of Biological Sciences* 34, 247–367. (16)

Sheldon, B.L. and Milton, M.K. 1972. Studies on the scutellar bristles of *Drosophila melanogaster*. II. Long-term selection for high bristle number in the Oregon RC strain and correlated responses in abdominal chaetae. *Genetics* 71, 567–95. (16)

Shields, D.C. and Sharp, P.M. 1987. Synonymous codon usage in *Bacillus subtilis* reflects both translational selection and mutational biases. *Nucleic Acids Research* 15, 8023–40. (4)

Shields, D.C., Sharp, P.M., Higgins, D.G. and Wright, F. 1988. Silent sites in Drosophila genes are not neutral: evidence of selection among synonymous codons. *Molecular Biology and Evolution* 5, 704–16. (4)

Shih, C.-C., Stoye, J.P. and Coffin, J.M. 1988. Highly preferred targets for retroviral integration. *Cell* 53, 531–7. (36)

Shrimpton, A.E. 1982. The isolation of polygenic factors controlling bristle score in *Drosophila melanogaster*. PhD Thesis, University of Edinburgh. (21, 34)

Shrimpton, A.E. and Robertson, A. 1988a. (See AR bibliography)

Shrimpton, A.E. and Robertson, A. 1988b. (See AR bibliography)

Sifers, R.N. Carlson, J.A., Clift, S.M., DeMayo, F.J., Bullock, D.W. and Woo, S.L.C. 1987. Tissue specific expression of the human alpha-1-antitrypsin gene in transgenic mice. *Nucleic Acids Research* 15, 1459–75. (36)

Silvey, S.D. 1980. *Optimal designs*. London, UK; Chapman and Hall. (24)

Simmons, G.M., Kreitman, M., Quattlebaum, W. and Miyashita, N. 1988. Molecular analysis of the alleles of alcohol dehydrogenase along a cline of *Drosophila melanogaster*. I. Maine, North Carolina and Florida. *Genetics* (in press). (5)

Simmons, M.J. and Crow, J.F. 1977. Mutations affecting fitness in Drosophila populations. *Annual Review of Genetics* 11, 49–78. (14, 16)

Simmons, M.J., Preston, C.R. and Engels, W.R. 1980. Pleiotropic effects on fitness of mutations affecting viability in *Drosophila melanogaster*. *Genetics* 94, 467–75. (13)

Simmons, M.J., Sheldon, E.W. and Crow, J.F. 1978. Heterozygous effects on fitness of EMS-treated chromosomes in *Drosophila melanogaster*. *Genetics* 88, 576–90. (13, 16)

Simons, J.P., McClenaghan, M. and Clark, A.J. 1987. Alteration of the quality of milk by expression of sheep beta-lactoglobulin in transgenic mice. *Nature* 328, 530–2. (30, 36)

Simons, J.P., Wilmut, I., Clark, A.J., Archibald, A.L., Bishop, J.O. and Lathe, R. 1988. Gene transfer into sheep. *Bio/Technology* 6, 179–83. (32, 34, 35, 36)

Simpson, E., Bulfield, G., Brenan, M., Fitzpatrick, W., Hetherington, C. and Blann, A. (1982). *H*-2-associated differences in replicated strains of mice divergently selected for body weight. *Immunogenetics* 15, 63–70. (32)

Simpson, S.P. and Webb, A.J. 1988. Growth and carcass performance of British Landrace pigs heterozygous at the halothane locus. *Animal Production* (in press). (28)

Sing, C.F., Boerwinkle, E., Moll, P.P. and Templeton, P.R. 1988. Finding genes affecting quantitative traits in humans. *Proceedings of the 2nd International Conference on Quantitative Genetics* (edited by B.S. Weir, E.J. Eisen, M.M. Goodman and G.Namkoong). Sunderland, Mass.; Sinauer, pp.250–69. (34)

Sing, C.F., Chamberlain, M.A., Block, W.D. and Feiler, S. 1975. Analysis of genetic and environmental sources of variation in serum cholesterol in Tecumseh, Michigan. I. Analysis of the frequency distribution for evidence of a genetic polymorphism. *American Journal of Human Genetics* 27, 333–47. (20)

Singh, M. and Lewontin, R.C. 1966. Stable equilibria under optimizing selection. *Proceedings of the National Academy of Sciences, USA* 56, 1345–8. (1)

Singh, R.S. and Rhomberg, L.R. 1987. A comprehensive study of genic variation in natural populations of *Drosophila melanogaster*. I. Estimates of gene flow from rare alleles. *Genetics* 115, 313–22. (11)

Skjervold, H. and Langholz, H.J. 1964. Factors affecting the optimum structure of A.I. breeding in dairy cattle. *Zeitschrift für Tierzüchtung und Züchtungsbiologie* 80, 25–40. (29)

Slatkin, M. 1985a. Gene flow in natural populations. *Annual Review of Ecology and Systematics* 16, 393–430. (11)

Slatkin, M. 1985b. Rare alleles as indicators of gene flow. *Evolution* 39, 53–65. (11)

Slatkin, M. 1987a. Heritable variation and heterozygosity under a balance between mutations and stabilizing selection. *Genetical Research* 50, 53–62. (15)

Slatkin, M. 1987b. Gene flow and the geographic structure of natural populations. *Science* 236, 787–92. (11)

Smith, C.A.B. 1975. A non-parametric test for linkage with a quantitative character. *Annals of Human Genetics* 38, 451–60. (20)

Smith, C. 1960. Efficiency of animal testing schemes. *Biometrics* 16, 408–15. (27)

Smith, C. 1964. The use of specialised sire and dam lines in selection for meat production. *Animal Production* 6, 337–4. (28)

Smith, C. 1967. Improvement of metric traits through specific genetic loci. *Animal Production* 9, 349–58. (34)

Smith, C. 1969. Optimum selection procedures in animal breeding. *Animal Production* 11, 433–42. (12, 27)

Smith, C. 1982. Estimates of trends in the halothane gene in pig stocks with selection. *Zeitschrift für Tierzüchtung und Züchtungsbiologie* 99, 232–40. (34)

Smith, C. 1984. Rates of genetic change in farm livestock. *Research and Development in Agriculture* 1, 79–85. (29)

Smith, C. 1985. Scope for selecting many breeding stocks of possible economic value in the future. *Animal Production* 41, 403–12. (27)

Smith, C. 1986. Use of embryo transfer in genetic improvement of sheep. *Animal Production* 42, 81–8. (29)

Smith, C. 1988a. Applications of embryo transfer in animal breeding. *Theriogenology* 29, 203–12. (29)

Smith, C. 1988b. Genetic improvement of livestock in developing countries using nucleus breeding units. *World Animal Review* 65, 2–10. (29)

Smith, C. and Bampton, P.R. 1977. Inheritance of reaction to halothane anaesthesia in pigs. *Genetical Research* 29, 287–92. (21)

Smith, C., Meuwissen, T.H.E. and Gibson, J.P. 1987. On the use of transgenes in livestock improvement. *Animal Breeding Abstracts* 55, 1–10. (30, 34, 35)

Smith, C. and Simpson, S.P. 1986. The use of genetic polymorphism in livestock improvement. *Zeitschrift für Tierzüchtung und Züchtungsbiologie* 103, 205–17. (34)

Smith, C. and Webb, A.J. 1981. Effects of major genes on animal breeding strategies. *Zeitschrift für Tierzüchtung und Züchtungsbiologie* 98, 161–9. (34)

Smith, S.P. and Graser, H.-U. 1986. Estimating variance components in a class of mixed models by Restricted Maximum Likelihood. *Journal of Dairy Science* 69, 1156–65. (23)

Soller, M. 1978. The use of loci associated with quantitative traits in dairy cattle improvement. *Animal Production* 27, 133–9. (34)

Soller, M. and Beckmann, J.S. 1982. Restriction fragment length polymorphisms and genetic improvement. *Proceedings of the 2nd World Congress on Genetics Applied to Livestock Production*. Madrid; Editorial Garsi, Vol. 6, 396–404. (34)

Sorensen, D.A. and Kennedy, B.W. 1986. Analysis of selection experiments using mixed model methodology. *Journal of Animal Science* 68, 245–58. (23)

Soriano, P., Cone, R.D., Mulligan, R.C. and Jaenisch, R. 1986. Tissue specific and ectopic expression of genes introduced into transgenic mice by retroviruses. *Science* 234, 1409–13. (36)

Sorribas, A. and Bartrons, R. 1986. Theoretical analysis of the flux control properties of a substrate cycle. *European Journal of Biochemistry* 158, 107–15. (31)

Specht, L.W. and McGilliard, L.D. 1960. Rates of improvement by progeny testing in dairy herds of various sizes. *Journal of Dairy Science* 43, 63–75. (29)

Sperlich, D. and Karlik, A. 1970. The genetic conditions in heterozygous and homozygous populations of Drosophila. I. The fate of alien chromosomes. *Genetica* 41, 265–304. (14)

Spickett, S.G. 1963. Genetic and developmental studies of a quantitative character. *Nature* 199, 870–93. (21)

Spickett, S.G. and Thoday, J.M. 1966. Regular responses to selection. 3. Interaction between located polygenes. *Genetical Research* 7, 96–121. (17, 21)

Sprague, G.F., Russell, W.A. and Penny, L.H. 1960. Mutations affecting quantitative traits in selfed progeny of doubled monoploid maize stocks. *Genetics* 45, 855–66. (15)

Spuhler, J.H. 1976. The maximum opportunity for natural selection in some human populations. In: *Demographic Anthropology* (edited by E.B.W. Zubrow). Albuquerque, New Mexico; University of New Mexico Press. (13)

Stam, P. 1986. The use of marker loci in selection for quantitative characters. In: *Exploiting New Technologies in Animal Breeding. Genetic Developments* (edited by C. Smith, J.W.B. King and J.C. McKay). Oxford, UK; Oxford University Press. (34)

Steinmetz, M. 1986. Polymorphism and recombinational hot spots in the murine MHC. *Current Topics in Microbiology and Immunology* 127, 279–84. (6)

Steinmetz, M., Malissen, M., Hood, L., Orn, A., Maki, R.A., Dastoornikoo, G.R., Stephan, D., Gibb, E. and Romanink, R. 1984. Tracts of high or low sequence divergence in the mouse major histocompatibility complex. *European Molecular Biology Organization Journal* 3, 2995–3003. (6)

Steinmetz, M., Minard, K., Horvath, S., McNicholas, J., Srelinger, J., Wake, C., Long, E., Mach, B. and Hood, L. 1982. A molecular map of the immune response region from the major histocompatibility complex of the mouse. *Nature* 300, 35–42. (6)

Stewart, C.L., Schuetze, S., Vanek, M. and Wagner, E.F. 1987. Expression of retroviral vectors in transgenic mice obtained by embryo infection. *European Molecular Biology Organization Journal* 6, 383–8. (36)

Stewart, T.A., Hollingshead, P.G. and Pitts, S.L. 1988. Multiple regulatory domains in the mouse mammary tumor virus long terminal repeat revealed by analysis of fusion genes in transgenic mice. *Molecular and Cellular Biology* 8, 473–9. (36)

Storb, U. 1987. Transgenic mice with immunoglobulin genes. *Annual Reviews of Immunology* 5, 151–74. (36)

Strobeck, C. 1983. Expected linkage disequilibrium for a neutral locus linked to a chromosomal arrangement. *Genetics* 103, 545–55. (5)

Sutherland, R.A., Webb, A.J. and King, J.W.B. 1984. Evaluation of overseas pig breeds using imported semen. 1. Growth and carcass performance. *Journal of Agricultural Science* Cambridge 103, 562–70. (28)

Sutherland, R.A., Webb, A.J. and King, J.W.B. 1985. A survey of world pig breeds and comparisons. *Animal Breeding Abstracts* 53, 1–22. (28)

Sved, J.A. 1968. The stability of linked systems of loci with a small population size. *Genetics* 59, 543–63. (6)

Sved, J.A. 1971. Linkage disequilibrium and homozygosity of chromosome segments in finite populations. *Theoretical Population Biology* 2, 125–41 (1)

Sved, J.A. 1975. Fitness of third chromosome homozygotes in *Drosophila melanogaster*. *Genetical Research* 25,197–200. (13)

Sved, J.A. 1983. Does natural selection increase or decrease variability at linked loci? *Genetics* 105, 239–40. (6)

Sved, J.A. and Ayala, F.J. 1970. A population cage for heterosis in *Drosophila pseudoobscura*. *Genetics* 66, 97–113.(13, 14)

Sved, J.A. and Feldman, M.W. 1973. Correlation and probability methods for one and two loci. *Theoretical Population Biology* 4, 129–32. (1)

Swain, J.L., Stewart, T.A. and Leder, P. 1987. Parental legacy determines methylation and expression of anautosomal transgene: a molecular mechanism for parental imprinting. *Cell* 50, 719–27. (36)

Swanson, L.W., Simmons, D.M., Arriza, J., Hammer, R., Brinster, R.L., Rosenfeld, G.M. and Evans, R.M. 1985. Novel developmental specificity in the nervous system of transgenic animals expressing growth hormone fusion genes. *Nature* 317, 363–6. (36)

Swift, G.H., Hammer, R.E. MacDonald, R.J. and Brinster, R.L. 1984. Tissue-specific expression of the rat pancreatic elastase I gene in transgenic mice. *Cell* 38, 639–46. (36)

Syvanen, M. 1984. The evolutionary implication of transposable elements. *Annual Review of Genetics* 18, 271–93. (8)

Tallis, G.M. 1959. Sampling errors of genetic correlation coefficients calculated from analyses of variance and covariance. *Australian Journal of Statistics* 1, 35–43. (24)

Tanksley, S.D., Medino-Filho, H. and Rick, C.M. 1982. Use of naturally occurring enzyme variation to detect and map genes controlling quantitative traits in an interspecific backcross of tomato. *Heredity* 49, 11–25. (34)

Taylor, B.A. 1978. Recombinant inbred strains. In: *Origins of inbred mice* (edited by H.C. Morse III). New York; Academic Press, pp.423–8. (32)

Taylor, C.E. and Powell, J.R. 1983. Population structure of *Drosophila*: Genetics and ecology. In: *The Genetics and Biology of Drosophila* (edited by M. Ashburner, H.L. Carson and J.N. Thompson, Jr). New York; Academic Press, Vol. 3d, pp.29–59. (11)

Taylor, J.F., Bean, B., Marshall, C.E. and Sullivan, J.J. 1985. Genetic and environmental components of semen production traits of artificial insemination Holstein bulls. *Journal of Dairy Science* 68, 2703–22. (23)

Taylor, P.D. and Jonker, L.B. 1978. Evolutionarily stable strategies and game dynamics. *Math. Biosciences* 40, 145–56. (10)

Taylor, P.D. and Sauer, A. 1980. A selective advantage of sex ratio homeostasis. *American Naturalist* 116, 305–10. (10)

Taylor, St C.S. 1976a. Multibreed designs. 1. Variation between breeds. *Animal Production* 23, 133–44. (24)

Taylor, St C.S. 1976b. Multibreed designs. 2. Genetic variation within and between breeds. *Animal Production* 23, 145–54. (24)

Taylor, St C.S. and Hnizdo, E. 1987. Multibreed designs. 3. Interbreed relationships. *Animal Production* 44, 39–53. (24)

Taylor, St.C.S., Thiessen, R.B. and Moore, A.J. 1986. Single sex beef cattle systems. In: *Exploiting New Technologies in Animal Breeding* (edited by C. Smith, J.W.B. King and J.C. McKay). Oxford, UK; Oxford University Press, pp. 183–93. (29)

Thoday, J.M. 1961. Location of polygenes. *Nature* 191, 368–70. (20, 21, 34)

Thoday, J.M. and Boam, T.B. 1961. Regular responses to selection. I. Description of responses. *Genetical Research* 2, 161–76. (16, 21)

Thoday, J.M., Gibson, J.B. and Spickett, S.G. 1964. Regular responses to selection. 2. Recombination and accelerated response. *Genetical Research* 5, 1–19. (16, 20, 21)

Thompson, R. 1973. The estimation of variance and covariance components with an application when records are subject to culling. *Biometrics* 29, 527–50. (23)

Thompson, R. 1976a. Design of experiments to estimate heritability when observations are available on parents and offspring. *Biometrics* 32, 283–304. (24)

Thompson, R. 1976b. The estimation of maternal genetic variances. *Biometrics* 32, 903–17. (24)

Thompson, R. 1988. Generalized linear models and application to animal breeding. In: *Proceedings of the International Symposium on Advances in Statistical Methods for Genetical Improvement in Livestock* (edited by D. Gianola and K. Hammond). Heidelberg, Germany; Springer Verlag (in press). (25)

Thompson, R. and Juga, J. 1989a. Selection differentials and realized heritability. *Animal Production* (in press). (24)

Thompson, R. and Juga, J. 1989b. A derivative-free approach for estimating variance and covariance components in bivariate animal models by Restricted Maximum Likelihood. *Acta agriculturae scandinavica* (submitted). (23)

Thompson, R., McGuirk, B.J. and Gilmour, A.R. 1985. Estimating the heritability of all-or-none and categorical traits by offspring-parent regression. *Zeitschrift für Tierzüchtung und Züchtungsbiologie* 102, 342–54. (25)

Thompson, R. and Meyer, K. 1986. Estimation of variance components: What is missing in the EM algorithm? *Statistical Computing and Simulation* 24, 215–30. (23)

Thompson, R. and Meyer, K. 1988. Estimating genetic parameters using an animal model with imaginary effects. *Génétique, Sélection, Évolution* (submitted). (23)

Thompson, S., Clarke, A.R., Pow, A.M., Hooper, M.L. and Melton, D.W. 1989. Germ line transmission and expression of a corrected HPRT gene produced by gene targeting in embryonic stem cells. *Cell* 56, 313–321. (36)

Thompson, V. 1977. Recombination and response to selection in *Drosophila melanogaster*. *Genetics* 85, 125–40. (12, 18)

Thomson, G. 1977. The effect of a selected locus on linked neutral loci. *Genetics* 85, 753–88. (6)

Throckmorton, L. 1975. The phylogeny, ecology and geography of *Drosophila*. In: *Handbook of Genetics* vol.3 (edited by R.C. King). New York; Plenum Press, pp.421–69. (4)

Togashi, K., Takeda, H. and Yokouchi, K. 1986. Selection response in open nucleus breeding system with overlapping generation. *Japanese Journal of Zootechnical Science* 57, 842–9. (27)

Togashi, K., Yokouchi, K. and Takeda, H. 1987. Inbreeding coefficients in open nucleus breeding schemes with overlapping generations. *Japanese Journal of Zootechnical Science* 58, 236–44. (27)

Toro, M.A. and Nieto, B.M. 1984. A simple method for increasing the response to artificial selection. *Genetical Research* 44, 347–9. (18)

Toro, M.A., Silió, L., Rodrigáñez, J. and Dobao, M.T. 1988. Inbreeding and family index selection for prolificacy in pigs. *Animal Production* 46, 79–85. (28)

Townes, T.M., Lingrel, J.B., Chen, H.Y., Brinster, R.L. and Palmiter, R.D. 1985. Erythroid-specific expression of human beta-globin genes in transgenic mice. *European Molecular Biology Organization Journal* 4, 1715–23. (36)

Tronik, D., Dreyfus, M., Babinet, C. and Rougeon, F. 1987. Regulated expression of the Ren-2 gene in transgenic mice derived from parental strains carrying only the Ren-2 gene. *European Molecular Biology Organization Journal* 6, 983–7. (36)

Tsubota, S. and Schedl, P. 1986. Hybrid dysgenesis-induced revertants of insertions at the 5′ end of the *rudimentary* gene in *Drosophila melanogaster*: Transposon-induced control mutations. *Genetics* 114, 165–82. (16)

Turelli, M. 1984. Heritable genetic variation via mutation-selection balance: Lerch's zeta meets the abdominal bristle. *Theoretical Population Biology* 25, 138–93. (15)

Turelli, M. 1985. Effects of pleiotropy on predictions concerning mutation-selection balance for polygenic traits. *Genetics* 111, 165–95. (15)

Turelli, M. 1988. Population genetic models for polygenic variation and evolution. In: *Proceedings of the 2nd International Conference on Quantitative Genetics* (edited by B.S. Weir, E.J. Eisen, M.M. Goodman and G. Namkoong). Sunderland, Mass.; Sinauer, pp.601–18. (15, 17)

Turner, H.N. and Young, S.S.Y. 1969. *Quantitative Genetics in Sheep Breeding*. Melbourne, Australia; Macmillan. (27).

Uematsu, Y., Ryser, S., Dembic, Z., Borgulya, P., Krimpenfort, P., Berns, A., von Boehmer, H. and Steinmetz, M. 1988. In transgenic mice the introduced functional T cell receptor beta gene prevents expression of endogenous beta genes. *Cell* 52, 831–41. (36)

Van Arendonk, J.A.M. 1989. Method to estimate genotype probabilities at individual loci in farm animals. *Theoretical and Applied Genetics*, (submitted). (34)

Van Raden, P.M. and Freeman, A.E. 1986. Computing Restricted Maximum Likelihood estimates of variances and covariances. *Journal of Dairy Science* 69, Suppl. 1, 208 (Abstr.). (23)

Van Raden, P.M., Freeman, A.E. and Rothschild, M.F. 1984. Maximizing genetic gain under multiple-stage selection. *Journal of Dairy Science* 67, 1761–6. (27)

Van Raden, P.M. and Jung, Y.C. 1988. A general purpose approximation to Restricted Maximum Likelihood: The tilde-hat approach. *Journal of Dairy Science* 71, 187–94. (23)

Van Vleck, L.D. 1964. Sampling the young sire in artificial insemination. *Journal of Dairy Science* 47, 441–6. (29)

Van Vleck, L.D. 1972. Estimation of heritability of threshold characters. *Journal of Dairy Science* 55, 218–25. (25)

Van Vleck, L.D. 1981. Potential genetic impact of artificial insemination, sex selection, embryo transfer, cloning, and selfing in dairy cattle. In: *New Technologies in Animal Breeding* (edited by B.G. Brackett, G.E. Seidel, Jr and S.M. Seidel). New York; Academic Press, pp.221–42. (29)

Varenne, S., Buc, J., Lloubes, R. and Lazdunski, C. 1984. Translation is a non-uniform process. Effect of tRNA availability on the rate of elongation of nascent polypeptide chains. *Journal of Molecular Biology* 180, 549–76. (4)

Verghese, M.W. 1974. Interaction between natural selection for heterozygotes and directional selection. *Genetics* 76, 163–8. (12)

Wade, M.J. and McCauley, D.E. 1984. Group selection: The interaction of local deme size and migration in the differentiation of small populations. *Evolution* 38, 1047–58. (11)

Wagner, E.F., Stewart, T.A. and Mintz, B. 1981. The human beta-globin gene and a functional thymidine kinase gene in developing mice. *Proceedings of the National Academy of Sciences, USA* 78, 5016–20. (36)

Wagner, E.F., Vanek, M. and Vennstrom, B. 1985. Transfer of genes into EC cells by retrovirus infection: efficient expression from an internal promoter. *European Molecular Biology Organization Journal* 4, 663–6. (36)

Wagner, T.E., Hoppe, P.C., Jollick, J.D., Scholl, D.R., Hodinka, R. and Gault, J.B. 1981. Microinjection of a rabbit beta-globin gene into zygotes and its subsequent expression in adult mice and their offspring. *Proceedings of the National Academy of Sciences, USA* 78, 6376–80. (36)

Walkley, J.R.W. and Smith, C. 1980. The use of physiological traits in genetic selection for litter size in sheep. *Journal of Reproduction and Fertility* 59, 83–8. (33)

Wallace, B. 1948. Studies on 'Sex Ratio' in *Drosophila pseudoobscura. Evolution* 2, 189–217. (14)

Walsh, K. and Koshland, D.E. 1985. Characterisation of rate controlling steps *in vivo* by use of an adjustable expression vector. *Proceedings of the National Academy of Sciences, USA* 82, 3577–81. (31)

Walters, J.R. and Sellwood, R. 1988. Genetic resistance to K88 *Escherichia coli* – pig performance trials. *Animal Production* 46, 486–7 (Abstr.). (28)

Ward, K.A., Franklin, I.R., Murray, J.D., Nancarrow, C.D., Raphael, K.A., Rigby, N.W., Byrne, C.R., Wilson, B.W. and Hunt, C.L. 1986. The direct transfer of DNA by embryo microinjection. In: *Proceedings of the 3rd World Congress of Genetics Applied to Livestock Production*. Lincoln, Nebraska; University of Nebraska, Vol. 12, 6–21. (30)

Ward, R.D. and Skibinski, D.O.F. 1985. Observed relationships between protein heterozygosity and protein genetic distance and comparison with neutral expectations. *Genetical Research* 45, 315–40. (2)

Warren, D.C. 1924. Inheritance of egg size in *D. melanogaster. Genetics* 9, 41–69. (21)

Watterson, G.A. 1975. On the number of segregating sites in genetical models without recombination. *Theoretical Population Biology* 6, 217–50. (3)

Watterson, G.A. 1978. The homozygosity test of neutrality. *Genetics* 88, 405–17. (6)

Weatherall, D.J. 1985. *The New Genetics and Clinical Practice*, 2nd Edition. Oxford, UK; Oxford University Press. (6)

Webb, A.J. and Bampton, P.R. 1988. Benefits of Selection for litter size in pig dam lines. *Animal Production* 46, 486 (Abstr.). (28)

Webb, A.J., Carden, A.E., Smith, C. and Imlah, P. 1982. Porcine stress syndrome in pig breeding. *Proceedings of the 2nd World Congress of Genetics Applied to Livestock Production*. Madrid; Editorial Garsi, Vol. 5, 588–608. (16, 34)

Webb, A.J. and Curran, M.K. 1986. Selection regime by production system interaction in pig improvement: a review of possible causes and solutions. *Livestock Production Science* 14, 41–54. (28)

Webb, A.J. and King, J.W.B. 1976. Development of a synthetic pig sire line by selection with immigration. 1. Results of selection and heritability estimates. *Animal Production* 22, 231–44. (28)

Webster, A.J.F. 1977. Selection for leanness and the energetic efficiency of growth in meat animals. *Proceedings of the Nutrition Society* 36, 53–9. (19)

Weinberg, W. 1909. Uber Vererbungsgesetze beim Menschen. *Zeitshrift für Abstammungs- und Vererbungslehre* 1, 377–92, 440–60. (21)

Weinberg, W. 1910. Weitere Beiträge zur Theorie der Vererbung. *Archiv für Rassen- und Gesellschaftsbiologie* 7, 35–49, 164–73. (21)

Weir, B.S. 1979. Inferences about linkage disequilibrium. *Biometrics* 35, 235–54. (7)

Weir, B.S. and Brooks, L.D. 1986. Disequilibrium on human chromosome 11p. *Genetic Epidemiology Supplement* 1, 177–83. (7)

Weir, B.S. and Cockerham, C.C. 1974. Behavior of pairs of loci in finite monoecious populations. *Theoretical Population Biology* 6, 323–54. (1)

Weir, B.S. and Cockerham, C.C. 1984. Estimating F-statistics for the analysis of population structure. *Evolution* 38, 1358–70. (11)

Weir, B.S. and Cockerham, C.C. 1989. Complete characterization of disequilibrium at two loci. In: *Mathematical*

Evolutionary Theory (edited by M.W. Feldman). Princeton, New Jersey; Princeton University Press, pp.86–110. (7)

Weir, B.S. and Hill, W.G. 1986. Nonuniform recombination within the human ß-globin gene cluster. *American Journal of Human Genetics* 39, 776–8. (6)

Weiss, G.H. and Kimura, M. 1965. A mathematical analysis of the stepping stone model of genetic correlation. *Journal of Applied Probability* 2, 129–49. (11)

Weller, J.I. 1986. Maximum likelihood techniques for the mapping and analysis of quantitative trait loci with the aid of genetic markers. *Biometrics* 42, 627–40. (20)

Weller, J.I. 1987. Mapping and analysis of quantitative trait loci in *Lycopersicon* (tomato) with the aid of genetic markers using approximate maximum likelihood methods. *Heredity* 59, 413–21. (20)

Westerhoff, H.V., Groen, A.K. and Wanders, R.J.A. 1984. Modern Theories of Metabolic Control and Their Application. *Bioscience Reports* 4, 1–22. (31)

Whitehead, C.C. and Griffin, H.D. 1985. Direct and correlated responses to selection for decreased body fat in broilers. In: *Poultry Genetics and Breeding* (edited by W.G. Hill, J.M. Manson and D.Hewitt). Harlow, UK; Longman, pp.113–23. (33)

Wijsman, E. 1984. The effect of mutagenesis on competitive ability in *Drosophila*. *Evolution* 38, 571–81. (13)

Wilkie, T.W. and Palmiter, R.D. 1987. Analysis of the integrant in Myk-103 transgenic mice which fail to transmit the integrant. *Molecular and Cellular Biology* 7, 1646–55. (36)

Willadsen, S.M. 1982. Micromanipulation of embryos of the large domestic species. In: *Mammalian Egg Transfer* (edited by G.E. Adams). Boca Raton, Florida; CRC Press, pp. 185–210. (35)

Willadsen, S.M. 1985. Micromanipulation of embryos of domestic species. *Journal of the Royal Agricultural Society* 146, 160–71. (35)

Willadsen, S.M. 1986. Nuclear transplantation in sheep embryos. *Nature* 320, 63–5. (35)

Wilson, A.C. 1976. Gene regulation in evolution. In: *Molecular Evolution* (edited by F.J. Ayala). Sunderland, Mass.; Sinauer, pp.225–34. (16)

Wilson, A.C. *et al.* (9 others) 1985. Mitochondrial DNA and two perspectives on evolutionary genetics. *Biological Journal of the Linnean Society* 26, 375–400. (6)

Wilson, A.C., Ochman, H. and Prager, E.M. 1987. Molecular time scale for evolution. *Trends in Genetics* 3, 241–7. (4)

Wolstenholme, D.R. and Thoday, J.M. 1963. Effects of disruptive selection. VII. A third chromosome polymorphism. *Heredity* 18, 413–32. (21)

Womack, J.E. and Moll, Y.D. 1986. Gene map of the cow: conservation of linkage with mouse and man. *Journal of Heredity* 77, 2–7. (34)

Wood, J.D., Kempster, A.J., David, P.J. and Bovey, M. 1987. Observations on carcass and meat quality in Duroc, Landrace and Duroc × Landrace pigs. *Animal Production* 44, 488 (Abstr.). (28)

Woolliams, J.A. 1989. The value of cloning in MOET nucleus breeding schemes for dairy cattle. *Animal Production* 48, 31–5. (29)

Woolliams, J.A. and Smith, C. 1988. The value of indicator traits in the genetic improvement of dairy cattle. *Animal Production* 46, 333–45. (29, 33)

Wray, N.R. 1988. Breeding value estimation for pigs in closed nucleus herds. *Animal Production* 46, 486 (Abstr.). (28)

Wright, F.G. 1988. Theoretical analysis of codon usage in DNA. PhD Thesis, Department of Genetics, University of Edinburgh. (1)

Wright, H.B., Pollak, E.J. and Quaas, R.L. 1987. Estimation of variance and covariance components to determine heritability and repeatability of weaning weight in American Simmental cattle. *Journal of Animal Science* 65, 975–81. (23)

Wright, S. 1921. Systems of mating. *Genetics* 6, 111–78. (12, 21)

Wright, S. 1922. The effects of inbreeding and crossbreeding in guinea pigs. III. Crosses between highly inbred families. *United States Department of Agriculture Bulletin* 1121, 1–60. (11)

Wright, S. 1931. Evolution in Mendelian populations. *Genetics* 16, 97–159. (11, 29)

Wright, S. 1934a. Molecular and evolutionary theories of dominance. *American Naturalist* 63, 24–53. (31)

Wright, S. 1934b. An analysis of variability in number of digits in an inbred strain of guinea pigs. *Genetics* 19, 506–36. (25)

Wright, S. 1935. Evolution in populations in approximate equilibrium. *Journal of Genetics* 30, 257–66. (1, 15)

Wright, S. 1939. Genetic principles governing the rate of progress of livestock breeding. *Proceedings of the American Society for Animal Production* 32, 18–26. (18)

Wright, S. 1951. The genetical structure of populations. *Annals of Eugenics* 15, 323–54. (1, 11)

Wright, S. 1952. The genetics of quantitative variability. In: *Quantitative Inheritance* (edited by E.C.R. Reeve and C.H. Waddington). London, UK; HMSO, pp.5–41. (20, 21)

Wright, S. 1959. On the genetics of silvering in the guinea pig with especial references to interaction and linkage. *Genetics* 44, 383–405. (20)

Wright, S. 1968. *Evolution and the Genetics of Populations*, Vol. 1. *Genetic and Biometric Foundations*. Chicago, Ill.; University of Chicago Press. (20)

Wright, S. 1969. *Evolution and the Genetics of Populations*, Vol. 2. *The Theory of Gene Frequencies*. Chicago, Ill.; University of Chicago Press. (11)

Wright, S. 1977. *Evolution and the Genetics of Populations*. Vol.3. *Experimental Results and Evolutionary Deductions*. Chicago, Ill.; University of Chicago Press. (17)

Wright, S. 1978. *Evolution and the Genetics of Population*, Vol. 4. *Variability Within and Among Natural Populations*. Chicago, Ill.; University of Chicago Press. (11)

Xu, K.P. and Greve, T. 1988. A detailed analysis of early events during *in vitro* fertilization of bovine follicular oocytes. *Journal of Reproduction and Fertility* 82, 127–34. (35).

Yamao,F., Muto, A., Kawauchi, Y., Iwami, M., Iwagami, S., Azumi, Y. and Osawa, S. 1985. UGA is read as tryptophan in *Mycoplasma capricolum*. *Proceedings of the National Academy of Sciences, USA* 82, 2306–9. (2)

Yamaguchi, O., Yamazaki, T., Saigo, K., Mukai, T. and Robertson, A. 1987. (See AR bibliography)

Yannopoulos, G., Stamatis, N., Monastirioti, M., Hatzopoulos, P. and Louis, C. 1987. *hobo* is responsible for the induction of hybrid dysgenesis by strains of *Drosophila melanogaster* bearing the male recombination factor 23.5 MRF. *Cell* 49, 487–95. (16)

Yoo, B.H. 1980a. Long-term selection for a quantitative character in large replicate populations of *Drosophila melanogaster*. I. Responses to selection. *Genetical Research* 35, 1–17. (16, 17)

Yoo, B.H. 1980b. Long-term selection for a quantitative character in large replicate populations of *Drosophila melanogaster*. II. Lethals and visible mutants with large effects. *Genetical Research* 35, 19–31. (15, 16, 17)

Yoo, B.H. 1980c. Long-term selection for a quantitative character in large replicate populations of *Drosophila melanogaster*. V. The inbreeding effect of selection. *Australian Journal of Biological Sciences* 33, 713–23. (18)

Young, M.W. and Schwartz, H.E. 1981. Nomadic gene families in *Drosophila*. *Cold Spring Harbor Symposia on Quantitative Biology* 45, 629–40. (6)

Young, S.S.Y. 1961. A further examination of the relative efficiency of three methods of selection for gains under less-restricted conditions. *Genetical Research* 2, 106–21. (12)

Young, S.S.Y. and Skavaril, R.V. 1976. Computer simulations of within family selection in finite populations. *Theoretical and Applied Genetics* 48, 45–51. (18)

Yuhas, J.M., Angel, C.R., Mahin, D.T., Faris, R.D., Woodward, K.T. and Storer, J.B. 1967. Plasma enzyme activities in inbred mice. *Genetics* 57, 618–24. (21)

Yuksel, E., Hill, W.G. and Roberts, R.C. 1981. Selection for efficiency of feed utilisation in growing mice. *Theoretical and Applied Genetics* 59, 129–37. (19)

Yukuhiro, K., Harada, K. and Mukai, T. 1985. Viability mutations induced by the *P* elements in *Drosophila melanogaster*. *Japanese Journal of Genetics* 60, 531–7. (8, 16)

Zeng, Z.-B. and Hill, W.G. 1986. The selection limit due to the conflict between truncation and stabilizing selection with mutation. *Genetics* 114, 1313–28. (15)

Zhuchenko, A.A., Samovolk, A.P., Korol, A.B. and Andryushchenko, V.K. 1979. Linkage between loci of quantitative characters and marker loci. II. Influence of three tomato chromosomes on variability of five quantitative characters in backcross progenies. *Genetika* 15, 672–83. (20)

Zimmer, A. and Gruss, P. 1989. Production of chimaeric mice containing embryonic stem (ES) cells carrying a homeobox *Hox1.1* allele mutated by homologous recombination. *Nature* 338, 150–3. (36)

Zuckerkandl, E. and Pauling, L. 1965. Evolutionary divergence and convergence in proteins. In: *Evolving Genes and Proteins* (edited by V. Bryson and H.J. Vogel). New York; Academic Press, pp.97–166. (3)